Materials for Springs

Y. Yamada (chief Ed.)

Materials for Springs

Translated from the Japanese original edition published by JSSE

Organized by JSSE translation committee

With 336 Figures

Springer

JSSE Japan Society of Spring Engineers

Dr. Eng. Yoshiro Yamada
Dr., Professional eng. Japan in Metallurgy
Yamada Research & Consultant Office
Lecturer, Setsunan Univ.
Japan

Mr. Toshio Kuwabara
General Affairs Director
Japan Society
of Spring Engineers (JSSE)
MH-KIYA Building 3F
12, KANDA-MIKURACHO
CHIYODA-KU, TOKYO
101-0038 JAPAN

Library of Congress Control Number: 2007933493

ISBN 978-3-540-73811-4 Springer Berlin Heidelberg New York

Springer is a part of Springer Science+Business Media
springer.com

Typesetting: by the authors and Integra using a springer LaTeXmacro package
Cover design: deblik, Berlin

Printed on acid-free paper SPIN: 11731405 89/Integra 5 4 3 2 1 0

Foreword

The Japan society of Spring Engineers, formerly called the Japan Society of Spring Research, was established in 1947. Thereafter, the society has endeavored to edit and publish many books on spring technologies. Among these publications, there have been such books as "Bane" in Japanese or "Springs" in English published in 1959, revised in 1964 and 1982, and "Spring Design" published in 1963, revised in 1982. Since then, various advanced materials and improved process technology have been developed. Reflecting the rapid development of PC hardware and software, we published "FEM for springs", in 1997 for advanced spring design analyses. The book was later translated into English and published from the Springer publishing company in 2003.

As for spring materials, "Spring Materials and their characteristics (in Japanese)" was published in 2000.

The book "Materials for Springs" is the English version of this book with some additional information currently available in the latest Japanese material industry.

I am confident that this book will provide a lot of advanced technical information on spring materials to engineers, researchers and students in worldwide spring industry.

April 2007

Kosuke Nagaya
Chairman, The Japan Society of Spring Engineers
Professor, Dr., Gunma University, Japan

Preface to Japanese Edition

Since the publication of the last edition of the book entitled "Springs" (written in Japanese) in 1982, extensive progresses have been made in all areas of spring materials technology. Publishing a new book on spring materials and the related technologies and adding information obtained in the twenty years after the last edition seem to be appropriate.

This book is basically intended for engineers related to spring materials and technologies who graduated from metallurgical or mechanical engineering course in technical high school, or in other higher engineering schools, as well as those who are related to purchases or sales of spring materials.

In the first chapter, the fundamental selection processes of spring materials and concomitant working processes including the information sources on materials database are given, followed by the basic mechanisms and theories of spring failures such as fatigue fracture, creep/stress relaxation and stress corrosion cracking of metallic materials.

In the second chapter, ferrous and non-ferrous metallic materials are the main topics. In the third and fourth chapters, polymer materials, FRP (Fiber Reinforced Plastics), ceramics and C/C composite materials are the main subject respectively. In each chapter from the second to the fourth chapter, material grades belonging to each material category, their characteristics, production processes, and special cares to be taken of when actually using the materials for springs, are given. Spring design technologies are little included in this volume however, because they are separately dealt with in the fourth volume (in Japanese) in this publication series.

In the fifth chapter, lists of Japanese spring material manufacturers and their material grades being produced, relative prices of some spring materials compared to standard materials in Japan, comparisons of spring materials in the Japanese Industrial Standards with some other foreign standards, etc, are summarized. It is to be remarked that the information on the relative prices of some spring materials in this chapter is not always applicable in any cases, since the price depends on various factors such as quantity of one time purchase, delivery condition, etc.

I express my feeling of great thanks to Mr. Yoshihide Nagai, a former director of the Japan Society of Spring Research and Mr. Kazuya Nakagawa, Nikkan Kogyo Shimbun Company, for their help for editing this book.

January 2000

Yoshiro Yamada, Dr.
Leader of the publication Working Group

Preface to the English Version

In editing this English version, efforts were made to update the contents in the original Japanese edition such as Japanese Industrial Standard (JIS) specifications and some other information that was felt better to up-date. For full particulars of any national or international standard, the up-to-date edition should always be consulted.

Since new information on delayed fracture of spring steels has been obtained, topics on delayed fracture and notch susceptibility of high strength steels in corrosive environments have been newly added at the end of the Chapter 1.

It is hoped that readers of this English version will gain knowledge from this book in which the most advanced spring and spring material technologies are contained.

Special thanks are due to Mr. Kanji Inoue, Senior Manager of NHK Spring Co. Ltd. for proof reading most part of the English translation manuscript and to Ms. Shimiko Shimamura, President of Plain Corporation for her efforts for preparing galley proofs. Acknowledgement is also given to Mr. Toshio Kuwabara, Director of the JSSE, for correspondence with the Springer Verlag Company for publishing this English edition and proof reading of Chapter 3, to Mr. Hiroshi Koyama for his important efforts to promote the activity throughout this project, and to all the engineers and researchers who participated in the English translation of this original Japanese book.

April 2007

Yoshiro Yamada, Dr., Professional eng. Japan in Metallurgy
Leader of the English Translation Committee in JSSE

Author Index

Japanese Edition Authors (Organization belonged when published.)

Tadasu Abumiya (Kobe Steel Ltd., Chap. 2.5)
Hiroaki Hayashi (Suzuki Metal Industry Co. Ltd., Chap. 2.1.3 (2), (4), and 2.3)
Hiroshi Horikawa (Furukawa Techno Materials Co. Ltd., Chap. 2.6)
Tomohito Iikubo (Daido Steel Co. Ltd., Chap. 2.2)
Hideaki Iwata (Tokai Rubber Industries Ltd., Chap. 3.2)
Kenji Kanazawa (Chuo University, Chap. 1.2.3)
Yutaka Maeda (MRC Techno Research Inc., Chap. 3.3)
Teruyuki Murai & Akio Ikenaga (Sumitomo Electric Industries, Ltd., Chap. 2.1.3 (1))
Takao Nakagawa (Across Corporation, Chap. 4.2)
Toshimasa Ochiai (NGK Industries, Chap 2.4.1, 2.4.2, 2.4.4, and 2.4.5)
Ken Okabe (Harada Metal Industries, Chap. 2.4.3)
Yasuhiro Oki & Nobuhiko Ibaraki (Kobe Steel Ltd., Chap. 2.1.1 (3) and (10) i)
Masao Sakamoto (National Research Institute for Materials Sciences, Chap. 1.2.2 (3) and (4))
Shigemi Satoh (NHK Spring Co. Ltd., Chap. 4.1)
Yatsuka Takata (Aichi Steel Corporation, Chap. 2.1.1 and 2.1.2)
Yoshinori Tanimoto (Nippon Seisen Co. Ltd., Chap. 2.1.3 (3))
Akio Tsuzuki (Togo Seisakusho Corporation, Chap. 5)
Katsuyuki Uchibori (Mitsubishi Steel Mfg. Co. Ltd., Chap 2.1.2 (11))
Isao Ueda (Horikiri Industries, Chap. 2.1.2 (10) ii)
Kyosuke Uemura (DuPont Kabushiki Kaisha, Chap. 3.1)
Yoshiro Yamada (Suncall Corporation, Chap 1.1, 1.2.1, 1.2.2 (1), (2), 2.1.3 (1) viii)

English Translation Committee:

Yoshiro Yamada, Committee Chief, Setsunan University
Hiroshi Koyama, Committee sub-Chief, Japan Society of Spring Engineers
Kanji Inoue, Translation Supervisor, NHK Spring Co. Ltd.
Toshio Kuwabara, Coordinator, Japan Society of Spring Engineers

Abraham Mehari, Aichi Steel Corporation
Nobuhiko Ibaraki, Kobe Steel Ltd.
Teruyuki Murai, Sumitomo Electric Industries Ltd.
Kenichi Shimizu, Sumitomo (SEI) Steel Wire Corp.
Fumiaki Nozoe, Sumitomo (SEI) Steel Wire Corp.
Akihiko Nishikawa, NHK Spring Co. Ltd.
Tomohito Iikubo, Daido Steel Co. Ltd.
Suguru Nomura, NHK Spring Co. Ltd.

Contents

1

A Guide to Spring Material Selection

1.1 Introduction

Spring material and its quality can be normally taken into consideration or highlighted in such cases as; (i) a spring installed in mechanical products, failed either by fracture or by significant deformation in use, (ii) a mechanical product newly designed or improved, where a new design of a spring is required, in this case, the spring with higher quality can be required, and (iii) a cost reduction requested for the spring having been used without any difference of the quality.

The case (i) can be divided into the following two cases:

(A) Quality requirements set up in the initial quality design stage, were not achieved in the actual products.
(B) Although the quality requirements set up at the design stage were satisfied in use, the springs were used in severer conditions than initially expected or some important quality requirement failed to be included in the initial quality requirements in the design stage.

When a spring has failed due to (A) or (B), an investigation for finding the failure cause is carried out, and the quality of the material used and the manufacturing process history are required to be studied. In the case of (ii) or (iii) described above, making the grade of material, dimensions for similar applications and the spring working processes clear, the search of substitute material grades can be started. Considering the availability, quality level, price, and the matching with working processes (conventional or new process), the most suitable material can be chosen.

As described above, the most suitable selection of spring material and its working processes can be said to play a very important role in quality and cost.

1.2 Functions and Qualities Required for Springs and the Spring Material Selection

1.2.1 Basic Items to be considered in Spring Material Selection

In the spring material selection and the working process designing, the following points shall be taken into consideration.

(1) Selected material and manufacturing process are such that the quality of finished springs satisfies customer's quality requirements.
(2) Availability of selected material.
(3) Economical feasibility of material and spring working processes (cost).
(4) The manufacturing processes, where the material quality should not be deteriorated.
(5) recycling.
(6) No pollution, safety, and regulations observance through spring manufacturing to disposal or recycling.

Springs play at least one of the following five basic roles or functions [1]:

(1) When unloaded, springs return to the original position or to the original shape.
(2) Absorption or utilization of vibration.
(3) Relaxation or absorption of impact force.
(4) Storage and/or release of energy.
(5) Measurement of force.

An unloaded spring often does not recover to its original shape, and this kind of shape change is called a permanent set of a spring. If a permanent set takes place in a spring, it may exert some deleterious effect on the measurement of force (5), or possibly on the functions (2) to (4) described above. In whatever basic spring functions are required for springs, springs loaded under repeated or varying stresses can sometimes fracture due to fatigue. In general, permanent set and fatigue fracture can be said to be the most important quality factors of springs to be paid attention to. In addition to these, failure of springs due to wear and/or corrosion is to be taken into consideration according to the application or the environments in use. Table 1.1 shows typical failures of springs, which occur in use.

Depending upon the load pattern of springs in use, quality requirements for the spring material vary. In Table 1.2 [2], relations between types of spring load pattern and the essential quality required for spring material are summarized. The quality requirements for spring materials change with such conditions as temperature or environments where the springs are used.

In Table 1.3, quality and performance characteristics of springs are listed.

As workability of spring materials, forming ability during spring forming, heat treatment performance (for example, hardenability of spring steel) to

Table 1.1. Principal types of spring failures

Fracture	Fracture with repeated stresses • Fatigue with no corrosion • Corrosion fatigue • Fatigue from fretting corrosion or wear Fracture with impact stress • Brittle fracture (Low temperature brittle fracture) • Ductile fracture Fracture with static stress • Stress corrosion cracking • Delayed fracture (Hydrogen embrittlement fracture)
Deformation (Permanent set)	• Yielding, plastic deformation (due to over stressing) • Static creep • Dynamic creep • Stress relaxation
Decrease of cross-sectional dimensions	• Wear • Fretting • General corrosion • Local corrosion • Errosion

Table 1.2. Types of load [2]

Types of load		Figures to be acquainted	Properties required for material
Static load	A constant and invariable load (permitted insignificant variation of load)	Load and deflection	High elastic limit
Repeated load	Constant loads repeatedly applied	Mean load Load amplitude Deflection Number of cycles	High fatigue strength
Impact load	A load applied abruptly at high speed	Impact force Deflection Number of cycles	High elastic limit High impact value
Load for measuring load	Accurate load being ensured for a wide range of deflection, like a spring balance	Spring constant Maximum load	High elastic limit High dimensional accuracy

obtain targeted quality, and the workability during material production processes, can be taken into consideration, since these workability characteristics are also very important in spring material selection.

Table 1.3. Quality and characteristic properties of spring materials

<u>Physical properties</u> Crystal structure Transformation temperature Melting temperature Density Electrical conductivity Electrical conductivity at 20°C (% IACS) Electrical resistivity Dielectric coefficient Magnetic permeability Magnetic saturation flux density Coercive force Magnetic hysteresis Magnetic striction Specific heat Thermal conductivity Linear thermal expansion coefficient <u>Microstructure</u> Metallurgical structure Crystal grain diameter Second phase particles Chemical composition Non-metallic inclusion Surface decarburization Internal oxidation <u>Mechanical properties</u> Elastic modulus (Young's modulus E) Shear modulus G Poisson's ratio Elastic limit (tension, compression, torsion, bend) Yield strength (ibid) Proof stress (ibid) Spring deflection limit	Residual stress Tensile strength Torsion strength Hardness Viscosity Viscoelasticity Fatigue characteristics *Fatigue strength *Fatigue limit *Fatigue strength at a particular life *Fatigue life *S-N curve (diagram) *Fatigue limit diagram *Crack growth rate *Lower limit threshold stress intensity factor range Fracture energy (Toughness) *Charpy impact value *Fracture bending stress *Fracture toughness <u>Environmental resistance</u> Corrosion Oxidization Delayed fracture Hydrogen embrittlement Stress corrosion cracking Stress cracking resistance Atmospheric corrosion Corrosion fatigue Wear Fretting corrosion Erosion <u>Workability at spring manufacturing processes</u> Material and spring workability *Drawability *Rolling ability	*Machinability *Grindability *Bendability *Torsion characteristics (wire) *Surface lubrication *Drawability (sheet) *Punching ability *Heat treatment ability *Hardenability *Isothermal transformation characteristics *Continuous cooling transformation characteristics *Temper softening characteristics *Decarburization characteristics *Temperature-time-austenitization characteristics *Low temperature annealing characteristics *Age hardenability (precipitation characteristics) <u>Shape characteristics</u> *Cross section *dimensions and its accuracy *Flatness *Camber *Wire cast *Twist *Bend *Notch shape *Surface flaw <u>Recycling ability</u> <u>Safety</u>

1.2.2 Spring Material Selection Method

Procedures of Spring Material Selection

In Sect. 1.2.1, basic factors to be considered in selecting spring material were described. In this section, more precise material selection procedures and special remarks can be discussed.

(1) The condition where the spring is used, such as volume, dimension, temperature and environmental atmosphere should be made clear.
(2) The quality requirements (load, deflection, fatigue life, etc.) should be made clear.
(3) The material grade, section size, elastic modulus, design stress, spring shape and its dimensions should be determined. Since in conventional usages, kinds of material used are normally known, see Table 1.4, it can be possible to make the material selection based on these kinds of information. It seems to be desired that the conditions where various kinds of springs are used, and the corresponding materials characteristics such as material grade, hardness, production process and dimensions, are collected and compiled as case-bases [4], and they can be utilized when necessary and can be updated.
(4) The testing methods and evaluation standards shall be determined in preparing material specifications.
(5) According to Japanese Industrial Standards (JIS) B 2704-2000 Helical compression and extension springs – Requirements for design, B 2709-2000 Helical torsion springs – Requirements for design, allowable stresses can be recommended for these kinds of springs used under static load with different wire diameters. In these specifications, fatigue life estimation methods based on the fatigue strength diagram between the maximum stress and minimum stress, are also described for reference. However, when higher quality springs than JIS specifications are required, spring materials other than JIS specification and spring manufacturing technologies such as shot-peening, should be studied.
(6) According to the JIS B 2704-2000 and B2709-2000 specifications, it can be recommendable to use figures listed in Table 1.5 as for elastic modulus or Young's modulus, E, and modulus of rigidity, G, for springs produced with various material grades and used at room temperature. It is noted, however, that cold rolled sheet material directionally varies elastic modulus. The extent of anisotropy caused by cold work changes with the degree of cold work and material chemical compositions.
(7) Material selection has to be made according to the temperature in use. Piano wire and hard drawn wire are the most popular material grades and their procurement is comparatively easy. Steel rope made of hard drawn wire can be used under dynamic stress even at 233 K (minus 40°C) without any problems. This means that piano wire springs and hard drawn wire springs can be used at such low temperatures.

Table 1.4. Types of springs, their applications and examples of material used

Type	Applications	Material grade (JIS)
Multi leaf spring	Suspension for automobile and railway train	SUP 3, SUP 6, SUP 9A, SUP 11A
Helical spring	Suspension for automobile and railway train	SUP 6, SUP 7, SUP 9, SUP 12
	Suspension for railway train and large spring for machine for general use	SUP 9A, SUP 10, SUP 11A, SUP 13
	Small spring for machine in general use	SW, SWP
	Machines in general use and automotive suspension	SWO, SWOSM, AWOSC-B
	Engine valve spring	SWO-V, SWOCV-V, SWOSC-V, SWP-V
	Mechanical spring with corrosion resistance	SUS-WP
Torsion bar	Torsion bar for automobile	SUP 9, S45C
	Stabilizer bar for automobile	SUP 9, S48C, SUP 11, STKM 15A equivalent
	Automobiles, Railway trains, OA appliances, Torsion bars with arm	SWP-A, SWP-B, SWO-V, SWOSC-V, SUS304WPB
Sheet spring (including Belleville, washers)	Industrial machine, Automobile, Electrical appliance	SK 4, SK 5, S45C-S70C, SUP 10
	Computer related equipment requiring corrosion resistance	SUS301CSP, SUS304CSP
	Engine part requiring heat resistance	SUS631CSP
	Electrical contact for switches etc.	C5210, C1700
Spiral spring	Industrial and construction machine, Electrical appliances etc.	SK4, SK5
	Automobile, Industrial machine, Electrical appliance etc.	SWRH62-82, SUP12 (Flat wire), SK4, SK5, SUS301CSP, SUS631CSP
Spring washer	Prevention of bolts and nuts unfastening	SWRH, SW, SWP, SXXCM, SKXM, C5191P-H, C5212P-H
Mesh spring	Exhaust gas catalyzer, Airbag	SUS304, SUS316, SUS310S, Inconel 601
Retaining ring	Eccentric C-clip, Concentric C-clip, Circlip etc.	SWRH, SWX, SWPX, SXXCM, SKXM

Table 1.5. Young's moduli and moduli of rigidity for spring material

Material	Modulus of rigidity, G GPa	Young's modulus E, GPa
Spring steel	78.5	206
Hard drawn wire Piano wire Oil tempered wire	78.5	206
Stainless steel wires for springs SUS302, SUS304, SUS304N1, SUS316	68.5	186
Stainless steel wire SUS631J1	73.5	196
Brass wire	39.0	98
Nickel-silver wire	39.0	108
Phosphorus bronze wire	42.0	98
Copper beryllium wire	44.0	127

For springs used from room temperature to 423 K (150°C), piano wire and hard drawn wire can be normally used under the dynamic stress conditions and the statically stressed conditions respectively. If fatigue fractures and/or creep problem(s) cannot be overcome by piano wire or hard drawn wire, oil tempered wire can be considered.

Stainless steel is sometimes used at more elevated temperature than oil tempered wire, because creep resistance of the stainless steel is generally superior to that of the oil tempered wire. For spring used at higher temperature than stainless steel springs, iron-base superalloy A286 (JIS SUH660), Nickel-base superalloy or ceramics (silicon nitride) are mainly used. Figure 1.1 is a result of a survey inquiring the relation between spring material grades and their temperatures being used [5]. Open circle marks, ○, in the Fig 1.1 indicate temperatures of springs in use. Figure 1.2 is a copy from a book entitled 'Spring Materials Selector' [6], indicating maximum temperature of use for various materials. These two figures can be usable as a reference when choosing the candidate material, although the selection of final material is difficult since the material's maximum temperature of use varies depending on stress condition of the spring.

(8) Since there are such cases as the endurance of a spring is deteriorated because of hazardous environment the spring is exposed to, the material selection to meet with the environment in use or a measure insulating the spring from the environment can be required. Table 1.6 shows the suitability of material to various media or environments [6]. This refers

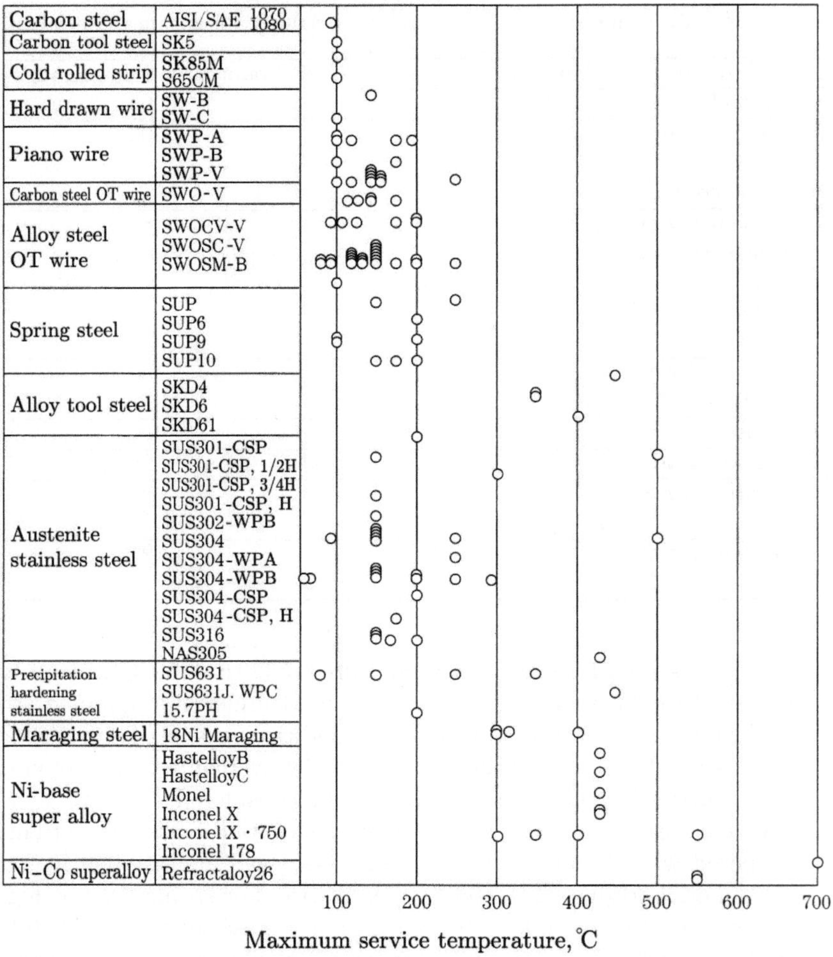

Fig. 1.1. Maximum service temperature of alloys [5]

to the original book "Spring Materials Selector, 2nd Edition" published by SRAMA (now, Institute of Spring Technology). This table should be regarded as a quick initial consideration, since small changes in temperature, concentration or impurities in the corrosive media can change behaviors remarkably.

(9) It can be normally said that the fatigue strength of metallic springs show the relationship proportionate to its hardness or tensile strength at least up to a certain level. As for steel springs, springs with too high hardness can often cause fatigue fracture in use, due to the notch sensitivity to small defect. Likewise, cracking troubles in spring cold-forming processes tend to occur more often as the spring material strength is in-

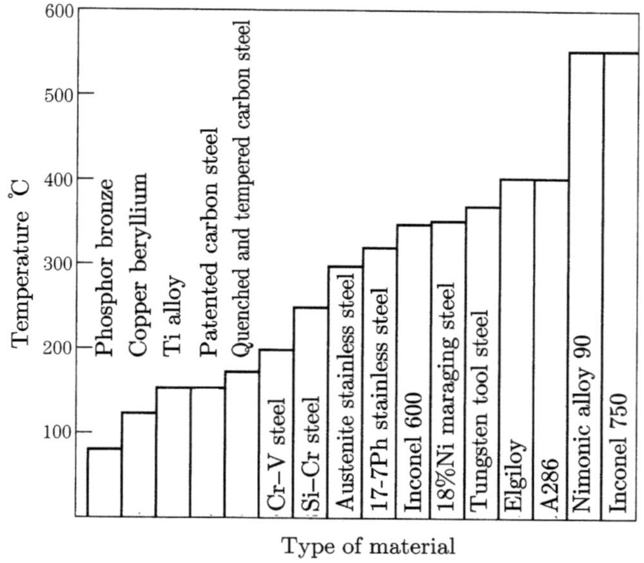

Fig. 1.2. Maximum temperature of use for various materials [6]

creased. Too high hardness also often can cause the fracture by hydrogen embrittlement or stress corrosion cracking, especially in steel springs. For example, in case of the steel spring manufactured by quenching and tempering, it should be cautious that temper at too low temperature may make the spring very susceptible to cause the delayed fracture under static load or the brittle fracture by impact.
Generally, piano wire and hard drawn wire have less susceptibility to hydrogen embrittlement than oil tempered wire and quenched-and-tempered steel. However, piano wire and hard drawn wire with excessively higher strength than specified can be susceptible to delamination (cracking along wire axis). Therefore, such high strength wire should not be used. Since austenitic stainless steel is susceptible to stress corrosion cracking in chloride containing environment, it is recommendable for austenitic stainless steel springs not to be used in such an environment. As for polymers, amorphous polymers such as polycarbonate, polystyrene, ABS, acrylic resins etc., tend to suffer from environmental stress cracking due to the exposure to ester compounds. In addition, the embrittlement due to ultra-violet light is to be prevented in polymer springs, as another example.

(10) Springs with comparatively large cross-section are normally made of spring steels through the process of hot-forming (or hot-working), followed by quenching and tempering. The spring steel grade actually used should be the one that has enough hardenability. This means that the material shall be hardened to the center of the cross-section and its

Table 1.6. Selection of spring material under chemical products [6]

✓= almost applicable
– = need to study
× = not applicable

chemicals	Density %	Temperature °C	Monel 400	Inconel 600	Inconel 625	Incoloy 825	Stainless 302	Stainless 316	Brass	Titan alloy
Acetaldehyde	99	40	✓	✓	✓	✓	✓	✓	–	×
Acetic Acid	0–99	30	–	–	✓	✓	✓	✓	×	✓
Acetic Anhydride	100	30	–	–	✓	✓	✓	✓	×	✓
Aceton	0–100	100	✓	✓	✓	✓	✓	✓	✓	✓
Acetylene	100	150	–	–	✓	✓	✓	✓	–	–
Acrolein	100	100	–	–	–	–	–	–	–	–
Air	100	–	✓	✓	✓	✓	✓	✓	✓	✓
Alcohol-Allyl	100	30	✓	✓	✓	✓	✓	✓	✓	✓
Ethyl	100	30	✓	✓	✓	✓	✓	✓	✓	✓
Mehyl	100	30	✓	✓	✓	✓	✓	✓	✓	✓
Allyl Chloride	100	30	✓	–	✓	✓	–	–	–	–
Aluminium Chloride	0–30	30	–	–	✓	–	–	–	×	✓
Aluminium Sulphate (Alum.)	100	30	–	×	✓	✓	×	✓	×	✓
Ammonia Liquid	0–100	30	×	–	✓	✓	–	–	×	–
Ammonium Bicarbonate	0–100	100	–	–	–	–	–	–	–	✓
Ammonium Carbonate	0–20	30	–	–	–	–	–	–	–	✓
Ammonium Chloride (dry)	0–20	20	–	–	✓	✓	–	✓	×	✓
	100	100	–	–	✓	–	–	✓	×	✓
Ammonium Hydroxide	0–30	70	×	–	✓	✓	✓	✓	×	✓
Ammonium Nitrate	0–40	80	–	–	✓	✓	✓	✓	×	✓
Ammonium Phosphate	5	100	–	–	✓	✓	✓	✓	×	✓
Ammonium Sulphate	0–40	100	–	–	✓	✓	✓	✓	×	✓
Amyl Acetate	100	30	✓	✓	✓	✓	–	–	–	–
Amyl Chloride	100	30	–	–	✓	–	–	–	–	–
Aniline	100	30	×	–	✓	✓	–	–	–	–
Barium Chloride	0–40	100	–	–	✓	–	–	–	–	✓
Barium Hydroxide	100	1040	–	–	✓	✓	–	–	–	✓
	0–50	100	–	–	✓	✓	–	–	–	✓
Beer	–	30	✓	✓	✓	✓	✓	✓	✓	–
Beet Suger Liquors	0–40	30	✓	✓	✓	✓	✓	✓	✓	–
Benzaldehyde	100	30	–	–	✓	✓	–	–	–	–
Benzene	100	30	✓	✓	✓	✓	✓	✓	✓	✓
Black Sulphate Liquor	100	100	–	✓	✓	✓	–	–	–	–
Boric Acid	0–20	100	–	✓	✓	✓	×	✓	×	✓
Bromine Dry	100	50	✓	✓	–	✓	–	–	–	×
Butadiene	100	30	✓	✓	✓	✓	✓	✓	✓	–
Butane	100	30	✓	✓	✓	✓	✓	✓	✓	–
Butyl Acetate	100	30	–	–	✓	✓	✓	✓	–	–
Butyric Acid	0–100	100	–	×	✓	✓	×	✓	–	✓
Calcium Chloride	0–25	30	✓	✓	✓	✓	✓	✓	–	✓
Calcium Hydroxide	0–30	100	–	–	✓	–	–	–	–	✓
Cane Suger Liquors	100	90	✓	✓	✓	✓	✓	✓	✓	–
Carbon Dioxide	100	450	✓	✓	✓	✓	✓	✓	✓	✓
Carbon Tetrachloride	100	30	–	✓	✓	✓	✓	✓	✓	✓
Caustic Soda	see Sodium hydroxide									
Chlorine, Dry	100	550	–	✓	–	×	✓	✓	✓	wet
Chlorobenzene	100	30	–	–	–	–	–	–	–	–
Chlorosulphonic Acid	100	30	–	–	–	–	–	–	×	–
Chloroform	100	100	✓	✓	✓	✓	–	–	–	✓
Chromic Acid	0–100	30	×	×	–	–	–	✓	×	✓
Citric Acid	100	30	–	–	✓	✓	–	✓	×	✓
Coffee	–	100	✓	✓	✓	✓	✓	✓	–	–
Copper Sulphate	0–30	100	×	×	✓	✓	✓	✓	×	✓
Cresylic Acid	100	30	–	–	–	–	–	✓	–	–

Table 1.6. Selection of spring material under chemical products [6]

✓= almost applicable
– = need to study
× = not applicable

chemicals	Density %	Temperature °C	Monel 400	Inconel 600	Inconel 625	Incoloy 825	Stainless 302	Stainless 316	Brass	Titan alloy
Dichloroethane	100	30	✓	✓	✓	✓	–	–	–	✓
Ethyl Acetate	100	30	–	–	✓	✓	✓	✓	–	–
Ethyl Cellulose		30	–	–	✓	✓	–	–	–	–
Ethyl Chloride	100	30	✓	✓	✓	✓	–	–	–	–
Ethylene Dichloride	100	30	✓	✓	✓	✓	–	–	–	–
Ethylene Glycol	100	30	–	–	✓	✓	✓	✓	✓	–
Fatty Acids	100	30	✓	✓	✓	✓	–	✓	–	–
Ferric Chloride	100	30	×	×	–	–	×	×	×	✓
Ferric Nitrate	100	30	×	×	✓	✓	–	–	–	–
Ferric Sulphate	0–30	30	–	×	✓	✓	✓	✓	×	✓
Fluoboric Acid	25	30	–	–	–	–	×	–	–	×
Fluosilicic Acid	20	30	×	–	–	–	–	–	–	×
Formaldehyde	0–100	30	✓	✓	✓	✓	✓	✓	✓	✓
Formic Acid	0–100	100	–	–	✓	✓	×	–	×	✓
Fuel Oil	100	30	✓	×	✓	✓	✓	✓	✓	✓
Furfural	30–100	100	–	–	✓	✓	–	–	–	–
Gelatine	0–40	50	–	✓	✓	✓	✓	✓	×	–
Glucose		30	✓	✓	✓	✓	✓	✓	–	–
Glutamic Acid		30	–	–	✓	✓	–	✓	–	–
Glycerine	100	30	✓	✓	✓	✓	✓	✓	×	✓
Glycerol	0–100	150	✓	✓	✓	✓	✓	✓	×	×
Hydraulic Oil	–	–	✓	✓	–	✓	✓	✓	✓	✓
Hydrazine	100	35	×	×	✓	×	✓	(1)	–	–
Hydrobromic Acid	40	30	×	×	×	–	–	–	–	✓
Hydrochloric Acid							×	×	×	×
Hydrocyanic Acid	100	35	×	✓	✓	✓	✓	✓	×	–
Hydrofluoric Acid	10–100	35	✓	–	–	–	×	×	×	×
Hydrogen Peroxide (acid free)		30	–	–	–	✓	✓	✓	×	–
Hydrogen Sulphide	0–20	150	–	–	✓	✓	✓	✓	×	✓
Hydroquinone		35	–	✓	✓	✓	✓	✓	–	–
Insulin	100	35	✓	✓	✓	✓	–	–	–	–
Lactic Acid	80	30	×	–	✓	✓	–	✓	–	✓
Lead Acetate	20	30	–	✓	✓	✓	–	–	–	✓
Lemon Juice		30	–	✓	✓	✓	✓	✓	–	–
Linseed Oil	100	30	✓	✓	✓	✓	–	–	–	✓
Lithium Chloride		30	✓	✓	✓	✓	✓	✓	–	✓
Lithium Hydroxide	10	30	✓	✓	✓	✓	–	–	–	–
Magnesium Carbonate		30	✓	✓	✓	✓	–	–	–	–
Magnesium Chloride	0–50	30	✓	✓	✓	✓	–	✓	×	✓
Magnesium Hydroxide		30	✓	✓	✓	✓	✓	✓	–	✓
Magnesium Nitrate		30	×	–	✓	✓	–	–	–	–
Magnesium Sulphate	30	30	✓	✓	✓	✓	✓	✓	–	✓
Maleic Acid		30	–	–	✓	✓	×	✓	–	✓
Mercuric Chloride		30	×	×	✓	×	×	×	×	✓
Mercuric Cyanide		30	×	×	–	–	–	–	–	✓
Mercuric Iodide		30	×	×	✓	×	–	–	–	–
Mercurous Nitrate		30	×	–	✓	✓	–	–	–	–
Methyl Alcohol	0–100	100	✓	✓	✓	✓	✓	✓	✓	✓
Methyl Chloride	100	30	✓	✓	✓	✓	✓	✓	–	–
Methyle Ethyl Ketone		100	–	–	✓	✓	✓	✓	–	–
Milk		30	–	✓	✓	✓	✓	✓	×	–
Mine Water		65	×	✓	✓	✓	✓	✓	✓	–
Molasses		30	✓	✓	✓	✓	✓	✓	✓	–
Mono (sodium, potassium or ammonium) Phosphate		30	✓	✓	✓	✓	–	–	–	–

Table 1.6. Selection of spring material under chemical products [6]

✓= almost applicable
– = need to study
× = not applicable

chemicals	Density %	Temperature °C	Monel 400	Inconel 600	Inconel 625	Incoloy 825	Stainless 302	Stainless 316	Brass	Titan alloy
Napthenic Acid	100	30	✓	✓	✓	✓	–	✓	–	–
Nickel Chloride		30	–	–	✓	✓	×	×	×	×
Nitric Acid	0–65	30	×	–	✓	✓	×	–	×	✓
	100	80	×	×	✓	✓	×	×	×	–
Nitrobenzene	100	100	–	–	–	–	–	–	–	–
Oils, Crude	100	30	✓	✓	✓	✓	✓	✓	–	✓
Oils, Essential	100	30	✓	✓	✓	✓	–	–	–	–
Oils, Mineral	100	30	✓	✓	✓	✓	✓	✓	✓	–
Oils, Palm	100	30	✓	✓	✓	✓	✓	✓	–	–
Oils, Peanut	100	30	✓	✓	✓	✓	✓	✓	–	–
Oils, Sulphanated	100	30	✓	✓	✓	✓	–	✓	–	–
Oils, Vegetable	100	30	✓	✓	✓	✓	✓	✓	✓	–
Oleic Acid	100	30	–	✓	✓	✓	✓	✓	×	–
Oleum	20	30	×	✓	✓	✓	–	–	–	–
Orange Juice		30	✓	✓	✓	✓	–	–	–	–
Oxalic Acid		30	–	–	✓	–	–	✓	×	×
Palmitic Acid	100	30	✓	✓	–	✓	–	✓	×	–
Paraffin	100	35	✓	✓	✓	✓	✓	✓	✓	–
Petrole	100	30	✓	✓	✓	✓	✓	✓	✓	–
Phenol	100	30	–	✓	✓	✓	✓	✓	–	✓
Phenol Sulphonic Acid	100	30	–	–	✓	✓	–	–	–	–
Phosphoric Acid	0–25	30	✓	✓	✓	✓	–	✓	×	✓
	25–85	85	✓	×	✓	✓	×	×	×	✓
Phthalic Anhydride	100	30	✓	✓	✓	✓	✓	✓	–	–
Picric Acid	100	30	×	×	–	–	✓	✓	×	–
Potassium Bicarbonate	0–30	30	✓	✓	✓	✓	✓	✓	✓	–
Potassium Carbonate		30	✓	✓	✓	✓	✓	✓	–	–
Potassium Chlorate		30	–	–	✓	✓	–	✓	–	–
Potassium Chloride		30	✓	✓	✓	✓	–	✓	×	✓
Potassium Chromate	0–30	30	–	✓	✓	✓	×	✓	–	–
Potassium Cyanide	0–30	100	–	–	–	–	–	–	–	–
Potassium Dichromate	0–20	30	–	–	✓	✓	✓	✓	–	✓
Potassium Fericyanide	0–30	30	–	–	–	–	–	–	–	✓
Potassium Hydroxide	0–50	30	✓	–	–	–	✓	✓	×	×
	0–50	100	✓	–	–	–	✓	✓	×	×
Potassium Nitrate		30	–	–	✓	✓	–	–	–	–
Potassium Sulphate	10	30	–	✓	✓	✓	✓	✓	✓	✓
Propane	100	100	✓	✓	✓	✓	✓	✓	✓	–
Salicyclic Acid		30	✓	✓	✓	✓	–	–	–	–
Sea Water	100	100	✓	✓	✓	✓	×	–	✓	✓
Silicon Tetrachloride	100	30	✓	✓	✓	✓	×	×	–	–
Silver Nitrate		30	×	–	–	–	–	–	–	✓
Soap	100	30	✓	✓	✓	✓	✓	✓	✓	✓
Sodium Acetate		30	–	–	✓	✓	✓	✓	–	✓
Sodium Bicarbonate		30	–	–	✓	✓	✓	✓	✓	–
Sodium Bisulphate	10	30	✓	–	✓	–	×	✓	×	✓
Sodium Bromide	0–50	30	–	–	–	–	–	–	–	✓
Sodium Carbonate	30	30	–	–	✓	✓	×	✓	–	✓
Sodium Chloride		30	✓	–	✓	✓	–	✓	×	✓
Sodium Hydroxide	0–50	30	✓	✓	✓	✓	✓	✓	×	✓
	50–75	30	✓	✓	✓	✓	×	×	×	✓

Table 1.6. Selection of spring material under chemical products [6]

✓= almost applicable
– = need to study
× = not applicable

chemicals	Density %	Temperature °C	Monel 400	Inconel 600	Inconel 625	Incoloy 825	Stainless 302	Stainless 316	Brass	Titan alloy
Sodium Metaphosphate		30	–	✓	✓	✓	✓	✓	×	–
Sodium Matascilicate	0–50	30	✓	✓	✓	✓	–	–	–	–
Sodium Nitrate	10	30	–	✓	✓	✓	✓	✓	×	✓
		30	–	✓	✓	✓	–	–	×	✓
Sodium Peroxide	100	100	–	–	–	–	✓	✓	×	–
Sodium Phosphate		30	✓	✓	✓	✓	✓	✓	×	✓
Sodium Sulphate		30	✓	✓	✓	✓	✓	✓	✓	✓
Sodium Sulphide		30	–	✓	✓	✓	30%	30%	×	✓
Steam	100	450	✓	✓	✓	✓	✓	✓	✓	✓
Stearic Acid		30	✓	✓	✓	✓	×	✓	×	✓
Sugar (Liquid)		30	✓	✓	✓	✓	✓	✓	–	–
Sulphuric Acid	0–15	30	✓	–	✓	✓	×	×	×	×
	15–75	30	–	×	✓	✓	×	×	×	×
Sulphurous Acid	0–60	100	–	–	–	–	×	–	×	$< 6\%$
Tall Oil	100	30	✓	✓	✓	✓	×	–	–	–
Tannic Acid	10	30	–	–	–	–	–	–	–	✓
Tartaric Acid	58	30	–	–	✓	✓	–	✓	×	✓
Tetraphosphoric Acid	100	30	×	–	✓	✓	–	–	–	–
Toluene	100	100	✓	✓	✓	✓	✓	✓	✓	–
Tricholorethylene	100	100	✓	–	–	–	✓	✓	✓	–
Turpentine	100	30	✓	✓	✓	✓	✓	✓	×	–
Urea	50–100	375	–	–	✓	✓	–	–	–	–
Vineger	100	30	✓	✓	✓	✓	×	✓	×	–
Vinyl Chloride	100	30	✓	✓	✓	✓	✓	✓	–	–
Water	100	100	✓	✓	✓	✓	✓	✓	✓	–
Xylene	100	100	–	✓	✓	✓	✓	✓	✓	–
Zinc Annonium Chloride	0–40	100	–	–	✓	–	×	×	×	–
Zinc Chloride	0–100	30	–	–	✓	–	×	×	×	✓
Zinc Nitrate	10	30	×	–	✓	✓	–	–	–	–
Zinc Sulphate	20	30	–	–	✓	✓	–	✓	×	✓

Table 1.6. Selection of spring material under chemical products [6]

chemical compositions (%)

	Fe	C	Si	Mn	Al	Cu	Ni	Co	Cr	Mo	Ti	Others
Monel 400	$\leqq 2.5$	$\leqq 0.3$	$\leqq 0.5$	$\leqq 2.0$		28.0/ 34.0	$\leqq 63.0$					
Inconel 600	6.0/ 10.0	$\leqq 0.15$	$\leqq 0.5$	$\leqq 1.0$		$\leqq 0.5$	$\geqq 72.0$		14.0/ 17.0			
Inconel 625	$\leqq 5.0$	$\leqq 0.10$	$\leqq 0.5$	$\leqq 0.5$	$\leqq 0.4$		Balance	$\leqq 1.0$	20.0/ 23.0	8.0/ 10.0	$\leqq 0.4$	Nb + Ta 3.15/4.15
Incoloy 825	Balance	$\leqq 0.05$	$\leqq 0.5$	$\leqq 1.0$	$\leqq 0.2$	1.5/ 3.0	Ni (+Co) 38.0/46.0		19.5/ 23.5	2.5/ 3.5	1.6/ 1.2	
Stainless 302B25	Balance	$\leqq 0.12$	0.20/ 1.0	0.50/ 2.00	$\leqq 0.030$		8.0/ 11.0		17.0/ 19.0			
Stainless 316	Balance	$\leqq 0.07$	0.20/ 1.0	0.50/ 2.00	$\leqq 0.030$		10.0/ 13.0		16.5/ 18.5	2.25/ 3.00		

metallurgical structure shall be hard and fine tempered martensite without including any pearlite or bainite. The hardenability of steel increases with adding alloy elements except for cobalt and aluminum. It can be estimated by temperature-time-isothermal transformation diagram (TTT diagram or S curve), continuous cooling transformation diagram (CCT diagram), Jominy end-quench test etc., for each steel grade.
Since the as-quenched steel springs are brittle and easy to fracture, the tempering is normally applied after quenching. A plotted curve of mechanical property at room temperature, against tempering temperature is called as the tempering characteristic curve. Too low temperature tempering should be avoided as explained above. Alloying elements (such as silicon, chromium, molybdenum, etc.) can keep the quenched steel more resistant to softening as tempering. For these reasons, alloy steel is often used for thick gauge steel springs. Since the fatigue strength and relaxation resistance can be also affected by alloy elements of steel, the spring material selection becomes important. The recent studies related with the development of new spring steel materials for the weight-saving of automobiles can be introduced in Sect. 2.1.2 (10) and Sect. 2.1.3 (2)

(11) For the applications where electrical conductivity is required, copper alloys are suitable. Adding alloy elements or impurities to pure metal can decrease the electrical conductivity. As for piano wire and hard drawn wire, the electrical conductivity decreases at the initial stage of wire drawing, then it recovers up to total reduction of 60 to 75% of wire drawing reduction, and decreases again as increasing the reduction. For the applications where the electrical current is relatively small, the plated piano wire and hard drawn wire can be sometimes used.

Information Sources on Spring Materials

Information sources on spring materials are classified into documents sources (data-books) and databases searchable in computers.

The most substantial document sources may be books, magazines, reports, etc., published by the Japan Society of Spring Research, JSSR (now, the Japan Society of Spring Engineers, JSSE), which has been established around sixty years before.

The society has been publishing Transactions of Japan Society of Spring Research (annually), preprint textbooks of spring or autumn meetings (twice a year), various technical committee reports, bulletins of the Japan Society of Spring Engineers (monthly), etc. Many of these publications supply technical information on spring materials. In the vol. 2, spring technology series, "Kinds of springs and their applications" [3], spring materials conventionally used for various kinds of springs are briefly described. Elevated temperature heat-resistant spring material data book [5] and its second volume [7] contain data on spring materials to be used over the room temperature.

In monthly Japanese Journal of the society of materials science, Japan, original papers dealing with a wide range of materials' fatigue properties and lectures related with materials failures have been carried. From the Japan Society of Mechanical Engineers, data books on fatigue fracture of metallic materials were published [8, 9].

As described earlier, elastic moduli of cold-worked polycrystalline metallic materials show anisotropy because of texture formation. A technical book that collected anisotropic properties of metallic materials was published from the Japan society of mechanical engineers [10]. National Research Institute for Metals (now, National Institute for Materials Science) in Japan published two data-sheet books on elastic moduli and Poisson's ratios for carbon steels, low alloy steels, spring steels, tool steels and stainless steels [11, 12].

As an introductory textbook for steel materials, there is, for example, a Japanese book "Iron and steel materials" published by the Japan Metals Society [13]. As for heat-treatment and properties of steels, there is a detailed and old handbook, "Heat-Treatment of Steels, fifth edition (in Japanese)" [14]. Information on stainless steels can be obtained from a book "Stainless Steel Handbook, third edition (in Japanese)" [15]. As for phosphorus bronze alloys for spring applications, there is a Japanese book written by Nishihata and Harada [16].

As English documents, Metals Handbook series (now Materials Handbook series) would be valuable ones; especially, Vol. 1 for steels and high-performance alloys [17], Vol. 2 for non-ferrous alloys and special purpose alloys [18], and Vol. 13 for corrosion resistance alloys and corrosion [19]. In Japan, "Metals Data Handbook (in Japanese)" [20] describes various properties of metallic materials.

There are many books on plastics or polymers. As an example, an English book, Materials Sciense of Polymers for Engineers, [21] can be referred to.

In Chap. 27 of the book "Engineering Materials 2, An Introduction to Microstructures, Processing and Design", relation between part design and material choice is described [23] though not specific to springs. Similar handbook that dealt with material selection process for design and data sources which supply material information, was published from ASM International [24].

JIS handbooks in which the Japanese Industrial Standards (JIS) on spring materials and their inspections are contained, are listed in the Sect. 5.3, Chap. 5. In Sect. 5.2, the same chapter, comparison tables are included, in which each JIS standard related to spring materials is compared with the corresponding similar foreign standards.

At the copying service center of the Japan Science and Technology Agency, service of copying science and technical articles can be available with some expense. Details are explained in the Sect. 5.3, in the Chap. 5.

In North America, the Spring Manufacturers Institute (or SMI) has been publishing quarterly journals "SPRINGS" and books on spring technology. In the magazine "SPRINGS", articles on spring materials and spring design have been taken.

In Sheffield, England, there was an organization called the Spring Research and Manufacturers' Association (or SRAMA). From this organization, a data book "SPRING MATERIALS SELECTOR" was published [6]. SRAMA was later reorganized to the Institute of Spring Technology (or IST) and the UK Spring Manufactures Institute (or UKSMI).

New and revised ISO international standards on springs and springs related technologies are now taken into consideration by the TC committee of springs. The European Spring Federation (or ESF) has developed and implemented European standards (EN norms) on spring technologies. In Japan, the Japan Spring Manufacturers Association (JSMA) can be responsible for considering spring related ISO and JIS standards.

Computer Assisted Information Source on Spring Materials

Searching spring materials information utilizing either CD-ROM or DVD databases, and on-line WEB search can be accessible.

BANEX [25] is a database specifically designed for cold wound helical springs and materials, where the data of fatigue strength and permanent set are included. Although this database was edited by the JSSR, in 1992, the computer operation system is not compatible with the Microsoft Windows. However, the data can be usable.

IST in UK has published a database 'Spring Materials Selector CD-ROM' in collaboration with the European Spring Federation, ESF. CD-ROM version (in Japanese) of the Transactions of JSSR, No. 1 (1952) to No. 48 (2003), was published in 2003.

As for the databases for materials in general though not specific to spring technology, there are many databases available from website. Japan Science and Technology Agency, JST runs database service named "JST Online Information System" [26]. Also in Japan, National Institute of Informatics runs similar service [27]. Through links from these websites, materials fact databases supplied by some Japanese organizations [28–30] can be available. At the data free way site of the National Institute for Materials Science in Japan [28, 31], pages on links to materials information and search-engines for materials information are operated. Other major websites are referred to as references [26–30, 32, 35, 36].

National Institute of Informatics made a survey on how databases are widespread in Japan. The survey was made against national or public universities and research institutes. The result suggested that around 250 databases existed in the field of engineering and technology, but materials databases were scarce. Internationally, activity of CODATA (Committee on Data for Science and Technology) is worthy of special mention [32].

Among websites dealing with materials strength information, Japan Science and Technology Agency has developed a multi-functions-integrated basal database system [33], in which spring related information is contained as the

Table 1.7. Data and databases on materials strengths

Metallic materials strength database (JST)
Pressure vessels databases (JST & JPVRC)
Databook on Fatigue Strength of Metallic Materials, Vol. 1–3 (Society of Materials Science, Japan)
Database on metallic materials fatigue crack propagation resistance (Society of Materials Science, Japan)
Stress Intensity Factors Handbook (Society of Materials Science, Japan)
Stress corrosion cracking and corrosion fatigue strength data sheets for metallic materials (Society of Materials Science, Japan)
Materials strength data sheets: Fatigue data sheets (National Institute for Materials Science, Japan)
Materials strength data sheets: Creep data sheets (National Institute for Materials Science, Japan)
Reliability design data book for metallic materials (Society of Materials Science, Japan)

functional alloy database. Preferred method of utilizing this system can be shown in another article [34].

Outside Japan, Alloys-DB [35] from the Joint Research Centre (JRC)'s Institute of Advanced Materials (IAM) in Petten, Netherlands and the database of ASM International [36] are usable. Information from these organizations can also be obtained in CD-ROM.

Some of fact-data books and databases on materials strength open to public in Japan are listed in Table 1.7. The database system based on the fatigue data book published from the Society of Materials Science, Japan, can be available from the same society for a fee. A data book carrying data on materials strength was published from the Society of Materials Science, Japan [37]. A design data book for designing with reliability was developed by the Society of Materials Science, Japan [38], based on the Material Strength Databases for Reliability Design (MSDRD) developed by the material strength probability model research committee in Japan. A materials strength data book containing fatigue and creep data sheets, published from the National Institute for Materials Science are also available [39].

Basic Characteristics of Spring Materials

Basic characteristics of spring materials to be taken in consideration in material selection process, are; (1) static mechanical properties, especially tensile strength, elastic limit, spring deflection limit, hardness and elastic modulus, (2) dynamic properties, especially, fatigue strength (fatigue life at a constant stress amplitude, or fatigue endurance limit), (3) creep (progressive deformation of material at constant stress) or stress relaxation (time-dependent decrease in stress under constant constraint), that causes permanent set, and

(4) corrosion resistance, in addition, for springs in which electric current flows or for applications where magnetic disturbance is undesirable, (5) electrical conductivity or magnetic properties.

Besides these characteristics, the elastic modulus which can greatly affect to spring characteristics, is discussed here:

An elastic modulus of metallic material under constant temperature has been regarded as a microstructure insensitive constant decided only by chemical compositions. However, the demands for more precise mechanical evaluation of parts have been increasing to apply more precise elastic modulus.

It seems to be based on the above background that the testing method of elastic modulus has been newly designated as a JIS standard. Figure 1.3 shows test results on 2% tin(Sn)-phosphor bronze sheets by Nishihata et al. [16]. The figure shows the relation between the modulus of elasticity, E, both in the parallel and transverse direction to rolling and the reduction of cold rolling. It can be seen from the Fig. 1.3, that the difference of the elastic modulus between the two directions increases and the anisotropy develops as the degree of cold rolling becomes high. In spring wire, the elastic modulus can change with the degree of drawing and also the stress relieving after drawing. Figure 1.4, as an example, shows that the shear modulus (modulus of rigidity) of SUS304 stainless steel wire changes with the reduction of area by cold drawing and the stress relief [40]. The Fig. 1.4 says, that the shear modulus tends to decrease with the wire drawing, and recover by the following stress relief, to some extent.

It is considered that the preferred alignment of crystal orientation called texture formation and multiplication of dislocations (dislocation is a linear crystal defect and their movements promote plastic deformation) can contribute to the change of elastic modulus with cold working (it is known that in the iron and stainless steel single crystals the elastic modulus depends

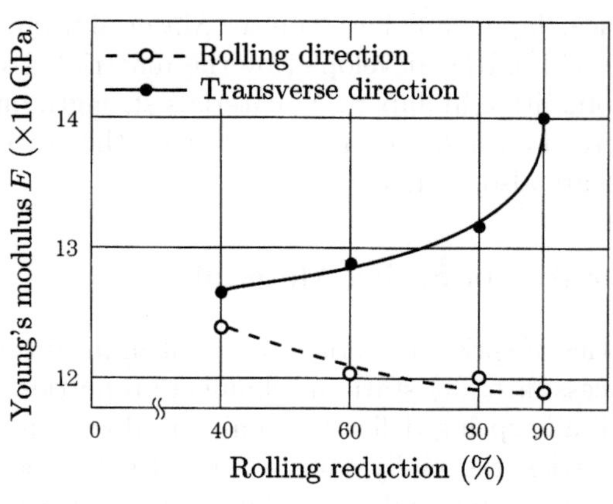

Fig. 1.3. Effect of rolling reduction on Young's modulus and its anisotropy for Sn 2% bearing phosphor bronze [16]

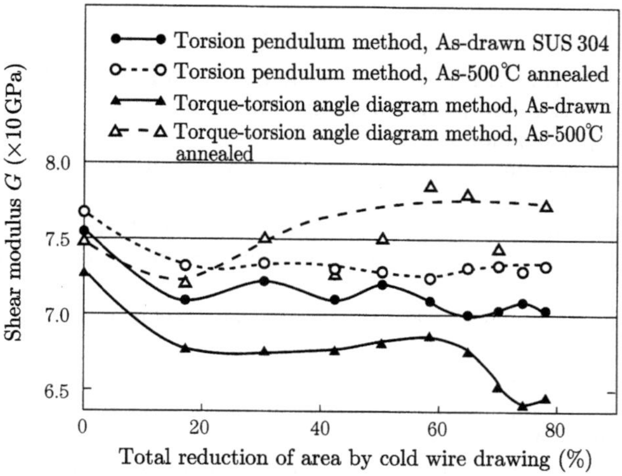

Fig. 1.4. Effect of reduction of area by wire drawing and 500°C annealing on shear modulus of SUS 304 stainless steel wire [40]

upon the crystal orientation). The modulus of elasticity (Young's modulus) and shear modulus are normally dependent upon the temperature and decrease with the temperature, except for some elinver type alloys. This point should be taken in consideration, when designing the springs to be used at the elevated temperature or at the sub-zero temperature.

Upon unloading shape memory alloy after loading to such a large strain seemingly in plastic strain region, there is a case where the spring recovers to the original shape. This is called super elasticity or pseudo elasticity. More details are described in the section of Ni–Ti shape memory alloys.

In composite materials such as fiber reinforced plastics (FRP), the elastic modulus can be determined by both the elastic modulus of matrix and fibers, and their volume or mass ratio. When all fibers are aligned parallel in one direction, it must be noted that the composite shows pronounced anisotropy in elastic and other mechanical properties.

When loading metallic materials within elastic limit, although the relationship between stress and strain is linear, more precise measurements may indicate that the slope of the stress divided by strain gradually decreases with strain. This deviation from the linearity can be caused by time-dependent phenomenon called as anelasticity or viscosity, and the stress relaxation and creep can be also the phenomena related to it. As for plastics materials, it must be also noted that the plastic materials show far pronounced viscous flow when loaded, compared with metallic materials.

References

1. M. Kodama, H. Sugisaka, S. Iwasaki and S. Watanabe: JIS How to use series, Spring utilization and points in design. July 1985, Japanese Standards Association, p. 1 (in Japanese)
2. A. Ono, H. Koyama, A. Tange: Springs, third edition, Dec., 1982, Maruzen, p. 324 (in Japanese)
3. Edited by the Japan Society of Spring Research: Vol. 2, Spring technology series, Kinds of springs and their applications, Sept., 1998, Nikkan Kogyo Simbun (in Japanese)
4. S. Kobayashi: Japanese Society for Artificial Intelligence, Vol. 7, No. 4, July 1992, p. 560 (in Japanese)
5. Edited by the Heat resistant spring materials survey committee, JSSR: Data book on high temperature strength of heat resistant materials, Japan Spring Manufacturers Association (in Japanese)
6. Spring Materials Selector, second edition: Institute of Spring Technology
7. Edited by the Heat resistant spring materials survey committee, JSSR: Data book on high temperature strength of heat resistant materials, second volume, Japan Spring Manufacturers Association (in Japanese)
8. Edited by the Japan Society of Mechanical Engineers: Data book of designing fatigue strength, Vol. I, revised edition, Feb., 1982 (in Japanese)
9. Ibid, Vol. II (in Japanese)
10. Edited by the Japan Society of Mechanical Engineers: Elastic coefficient of Metallic Materials, Japan Society of Mechanical Engineers, Oct., 1980 (in Japanese)
11. Data book of elastic coefficient of steels, Fatigue data book No. 83, National Research Institute for Metals, 1994 (in Japanese)
12. T. Abe, S. Nishijima and K. Kanazawa: Fatigue data sheets No. 15, data sheets of elastic coefficients of steels, National Research Institute for Metals, 1997
13. Lecture of current metallurgy, material parts, Vol. 4, Iron and Steel materials, Japan Society of Metals (in Japanese)
14. Edited by the Iron and Steel Institute of Japan: Heat-treatment of steels, revised fifth edition, Maruzen, Oct., 1969 (in Japanese)
15. Edited by the Japan Stainless Steel Association: Stainless Steel Handbook third edition, Nikkan Kogyo Shimbun, Jan., 1995 (in Japanese)
16. M. Nishihata and H. Harada: Basics and applications of phosphorus bronze, Chap. 6, Jan., 1997
17. Edited by ASM International: Metals Handbook, Vol. 1, Properties and Selection: Irons, Steels and High-Performance Alloys, 1990
18. Edited by ASM International: Metals Handbook, Vol. 2, Properties and Selection: Nonferrous Alloys and Special-Purpose Materials, 1990
19. Edited by ASM International: Metals Handbook, Vol. 13: Corrosion, 1987
20. Edited by the Japan Institute of Metals: Metals Data Book, Third revised edition, Maruzen, March, 1993 (in Japanese)
21. Tim A. Osswald and H. Georg L. Menges: Materials Science of Polimers for Engineer, Carl Hanser Verlag, 1996
22. I. Narusawa: Plastics Materials Strength Series Volume I, Mechanical Properties of Plastics, Sigma Publishing, April, 1994 (in Japanese)
23. M. F. Ashby and D. R. H. Jones: ENGINEERING MATERIALS 2, An introduction to Microstructures, Processing and Design, Pergamon Press, 1986

24. G. E. Dieter et al.: ASM Handbook Vol. 20, Materials Selection and Design, ASM International, Dec., 1997
25. Edited by the Japan Society of Spring Research: BANEX Illustration Book, Japan Spring Manufacturers Association 1991 (in Japanese)
26. Japan Science and Technology Agency: JST Online Information System
27. National Institute of Informatics: http://www.niiac.ac.jp/index-j.html
28. National Institute for Materials Science (NIMS): http://www.nims.go.jp/eng/index.html
29. National Institute of Advanced Industrial Science and Technology (AIST): http://www.aistgo.jp/RIODB/hm/Material.html
30. Institute for Materials Research, Tohoku University: http://www-db1.imr.tohoku.ac.jp/database/
31. M. Fujita: A Trial on Fact Database for Materials, Materia Japan, Vol. 38, No. 1, 1999, p. 24 (in Japanese)
32. CODATA: http://www.cisti.nrc.ca/codata/
33. M. Obara: New Databases for Materials Research – From Microscopic to Macroscopic Model, Materia Japan, Vol. 37, No. 6, 1998, p. 449 (in Japanese)
34. K. Iizima, K. Nagahashi, H. Ishiguro and H. Kaizuka: Functional Alloys Database – WWW-based full text retrieval and browsing system –, Journal of Information Processing and Management, Vol. 39, 8, 1996, p. 599 (in Japanese)
35. JRC Petten Materials Data Bank: http://mnemonyse.jrc.nl
36. ASM International: http://www.asm.intl.org/
37. Edited by the Society of Materials Science, Japan: Data Book on Fatigue Strength of Metallic Materials (Vol. 1–3, in Japanese)
38. Edited by the Society of Materials Science, Japan: Reliability design data book for metallic materials, 1982 (in Japanese)
39. Materials strength data sheet book, National Research Institute for Metals (in Japanese)
40. Y. Kawabata, T. Wakamiya, Y. Yamaoka and S. Yamashita: Effect of degree of wire drawing on elastic moduli of spring wires, Transactions of Japan Society of Spring Research, Vol. 23, 1978, p. 1 (in Japanese)

1.2.3 General Considerations of Spring Durability

The damage of spring durability means that springs cannot maintain the initial characteristics in use. As for the causes of spring durability damage, the environments where springs are exposed in use are to be considered. The environments can be expressed in terms of mechanical condition (mechanical environment) and atmospheric condition (atmospheric environment) where springs are exposed. The former can be further divided into the static and dynamic, according to the exerted force. The atmosphere is comprised of the corrosion and temperature. As shown in Fig. 1.5, according to the combined environments where springs are exposed, various problems which can damage the spring can arise.

In this section, from the described point of view, fundamental considerations of fatigue, creep, stress relaxation and stress corrosion cracking are described.

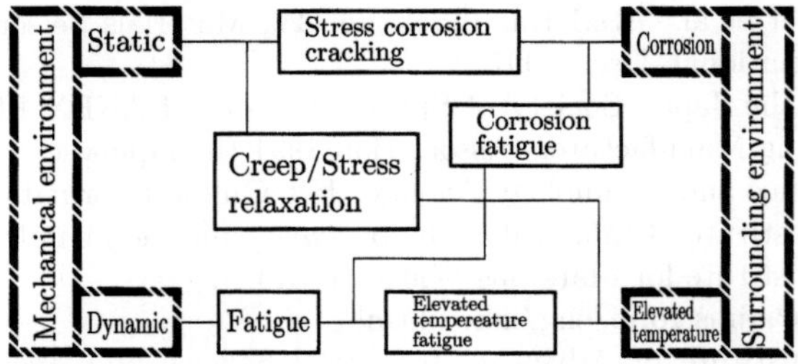

Fig. 1.5. Mechanical and environmental conditions related to spring durability

Fatigue Characteristics

(1) S-N curve (High cycle fatigue)

Fatigue characteristics of material can be expressed by the relation between the magnitude of repeated stress or stress amplitudes (S) and the number of cycles to failure (N), so-called S-N curve (S-N diagram). In general, the lower the stress is, the longer the number of cycles to failure becomes. Steel and some other metals or alloys show upper limit stress where the fracture never happens with prolonged stress repetitions, normally over 10^7 cycles. This stress is called as fatigue limit (endurance limit). The stress corresponding to a certain number of cycles is called fatigue strength at that number of cycles. The fatigue strength at 10^7 cycles or at 10^8 cycles can be also called as fatigue limit. The fatigue fractures below 10^4 cycles can be called as low cycle fatigue fracture, which can be distinguished from high cycle fatigue fracture over 10^4 cycles.

The fatigue life under S-N curve normally shows many scattering, which can be shown in Fig. 1.6. The Fig. 1.6 shows the results of quenched and tempered steel on the rotating bend fatigue tests, in the form of P-S-N curves taking fracture probability, P, as one parameter [41]. In many cases, samples of more than ten pieces are fatigue tested in total at different stresses to have a S-N curve. The mean line at each tested stress can be said as the S-N curve at 50% fracture probability.

(2) Mechanism of fatigue fracture

The fatigue fracture is not the phenomenon which occurs by one loading cycle. The fatigue starts by the generation of a small crack, which grows and propagates gradually to final fracture. Fatigue crack can normally initiate by the repetition of slip deformation around the surface. The slip band which occurs at the early stage of stress cycles grows gradually to the persistent slip band which is observable microscopically even on polished surface. These persistent slip bands soon form rough surface with extrusions and intrusions, where the stress can be concentrated to generate micro-cracks [42], see Fig. 1.7. Figure 1.8 [43] can show a model for the propagation of a fatigue crack, where a crack is repeatedly opened and closed under normal stress in the perpendicular direction to the crack surface, and propagate gradually. In fatigue fracture

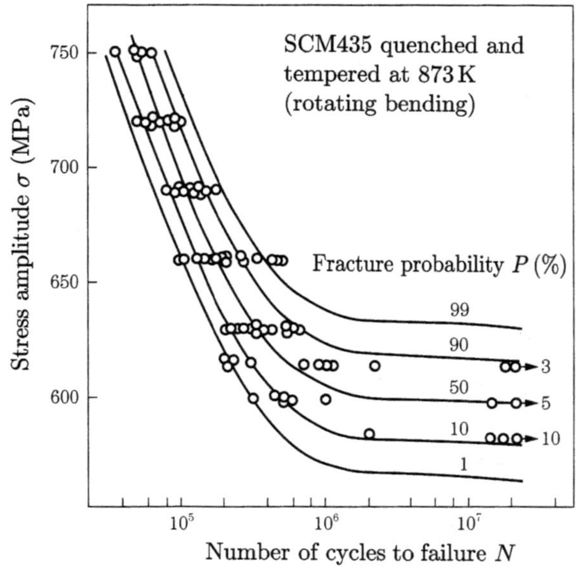

Fig. 1.6. An example of S-N curves (P-S-N curves)

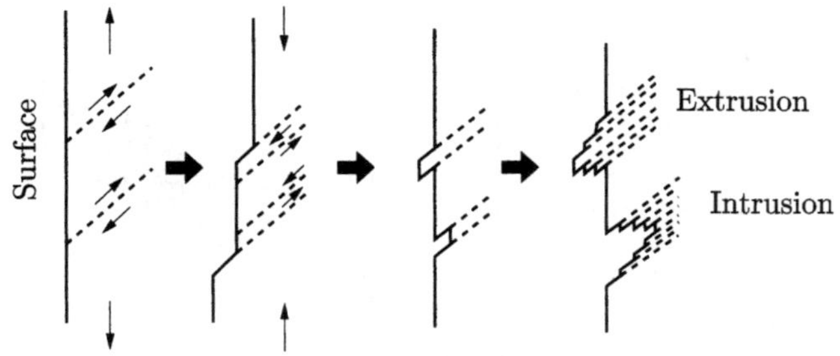

Fig. 1.7. A crack initiation model [42]

surface generated by such a process, striped pattern called striations can be sometimes observed.

Thus, it can be said, in fatigue, that the generation of cracks can be made by shear stress and its propagation can be made by normal stress. The cracks generated along slip bands are called Stage I crack, and cracks propagated in the perpendicular direction to normal stress are called Stage II cracks [44].

(3) Factors on fatigue strength

The fatigue strength is affected by many factors, which can be classified as (i) material, (ii) shape, (iii) stress or (iv) atmosphere. It should be noted that these factors affect fatigue properties not independently but interdependently.

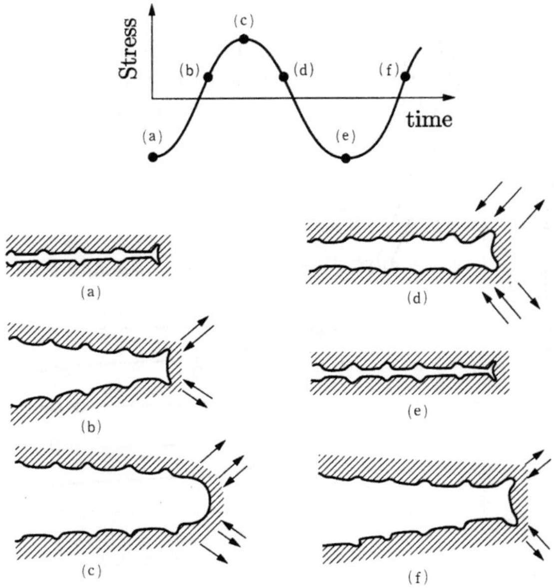

Fig. 1.8. A model of fatigue crack propagation [43]

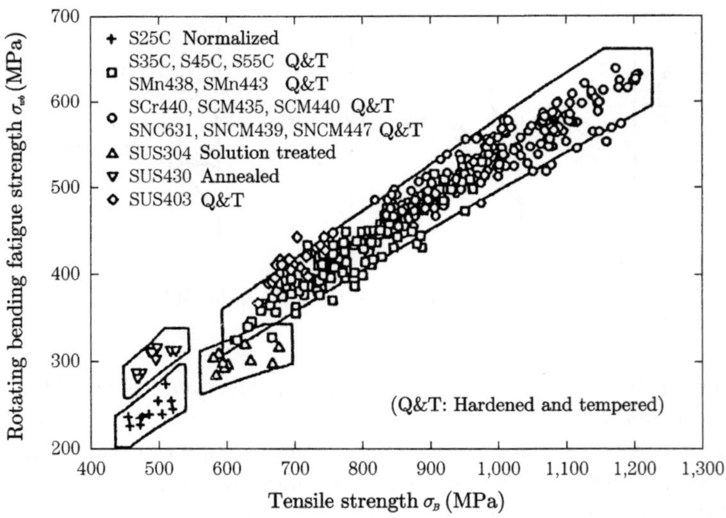

Fig. 1.9. Relationship between rotating bending fatigue strength and tensile strength [45]

(A) Material and heat treatment

Figure 1.9 shows the relations between the rotating bending fatigue strength and the tensile strength for various grades of heat-treated steel [45]. The figure indicates that fatigue strength is proportional to tensile strength as a whole. However, for steel with the tensile strength of more than 1,500 MPa, such as spring steel and tool steel, fatigue strength does not always increase with tensile strength, but rather tends to show scattering. It can be understood that the defects outside and inside the steel such as non-metallic inclusions tend to become sensitive to fatigue failure when the tensile strength exceeds around 1,500 MPa [46].

(B) Notch effect and size effect

Components or fatigue testing specimens with notches have lower fatigue strength than those with no notches, due to the stress concentration. The ratio of the fatigue strength of unnotched specimens, to the corresponding fatigue strength of notched specimens is known as the fatigue strength reduction factor or the fatigue notch factor, K_f.

$$K_f = \frac{\text{fatigue strength of un-notched specimen}}{\text{fatigue strength of notched specimen}} \tag{1.1}$$

While the stress concentration factor K_t is a function of shape and dimensions of a component, the fatigue strength reduction factor K_f can be decided by the strength of material and stress-strain relation at the notched root. Therefore, the K_f does not coincide with the K_t. The K_f is usually smaller than K_t, although the K_f comes closer to the K_t for the material with high sensitivity to notches. The notch sensitivity index q expressed by the equation (1.2), is widely used to represent fatigue strength sensitivity of notched specimens or components.

$$q = \frac{K_f - 1}{K_t - 1} \quad 0 \leqq q \leqq 1 \tag{1.2}$$

The notch sensitivity generally becomes higher with the higher tensile strength (higher hardness). Figure 1.10 shows the relation between fatigue strength and K_t [47], where σ_{w0} is the fatigue strength for plain specimens (without notches, $K_t = 1$), and σ_{w1} is the minimum stress where the fatigue crack is generated and σ_{w2} is the stress where the final fracture can occur, both in the corresponding notched specimens. As the notch becomes sharper, the fatigue crack can be easily generated. The cracks generated sometimes stay as non-propagating cracks when the stress is below a critical level.

Fatigue strength tends to be reduced with increasing the test specimen sizes. This is known as the size effect. As the causes of size effect, it can be considered that the homogeneity of materials becomes worse as the size increases, that the volume exposed under high stress increases, as the size increases, and that the increased volume exposed under high stress raises the possibility where the crack initiating defects can be included.

(C) Type of applied stress

As for the types of applied stress in fatigue testing, besides the rotating bending stress, there are the uni-axial stress, bending stress, torsional stress etc. Both the S-N curve and the fatigue strength vary according to the loading

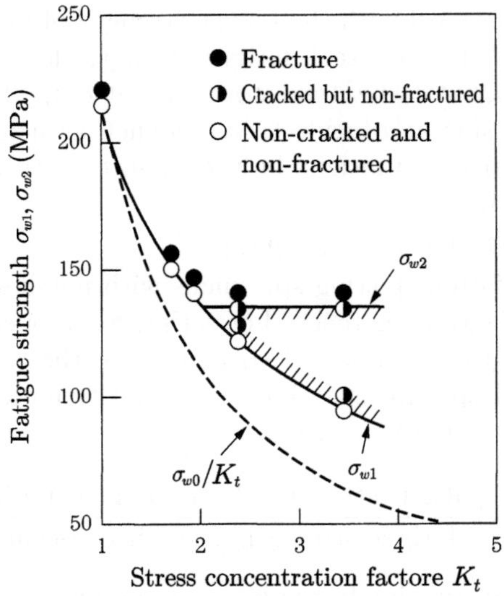

Fig. 1.10. Dependence of fatigue strength on stress concentration factor

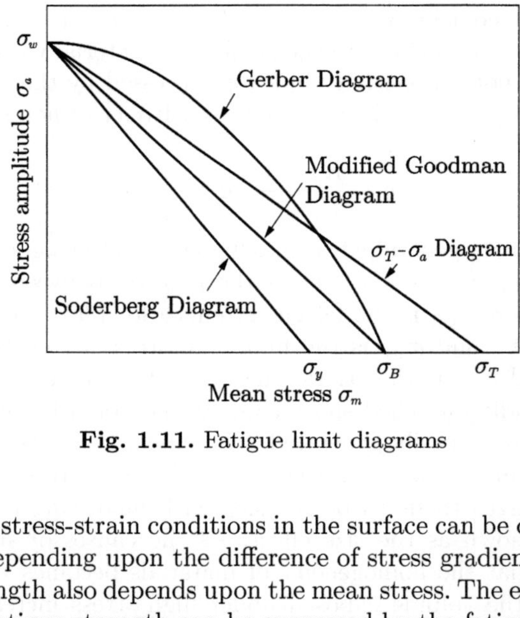

Fig. 1.11. Fatigue limit diagrams

types, since the stress-strain conditions in the surface can be different for each loading type, depending upon the difference of stress gradient etc.

Fatigue strength also depends upon the mean stress. The effect of the mean stress, σ_m, on fatigue strength can be expressed by the fatigue limit diagram where the mean stress is on the horizontal axis and the stress amplitudes (fatigue strength) are on the vertical axis, as shown in Fig. 1.11.

Following equations obtained experimentally, have been proposed.

$$\sigma_a = \sigma_w \left(1 - \frac{\sigma_m}{\sigma_B}\right) \qquad \text{Modified Goodman Diagram} \qquad (1.3)$$

$$\sigma_a = \sigma_w \left\{ 1 - \left(\frac{\sigma_m}{\sigma_B} \right)^2 \right\} \quad \text{Gerber Diagram} \tag{1.4}$$

$$\sigma_a = \sigma_w \left(1 - \frac{\sigma_m}{\sigma_y} \right) \quad \text{Soderberg Diagram} \tag{1.5}$$

$$\sigma_a = \sigma_w \left(1 - \frac{\sigma_m}{\sigma_T} \right) \quad \sigma_T\text{–}\sigma_a \text{ Diagram,} \tag{1.6}$$

where, σ_B is the tensile strength, σ_y is the yield strength, σ_T is the true fracture strength, and σ_w is the fatigue strength under completely reversed stresses.

The effect of compressive residual stress in the surface layer, which can be obtained by carburizing, nitriding, shot-peening etc., can be said to be similar with that of reducing the mean stress, to improve fatigue characteristics.

Figure 1.12 shows the effect of surface residual stress on fatigue strength of notched S35C (AISI 1035) steel specimens with $K_t = 3$ [48]. It can be seen from the Fig. 1.12 that although the fatigue strength σ_{w2} increases with the compressive residual stress, the strength σ_{w1}, where the crack is generated, is not affected by the compressive residual stress. Therefore, it can be said that the compressive residual stress in the surface layer does not affect the crack initiation, but can effectively suppress the propagation of cracks, to increase the fatigue strength σ_{w2}. It can be also understood that the fatigue crack generation depends upon the magnitude of repeated shear stress and the fatigue crack propagation depends on the magnitude of normal stress which brings about the opening and closing of cracks.

When the reversed torsion stress and normal stress (tensile and compressive, or bending) are combined, the fatigue strength shall be expressed in terms of both normal stress and shear (torsion) stress cpmponent. Figure 1.13 gives such an example of SNCM439 (AISI4340) steel under combined stress conditions. Figure 1.13 also shows that the fatigue strength in torsion decreases with the combined reversed normal stress. The curves based on the maximum

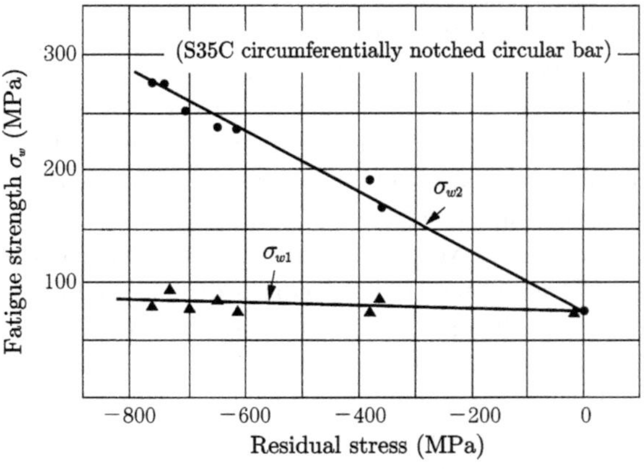

Fig. 1.12. Effect of compressive residual stress on fatigue strength [48]

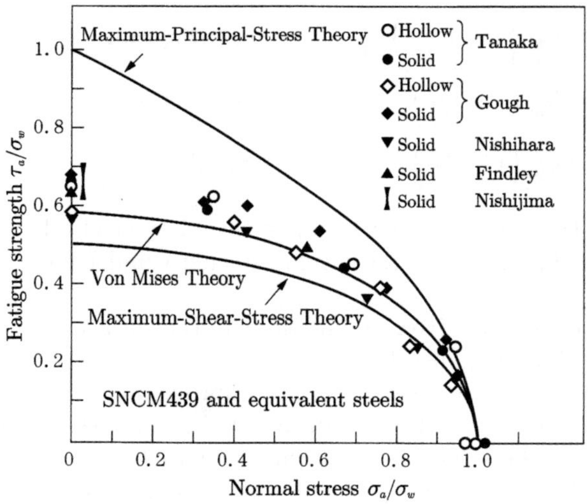

Fig. 1.13. Fatigue strength under combined stress [9]

shear strain (or distortion) energy criterion by Von Mises, maximum-shear-stress criterion by Tresca, and maximum principal stress criterion can be seen in the Fig. 1.13. These curves are based on the usually accepted criteria for predicting the onset of plastic deformation under a static combined stress loading, the static yield stress being replaced by the corresponding limiting cyclic stress amplitudes. The curve by Von Mises seems to have reasonable agreements with the actual data although the curve shows slightly safer results than the actual data.

The equivalent stress σ_{eq} can be given by the following equation (1.7).

$$\sigma_{\mathrm{eq}} = \sqrt{3\tau_a^2 + \sigma_a^2} \tag{1.7}$$

(D) Atmospheric environment

Fatigue strength of metallic materials in aqueous solutions depends on the condition of the aqueous solution. The fatigue strength sometimes decreases drastically in aqueous solution than in air. Figure 1.14 shows S-N diagrams for carbon steel, tested in clean water, in 1–10% salt water and in air, [50]. Under higher stress conditions, the effect of environment is rather small. However, under low stress conditions, the fatigue life in salt water is reduced significantly compared with that in air. No fatigue limit can be realized under the fatigue tests in salt water. It is known that the corrosion fatigue life can depend upon the speed of stress repetition. Many scatterings in the Fig. 1.14 can be partly due to the different speed of stress repetition.

The temperature is also an essential factor for the fatigue strength. Figure 1.15 shows the temperature dependence of rotating bending fatigue strength for various steel grades [51]. Generally speaking, the fatigue strength decreases with the rise of temperature. However, in carbon steels and low alloy

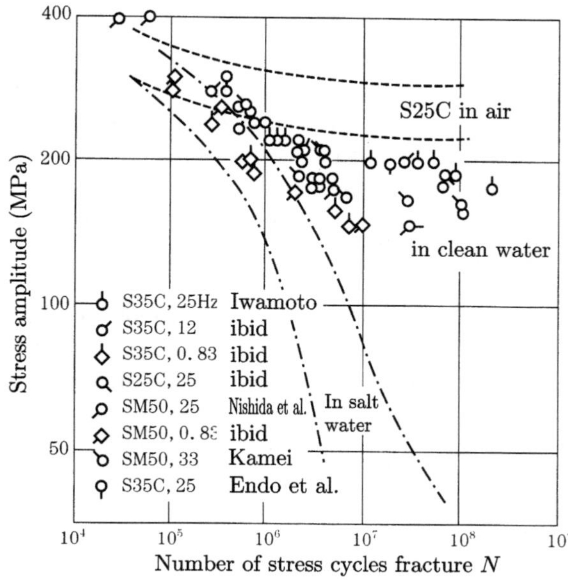

Fig. 1.14. S-N curves of carbon steels in corrosion environments [50]

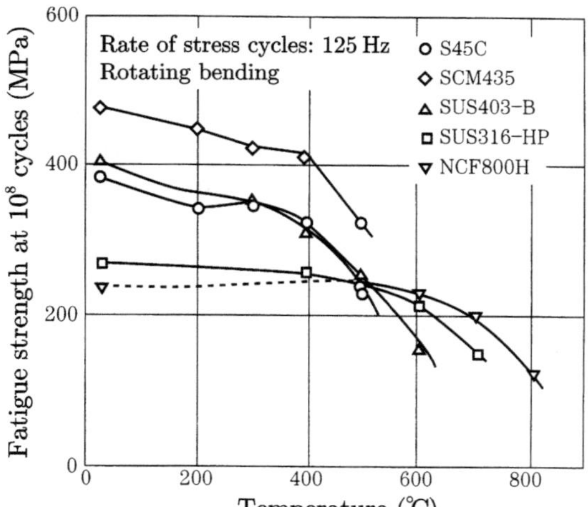

Fig. 1.15. Temperature dependence of fatigue strength [51]

steels, a peak of fatigue strength appears around 573 K (300°C), due to the blue brittleness. The relation between fatigue strength and tensile strength at each elevated temperature, can be in good correlation, as shown in Fig. 1.9 at room temperature. The ratio of fatigue strength to tensile strength at elevated temperature can be said to be around 0.5. The fatigue strength and fatigue life at elevated temperature also depend upon the speed of stress repetition.

(4) Low cycle fatigue

The fatigue fractures that take place below 10^4 cycles, are called as low cycle fatigue. This kind of fracture occurs when the component of machines or structures has been loaded over its yield strength. For example, in case of the structural material which can not be overloaded in normal use, but be possibly overloaded in earthquake, and the parts of instruments in high temperature use where the temperature fluctuation from the room temperature can be repeated when turning on and off, the low cycle fatigue becomes important.

Low cycle fatigue characteristics in metallic materials can be normally obtained under completely reversed stress conditions with the strain (deflection) controlled testing. As a typical example, the stress-strain behavior can be shown in Fig. 1.16. In this figure, ε_{ta} is total strain amplitude, ε_{pa} is plastic strain amplitude, ε_{ea} is elastic strain amplitude, and σ_a is stress amplitude. The relations between these strains and stresses can be expressed by

$$\varepsilon_{ta} = \varepsilon_{pa} + \varepsilon_{ea} = \varepsilon_{pa} + \frac{\sigma_a}{E} \tag{1.8}$$

where, E: Young's modulus.

The repeated strains including the plastic strain make the material hardened or softened. The relations between the total strain amplitude and the stress amplitude, say repeated stress-strain curve, is the characteristic which can be applied for strain analysis at the root of a notch. As shown in Fig. 1.17, the repeated stress-strain curve of material hardened with strain (age-hardening type aluminum alloy, 2024-T4) stays at higher stress side than the stress-strain curve obtained by tensile test. To the contrary, for the material softened by repeated strain (super high-strength steel, SAE4340, in Fig. 1.17), this relations become reversed [52].

In case of low cycle fatigue test, the relations between the strain amplitude and the number of cycles to fracture, N_f can be expressed by logarithmic coordinates (Fig. 1.18). The relations between ε_{pa} and N_f, and between ε_{ea}

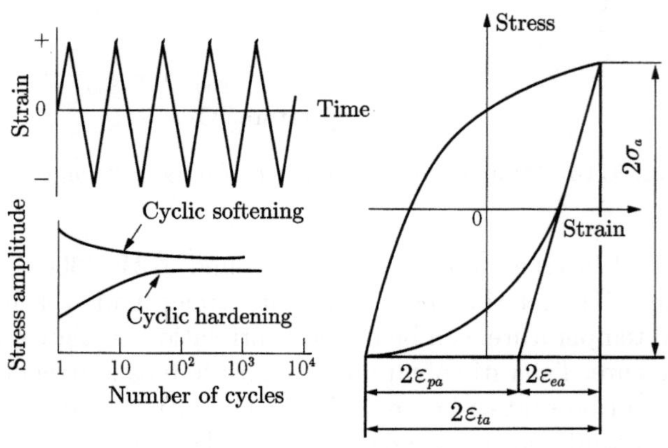

Fig. 1.16. Stress-strain behavior in low cycle fatigue

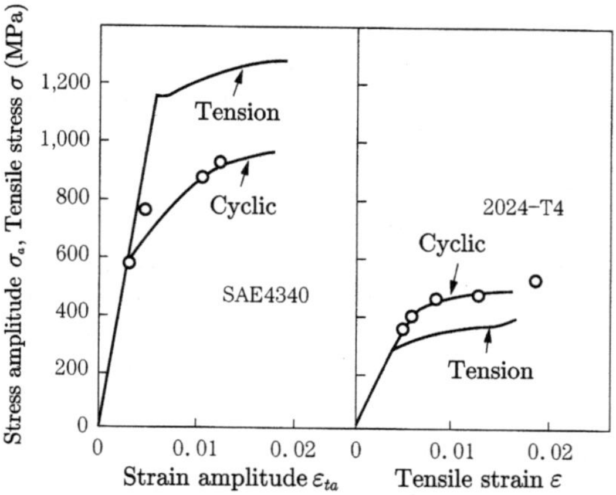

Fig. 1.17. Cyclic stress-strain curves [52]

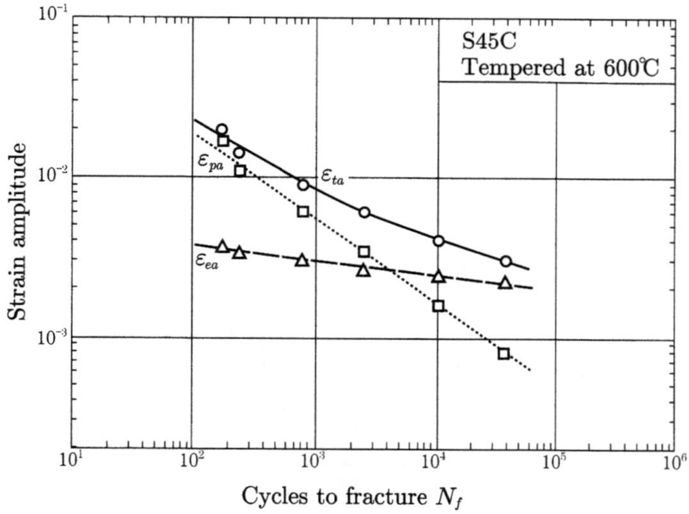

Fig. 1.18. Low cycle fatigue curves

and N_f appear to be linear, as expressed by the following equations.

$$\varepsilon_{pa} N_f^{k_p} = C_p \tag{1.9}$$

$$\varepsilon_{ea} N_f^{k_e} = C_e \tag{1.10}$$

The relations between ε_{pa} and N_f or equation (1.9) is called Coffin-Manson's relation. From the above equations, the relations between ε_{ta} and N_f can be written as

$$\begin{aligned} \varepsilon_{ta} &= \varepsilon_{pa} + \varepsilon_{ea} \\ &= C_p N_f^{-k_p} + C_e N_f^{-k_e} \end{aligned} \tag{1.11}$$

where, C_p, C_e, k_p and k_e are material constants, and the values of k_p and k_e are normally around 0.12 and 0.6 respectively.

The low cycle fatigue life at elevated temperature depends on testing temperature, strain rate, strain waveform, etc.

(5) Fatigue crack growth rate

Fatigue crack growth characteristics are normally expressed by the relation between stress intensity factor range ΔK $\{= A\Delta\sigma\sqrt{\pi a}\}$ and crack growth rate da/dn, where a is crack length, $\Delta\sigma$ $(= 2\sigma_a)$ is stress range applied, and A is a shape factor determined by the shape of test piece including the crack shape. Figure 1.19 shows some examples. With decreasing the ΔK, the crack growth rate also decreases drastically. It can be finally found that below a critical lower limit of stress intensity factor range ΔK_{th}, the fatigue crack does not propagate. The relation between the ΔK and da/dn can be shown to be linear in log-log plot, see equation (1.12), except for the stress intensity range near ΔK_{th}. This relation is called as Paris' law.

$$\frac{da}{dn} = C\Delta K^m \tag{1.12}$$

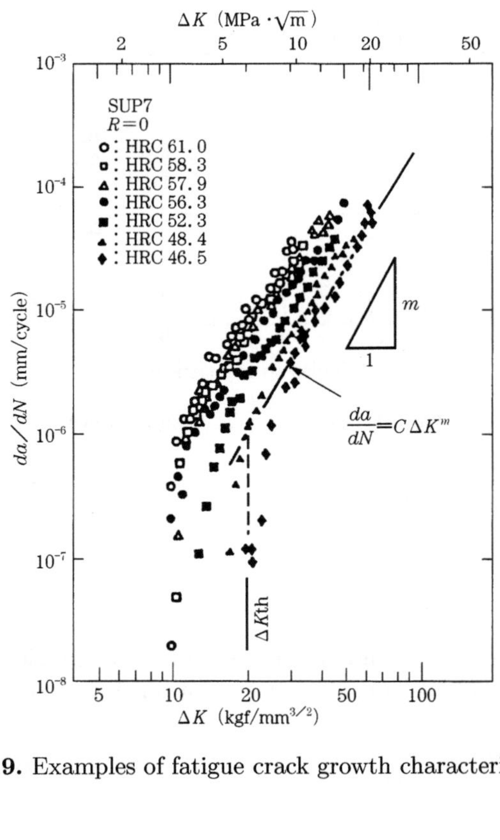

Fig. 1.19. Examples of fatigue crack growth characteristics [53]

(6) Estimation of fatigue strength and fatigue life

(A) Relation with mechanical properties

When high cycle fatigue fracture is a problem in designing machines and structures, fatigue limit (fatigue strength) becomes more important than fatigue life.

In order to estimate the fatigue strength of material without the actual fatigue testing, the figures shown in the Fig. 1.9 for steel, as an example, can be one solution. There is a good correlation between the hardness and fatigue strength.

On the other hand, for the machines and structures where low cycle fatigue fracture is a problem, it becomes important to estimate the fatigue life under the expected strain and cycles of those products within their commercially using period.

There are following equations to estimate low cycle fatigue life from the mechanical properties.

$$\text{Langer's equation: } \varDelta\varepsilon_t = \frac{\varepsilon_f}{2\sqrt{N_f}} + \frac{2\sigma_w}{E} \tag{1.13}$$

Manson's method of universal slopes

$$\varDelta\varepsilon_t = 3.5\left(\frac{\sigma_B}{E}\right)N_f^{-0.12} + \varepsilon_f^{0.6}N_f^{-0.6} \tag{1.14}$$

where, $\varDelta\varepsilon_t$ is total strain range $(= 2\varepsilon_{ta})$, $\varepsilon_f = \ln\{1/(1-\phi)\}$, ϕ is the reduction of area in tensile tests, E is Young's modulus, σ_w is completely reversed fatigue strength, and σ_B is tensile strength.

(B) Fatigue life under varying stress

Although the fatigue test can be normally carried out under constant repeated stress amplitudes, most structural members and machine components such as springs etc., in use, are not necessarily applied to constant stress conditions, and mostly applied to the varying stress conditions. The fatigue life of a component subjected to varying stress cycles, can be estimated by the linear cumulative damage rule, also known as Miner's rule. Assuming the fatigue damage $D = n_i/N_i$, where the n_i is number of cycles under the stress amplitude, σ_{ai}, and N_i is number of cycles to fracture under the σ_{ai}, in a S-N diagram, the total life of a specimen subjected to varying stress cycles can be estimated by adding up the damage accumulated by each stressing $D = \sum(n_i/N_i)$ until reaching to $D = 1$. Since in actual use, the stress below the fatigue limit can be frequently applied, the slope of S-N curve is extrapolated to longer life side and this modified S-N curve may be used for the estimation of fatigue life based on the Miner's rule. This method is called the modified Miner's rule. Also in designing, the D can be sometimes taken as 0.2–0.3, considering the safety factor.

Further, in applying the Miner's rule for complicated varying stress cycles, stress frequency distribution should be calculated at first. As methods to read out stress waveforms, there are the range-pair method, the rain-flow

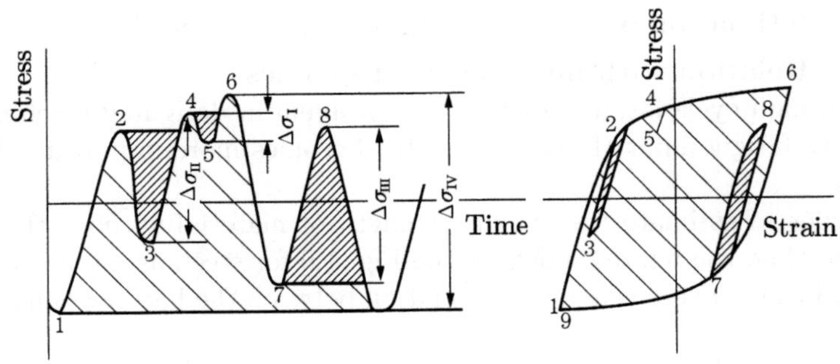

Fig. 1.20. Range-Pair counting method

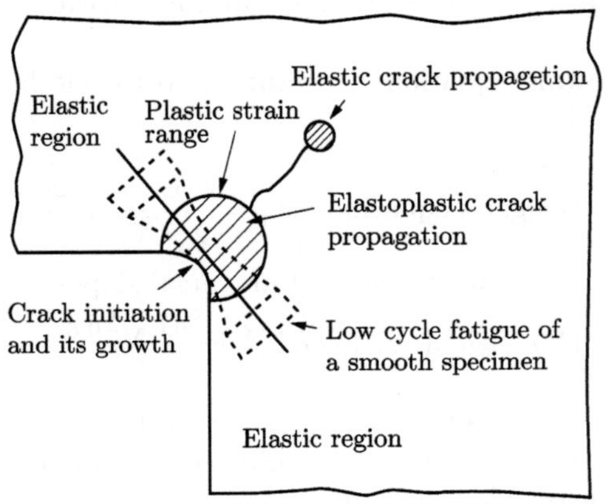

Fig. 1.21. Fatigue of a structural member at the notch root

method, etc. The range-pair method is a method to make pairs of upstream and downstream stress range or strain range with equal magnitude. As shown in Fig. 1.20, this method counts each stress-strain hysteresis loop consecutively, and is reputed to be the most rational counting method in analyzing fatigue failures.

(C) Crack initiation life and crack growth life

Fatigue fracture of metal progresses through the process of crack initiation and its growth (propagation). When the crack growth process of machine components or structure members can be checked at periodical inspections, the crack initiation may be allowed. Of course, there can be many structures where any crack initiation is never permitted.

The methods of estimating crack initiation life from a notch root and growth life are described. Figure 1.21 gives a schematic illustration of fatigue process starting from a notch root of a component. Even if repeated stress might be within elastic stress range in nominal stress, it can be sometimes realized that a plastic zone due to stress concentration can be actually formed around the notch root, where plastic strain occurs repeatedly. Then, the fa-

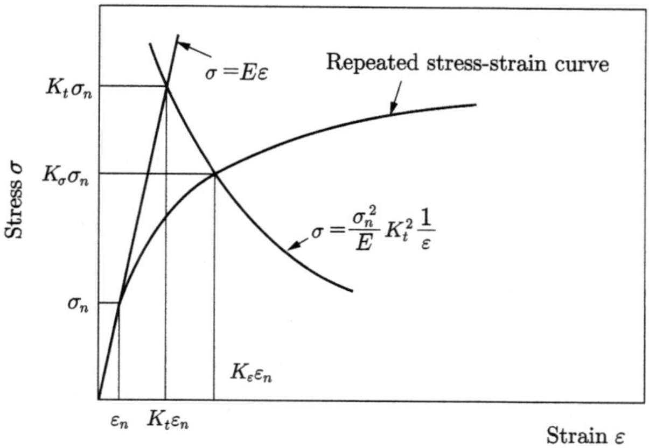

Fig. 1.22. Stress and strain estimation based on Neuber's rule

tigue crack initiation life at the notch root can be regarded as the low cycle fatigue life of plain specimens with no notches. However, at this time, the actual stress σ, and strain ε, at the notch root should be estimated from the elastic stress concentration factor K_t, nominal stress σ_n and nominal strain ε_n $(= \sigma_n/E)$ of the notched specimen. The stress and strain estimation method, based on Neuber's rule, is given schematically in Fig. 1.22.

Although stress concentration factor under plastic strain condition K_σ is smaller than K_t, strain concentration factor K_ε is larger than K_t. Among these factors, there is a following relation based on Neuber's rule.

$$K_t^2 = K_\sigma K_\varepsilon \tag{1.15}$$

Under the condition of nominal stress σ_n and nominal strain ε_n, the relationship between σ and ε is expressed by the following hyperbolic equation.

$$\sigma = \frac{\sigma_n^2}{E}\frac{K_t^2}{\varepsilon} \tag{1.16}$$

Since the stress-strain relationship under repeated strain is under repeated stress-strain relationship, the actual stress and strain are obtained as the crossing point of the equation (1.16) and the repeated stress-strain curve, as shown in Fig. 1.22. Fatigue life on the low cycle fatigue curve corresponding to the strain obtained by the method described above, can be regarded as the estimated crack initiation life.

Crack propagation behavior after a fatigue crack is found at a periodical inspection can be estimated in terms of fatigue crack growth (propagation) characteristics.

When a crack length is a_0 and the acting stress range anticipated is $\Delta\sigma$, the stress intensity factor range ΔK_0 can be expressed as follows.

$$\Delta K_0 = A\Delta\sigma\sqrt{\pi a_0} \tag{1.17}$$

If ΔK_0 is smaller than ΔK_{th}, the crack does not propagate and remains as a non-propagating crack. However, when ΔK_0 is larger than ΔK_{th}, the crack continues to propagate. Assuming that the crack propagation characteristic follows the Paris' law, the crack length increment per certain cycles Δn, or the stress cycle number needed to reach to certain crack length, can be obtained by integrating the equation (1.12) using the equation (1.17).

Assuming that the constant A in the equation (1.17) is not dependent upon crack length and m is not equal to 2, following equation can be obtained.

$$\Delta n = \frac{2}{m-2} \frac{a_0^{1-m/2} - a^{1-m/2}}{CA^m \pi^{m/2} (\Delta\sigma)^m}, \tag{1.18}$$

where, a_0 is the initial crack length, and a is the crack length after stress cycles Δn.

Creep and Stress Relaxation

The gradual deformation of metallic material that is kept under the constant load smaller than its tensile strength is known as creep. When material is sustained under constant strain, the initially introduced stress gradually decreases with the elapse of time. This characteristic is also known as stress relaxation. Schematic illustration of creep and stress relaxation is shown in Fig. 1.23. The characteristics of both creep and stress relaxation are sensitive to temperature and accelerated with increasing temperature.

(1) Creep

Creep characteristics have been studied with plain round bar specimens, applying them constant load under fixed temperature. Creep curve that ex-

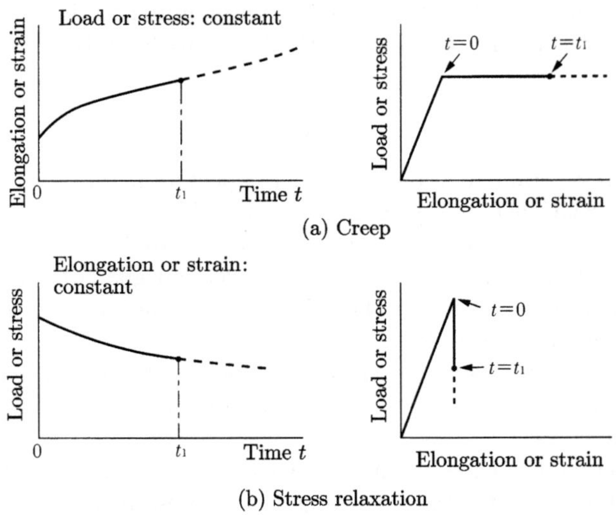

Fig. 1.23. Creep and stress relaxation

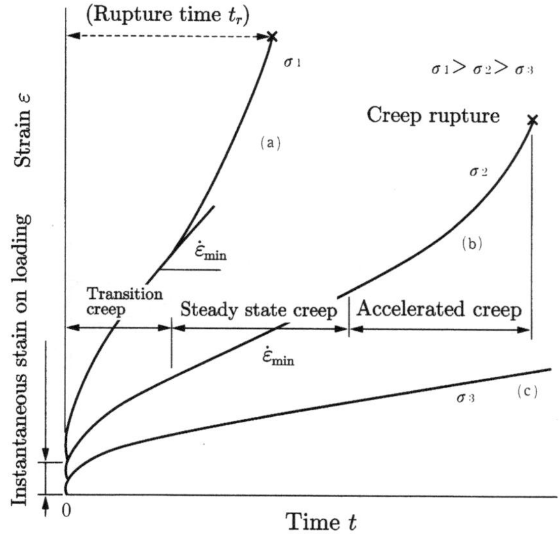

Fig. 1.24. Schematic illustration of creep curves

presses the relationship between the strain ε and the time t, taking the stress σ as parameter, is shown schematically in Fig. 1.24.

Creep deformation except the instantaneous strain that takes place just after loading, can be divided into three stages; the initial transition creep stage (primary creep) in which strain rate ($\dot{\varepsilon} = d\varepsilon/dt$) gradually increases with time, the following steady-state creep stage (secondary creep) in which strain rate is nearly constant, and the final accelerated creep stage (tertiary creep) in which strain rate increases with time and reaches to fracture. Since the strain rate at the steady-state creep stage is the minimum among the three stages, it is called the minimum creep rate $\dot{\varepsilon}_{\text{min}}$. The Figure presenting the relationship between stress and fracture time, taking testing temperature as parameter, is called stress-rupture curve. Figure 1.25 shows stress-rupture curves for SUS304 stainless steel [54]. Each curve in the figure was obtained with the time-temperature parameter (TTP) method. According to this method, a master curve called the main rupture curve is obtained by plotting numerical figures of a parameter containing temperature (absolute temperature T) and rupture time (t_r) against the stresses applied, based on the relationships between temperature and time contributing to creep deformation and stress-rupture. By utilizing a master curve, creep rupture time is estimated. Typical parameters are given below.

Larson–Miller (LM) parameter, $$P = T(\log t_r + C) \tag{1.19}$$

Orr–Sherby–Dorn (OSD) parameter, $$P = \log t_r - \frac{Q}{4.57239} \tag{1.20}$$

Manson–Haferd (MH) parameter, $$P = \frac{\log t_r - \log t_a}{T - T_a} \tag{1.21}$$

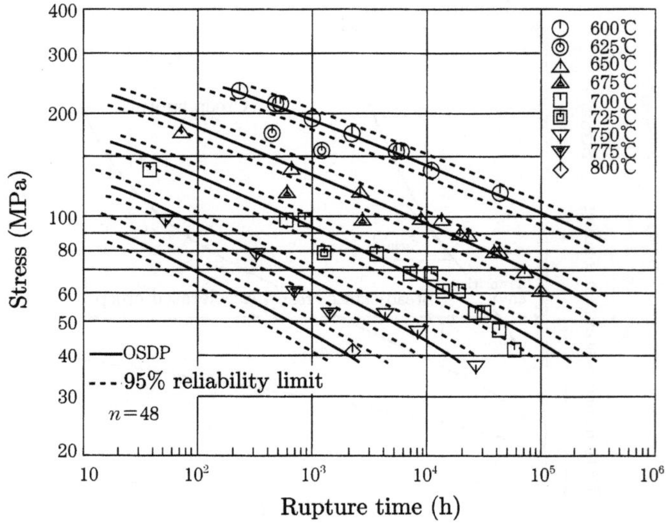

Fig. 1.25. Creep rupture curves [54]

where, C, Q, t_a and T_a are constants.

Creep rupture-curves at each temperature in the Fig. 1.25, were obtained from the main rapture curve by the OSD parameter. It should be noted that an extraordinary extrapolation of such a main rapture curve to longer time side may possibly make the estimation error significant since the real rupture mechanism under long time periods may change from the calculated one.

(2) Stress relaxation

Phenomena such as the repulsive force of a spring decreases progressively with the elapse of time (called spring permanent set) and the fastening force of a fastened bolt decreases with time at elevated temperature, can be caused by stress relaxation. Stress relaxation takes place due to non-elastic deformation of material. Assuming that the total strain ε_t is composed of elastic strain ε_e and time-dependent non-elastic strain ε_p, the following equation is obtained.

$$\begin{aligned}\varepsilon_t &= \varepsilon_e + \varepsilon_p \\ &= \frac{\sigma}{E} + \varepsilon_p \end{aligned} \tag{1.22}$$

where, E is the modulus of elasticity. Differentiating the equation (1.22) with time under the condition of ε_t being a constant, the following equation is obtained;

$$\dot{\sigma} = -E\dot{\varepsilon}_p \tag{1.23}$$

This equation indicates that with the increase of non-elastic strain due to the elapse of time the stress decreases. Figure 1.26 gives stress relaxation curves for bolts used for steam turbine.

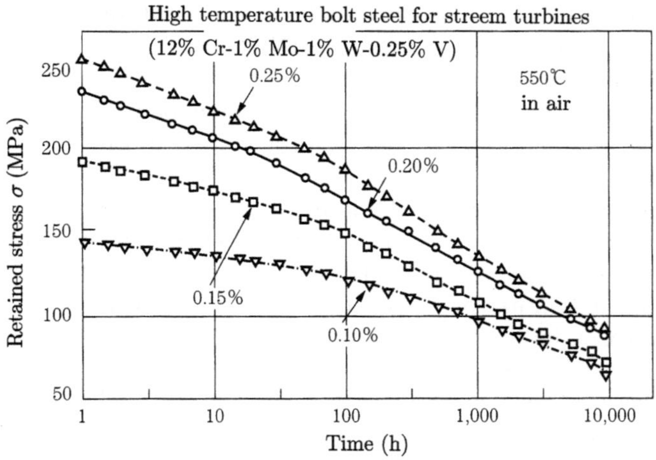

Fig. 1.26. Examples of stress relaxation curves (% in the figure indicates total strain) [55]

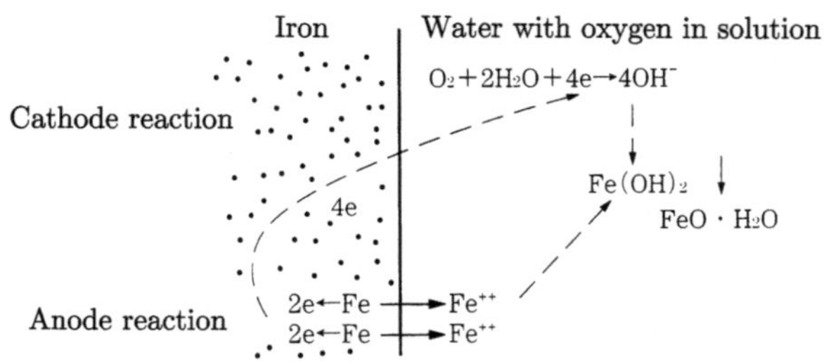

Fig. 1.27. Corrosion process

Stress Corrosion Cracking

For example, corrosion of steel in aerated aqueous solutions proceeds with two electro-chemical reactions shown in Fig. 1.27. An iron atom in steel goes into aqueous solution as ferrous ion Fe^{++}, leaving two electrons in the steel. This is called as anode reaction. The electrons left in the steel reduce oxygen dissolved in the aqueous solution, and as the result, OH^- ion is generated. This is called as cathode reaction. Further, OH^- ions and a Fe^{++} ion in the aqueous solution combine to form $Fe(OH)_2$ or $FeO–H_2O$ that is so called water rust. These reactions take place in solution with plenty of dissolved oxygen. In aqueous solution with insufficient dissolved oxygen, such as the aqueous solution in corrosion pit, the following reaction and other reactions take place as cathode reaction.

$$Fe^{++} + 2H_2O \rightarrow Fe(OH)_2 + 2H^+$$

As a result, hydrogen ion concentration is enriched and some of hydrogen ions are adsorbed on the surface of steel as cathode and further absorbed in steel. Absorbed hydrogen atoms diffuse to defects or to stress concentrated portions in steel and cause cracking under tensile or torsion load. This kind of fracture is called hydrogen induced cracking, hydrogen assisted cracking or hydrogen embrittlement.

Stress corrosion cracking (SCC) in broad sense is a kind of brittle fracture that occurs for metallic material in corrosive environment under sustained loading after the elapse of time. The SCC of steel is classified into (i) stress corrosion cracking in narrow sense due to active pass corrosion at crack front and (ii) hydrogen induced cracking or hydrogen assisted cracking due to hydrogen occluded in steel. The latter is also called delayed fracture or static fatigue. Whether SCC occurs or not is related to the quality of material, environment the material is exposed to, and the magnitude of stress applied. Namely, SCC occurs only in environment specific to each material and under tensile load. As factors to increase stress, there are following factors, i.e., stress concentration caused by irregular shape such as notches, shoulders and material defects, and larger component size.

Although SCC test is usually conducted under a constant load, it should be noted that SCC test under repeating load often accelerates the failure of component tested.

It is known that SCC in narrow sense may occur for α brass and some other copper alloys in ammonium solutions, for austenite stainless steels in chloride solutions, and for mild steels in alkaline solutions. SCC can sometimes take place under stress even if corrosion on the test piece surface is not visible. Fracture surfaces by SCC can be either inter-granular or trans-granular.

The SCC of austenite stainless steels can be prevented by cathodic polarization. On the contrary, some ferrite stainless steels and martensite stainless steels may fail by delayed fracture under rather high tensile load in cathodic polarization condition due to hydrogen assisted cracking.

Delayed fracture by hydrogen embrittlement (hydrogen induced cracking or hydrogen assisted cracking) occurs for high-strength low alloy steels in various environments. Although hydrogen solubility in tempered martensite steels with body-centered cubic structure is rather low at room temperature, low alloy steel tends to become more susceptible to delayed fracture due to hydrogen occluded, with increasing tensile strength or hardness. According to hydrogen permeability and delayed fracture tests for HP9-4-4.5 steel and 4340 steel (NiCrMo steel) by Barth et al. [56], delayed fracture occurs when hydrogen permeability is experimentally observed and it does not occur when hydrogen does not permeate steel, irrespective of polarized condition. They also reported that specimens generated surface pitting show hydrogen permeability and those with no surface pitting do not show hydrogen permeability. Austenite stainless steel and austenite high-manganese non-magnetic steel have higher hydrogen solubility than low alloy steel and can be resistant to hydrogen embrittlement even if relatively high amount of hydrogen is occluded. However, it should be noted that if austenitic steel with high carbon content transforms to martensite structure with high hardness under loading there is the possibility where delayed fracture takes place.

The temperature that the delayed fracture occurs for steels is ranged from 100 to −50°C. Delayed fracture also may occur in an environment apparently non corrosive, for example, in hydrogen gas and other hydrogen containing gas. Generally, low alloy steel hardened and tempered are more susceptible to delayed fracture as the tensile strength becomes higher and as the stress concentration and the sustained load or applied stress are larger. It is also known that chemical compositions and prior austenite grain size affect the susceptibility of steel to delayed fracture. Delayed fracture surface morphology of steel changes with the stress level: At relatively low stress applied, delayed fracture surface can be sometimes brittle inter-granular type or quasi-cleavage type. In contrast, at a higher stress applied, it usually occurs with ductile mode of dimple fracture.

It should be also noted that the hydrogen that promotes delayed fracture in steel springs can be absorbed either during acid cleaning and the following electro-plating, or during the service with the progress of corrosion reaction with environments.

References

41. Nishijima S., et al.: "Evaluation of Statistical Fatigue Properties and Their Heat-to-Heat Variations of S25C, S45C, and SCM3 Steels", Trans. NRIM, 19 (1977), p. 119
42. Wood W. A.: "Recent Observations on Fatigue Fracture in Metals", ASTM STP, 237 (1958), p. 110
43. Laird C: "The Influence of Metallurgical Structure on Mechanism of Fatigue Crack Propagation", ASTM STP, 415 (1967), p. 131
44. Forsyth P. J. E.: "Fatigue Damage and Crack Growth in Aluminum Alloys", Acta Met. 11 (1963), p. 110
45. Nishijima S. et al.: "Standard Fatigue Characteristics of JIS Machine Structural Steels", National Research Institute for Metals Fatigue Data Sheet, Vol. 5 (1989), p. 53 (in Japanese)
46. Kanazawa K. et al.: "Fatigue Characteristics of High Strength Steels", National Research Institute for Metals Material Strength Data Sheet, Vol. 9 (1995) p. 20 (in Japanese)
47. Nishitani H.: "Size effect of diverging point and fatigue limit at rotating bending test (Study on small specimens)", Monthly magazine, Japan Society of Mechanical Engineers, Vol. 34A, (1968) p. 371 (in Japanese)
48. Nishioka K., "Recent problems on fatigue", Symposium of the Society of Materials Science, Japan, (1965) p. 1 (in Japanese)
49. Tanaka K., et al.: "Fatigue characteristics of SNCM 8 steels under combined load (First report, Experimental method and the result)", Monthly magazine, Japan Society of Mechanical Engineers, Vol. 45A (1979) p. 195 (in Japanese)

50. Nishijima S. et al.: Preprint of the Japan Society of Mechanical Engineers, No. 822-12 (1982), p. 237 (in Japanese)
51. Kanao M. et al.: "Preparing data sheets for fatigue characteristics for domestic materials in service", Technical reports of the National Research Institute for Metals, Vol. 3, (1982), P. 66 (in Japanese)
52. Landgraph, R.W. et al., J. of Mater., 4 (1969) p. 176
53. Koyama H. et al.: "Fatigue crack propagation characteristics of spring steel" Transactions of Japan Society for Spring Research, Vol. 29 (1984) p. 30
54. Nishijima S. et al.: "Development of data evaluation method for constructing metallic materials strength database", Research reports of the National Research Institute for Metals, 11 (1990), p. 419
55. Tanaka C. et al.: "Long-time Stress Relaxation Properties of 12Cr–Mo–W–V and 19Cr–9Ni–Mo–W–Nb Steels" Trans. ISIJ, 18 (1978) p. 412
56. Birth C. F. et al.: Corrosion 25 (1969) p. 353

1.3 Supplements: Delayed Fracture of High Strength Steel and the Testing Methods

1.3.1 Delayed Fracture Tests of Straight Bar

Delayed fracture of high strength steel has been studied mostly under a constant tensile load or bending load by many researchers. The models of delayed fracture mechanism of carbon and low alloy steel have been proposed by many researchers and the delayed fracture of these steels is now believed to occur by the following hydrogen assisted damage mechanisms: They are classified as hydrogen enhanced decohesion and hydrogen affected localized plasticity. The latter is further classified as hydrogen enhanced localized plasticity and adsorption induced dislocation emission [1]. Delayed fracture testing methods can be also classified into two types: The first type is a delayed fracture test caused by internal hydrogen, where a constant load is applied to a specimen until the final fracture under a non-corrosive environment after introducing hydrogen into the specimen by dipping it in acid solution or by cathodic electrochemical charging [1, 2]. The cracking process in this type of delayed fracture can be classified as Internal Hydrogen Assisted Cracking (IHAC) or as Internal Hydrogen Embrittlement (IHE). The second type is a delayed fracture test caused by external hydrogen, where a specimen under a constant load and a corrosive environment simultaneously absorbs hydrogen from the surface to occur delayed fracture. The cracking process in this type of delayed fracture can be also classified as Hydrogen Environment Assisted Cracking (HEAC) or Hydrogen Environment Embrittlement (HEE). The first type (IHAC) of delayed fracture testing is generally important for investigating the quality of springs electro-plated. The second type (HEAC) evaluation is generally important for springs to be used in corrosive environments.

Springs normally have no notch on the surface. However, steel springs exposed to corrosive environments such as salt water and rain, sometimes generate corrosion pits, which the stress concentration can occur around.

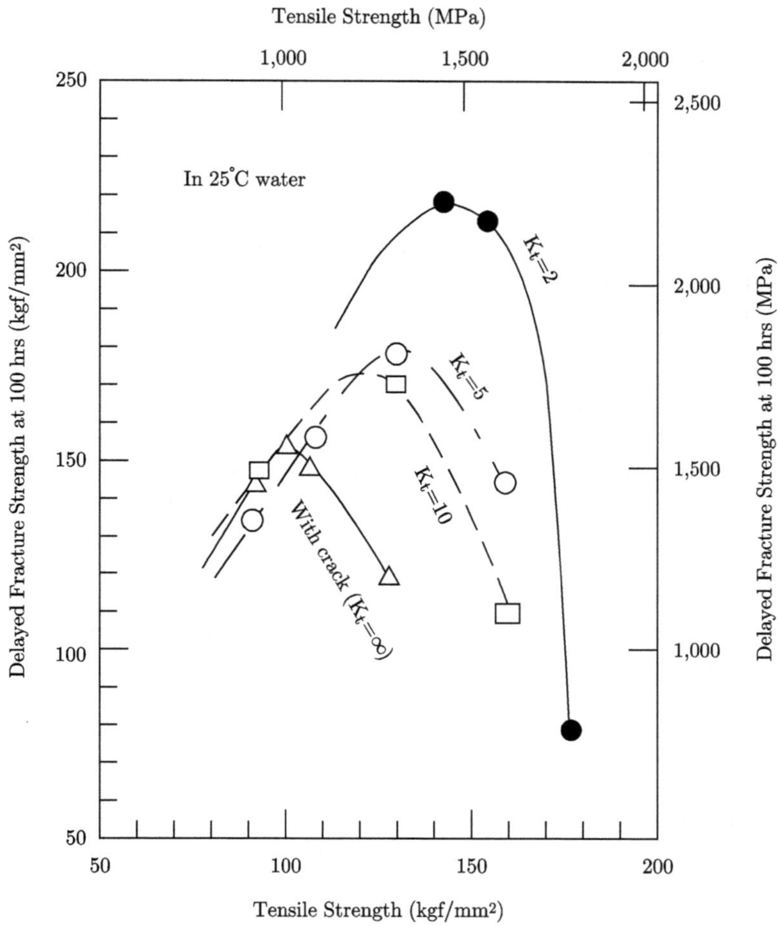

Fig. 1.28. Effect of stress intensity factor K_t of AISI 4340 steel specimen hardened and tempered to various tensile strengths on delayed fracture strength.

Figure 1.28 shows the effect of stress concentration factor on delayed fracture strength for hardened and tempered alloy steel (AISI 4340) [3]. The delayed fracture tests were carried out in water under constant tensile loads. It can be seen from the Fig. 1.28 that as the tensile strength is higher and the stress concentration factor is larger, the specimen becomes more susceptible to the delayed fracture. In the Fig. 1.28, each tensile strength was obtained by using a straight smooth specimen, and the delayed fracture strength at 100 hrs was determined by the delayed fracture load, at which delayed fracture took place after 100 hrs loading, divided by the circumferentially notched area before the test.

Helical springs (coil springs) are mainly used under torsional stresses. Therefore, it is important to know delayed fracture properties of helical springs and materials under torsional stresses.

Wu-Yang Chu et al. have made delayed fracture tests for ultra high strength hardened and tempered steel bar specimens with a circumferential notch under tension (Mode I) and/or torsion (Mode III) [4]. They found that hydrogen pre-charged specimens can cause delayed fracture under constant torsion stresses in air. They also reported that for both smooth and circumferentially notched specimens, the delayed fracture cracks propagated on the planes inclined at 45 degree to the longitudinal axis of the specimens and the fracture surface was inter-granular [4, 5]. The crack initiation site was at the notch root for notched specimens. After outgassing of the charged specimens, it was reported that the delayed cracking along the slanted plane did not occur, suggesting the effect of hydrogen for the occurrence of slanted cracking [4]. They also made delayed fracture tests in aqueous solutions under torsional stresses for ultra high strength steel with a circumferential notch [6]. It was reported that the fracture initiated and propagated along the planes inclined at 45 degree to the original notched plane.

T. Okamura et al. made similar delayed fracture tests under constant torsional stresses in corrosive solution (20% ammonium thiocyanate solution at 323 K) for smooth straight spring steel wire induction-hardened and tempered to various strengths [7]. The schematic illustration of the testing apparatus and the test specimen is shown in Fig. 1.29. It was reported that the smaller prior austenite grain size gave better resistance to delayed fracture when the tensile strength was the same level. In this paper, hydrogen amount evolved during heating at a constant rate was also reported. The apparatus for the measurement of hydrogen evolution during heating is shown schematically in Fig. 1.30. In many papers related with hydrogen evolution analyses, it was reported that the smaller amount of hydrogen evolved below 500 K show higher resistance to delayed fracture [2]. However, T. Okamura et al. [7] reported that as the prior austenite grain was finer, the amount of hydro-

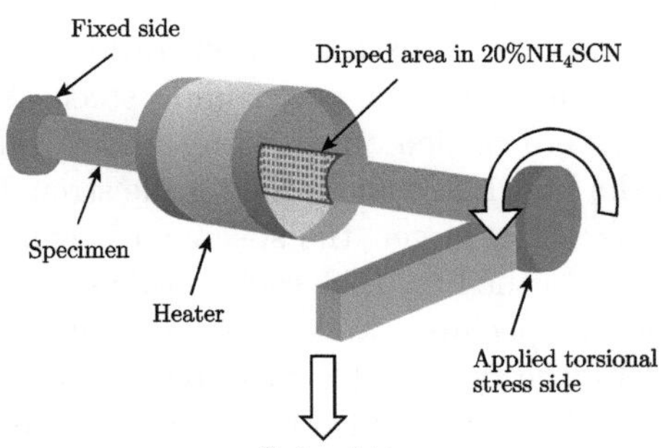

Fig. 1.29. Schematic figure of torsional delayed fracture testing apparatus

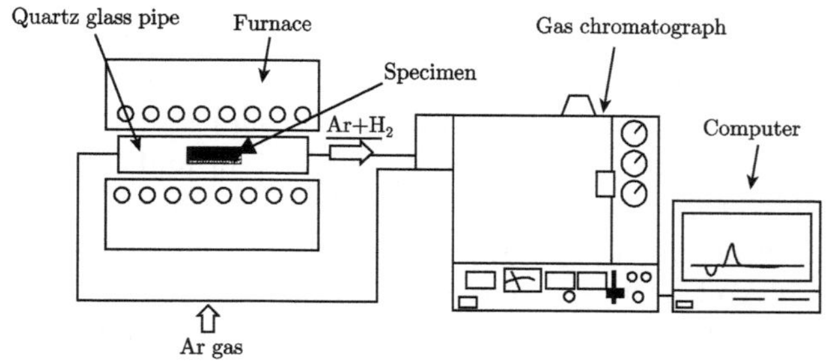

Fig. 1.30. Schematic diagram of hydrogen measuring apparatus

gen evolved during heating became larger, despite the fact that the delayed fracture characteristic was improved.

Slow strain rate tensile technique (SSRT) using a straight bar as a specimen in corrosive environment is a relatively simple method of evaluating delayed fracture strength of steel. In this method, the tendency that the fracture strength of high strength material decreases with the drop of tensile strain rate in tensile testing when the specimen is immersed in a corrosive environment, is utilized.

1.3.2 Testing for Helical Springs

K. Tango et al. developed a new method of evaluating delayed fracture properties of automotive suspension coil springs in corrosive solution [8]. The system of this method is shown in Fig. 1.31. According to this paper, a sensed acoustic emission can correspond to a crack generation or propagation at a certain location speculated from the arrival time difference of the acoustic signal. They also reported the effect of residual stress distributions given by shot-peening on delayed fracture of suspension coil springs (steel grade: JIS SUP7 or AISI 9260) [9] and that the delayed fracture was initiated by a trans-

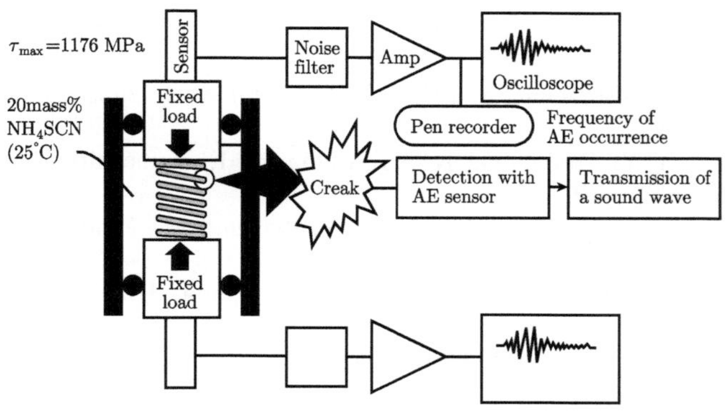

Fig. 1.31. Schematic illustration of system measuring delayed fracture of coil spring

granular (quasi-cleavage) crack generated along the maximum torsional stress plane, to be propagated along grain boundary and to the perpendicular direction to the applied principal (maximum tensile) stress. According to the experimental results by Wu-Yang Chu et al. [4–6] and Tango et al. [9], it can be presumed that delayed fracture cracks normally grow with the help of tensile stress acting on the cracked surface rather than shear stress.

Cold wound coil springs have tensile residual stresses at their inner spring surfaces. T. Aoki et al. made experiments where cold wound coil springs made of oil-tempered wire compressed slightly were dipped in hydrochloric acid solution, and their fracture initiations were checked by strain gauge [10]. It was reported that a generated crack was perpendicular to the wire axis and the crack depth was limited within the depth of tensile residual stress.

Generally, delayed fracture time in HEAC test becomes shorter when stronger acid are used as solutions. It should be noted that too strong an acidic environment may make it difficult to compare correctly the delayed fracture characteristic among different springs used in mild environment. In this sense, proper selection of testing environment in accordance with each spring application shall be studied further.

$*$ Section 1.3 was written by Yoshiro Yamada.

References

1. R. P. Gangloff: Hydrogen assisted cracking of high strength alloys, in Comprehensive Structural Integrity, Vol. 6, Environmentally-Assisted Fracture, edited by I. Miline et al., Elsevier Science, New York, 2003
2. T. Hasegawa, T. Nakahara, Y. Yamada and M. Nakamura: Wire Journal International, 25, 8 (1992) p. 49
3. T. Fujita and Y. Yamada: International Conference on Stress Corrosion Cracking and Hydrogen-Embrittlement of Iron Base Alloys, France, 1973, or NACE-5 (1977) 736, or Doctor Thesis of T. Fujita
4. W.-Y. Chu, C.-M. Hsiao, S.-Y. Ju, and C. Wang: Corrosion-NACE, 38, 8, August (1982) p. 446
5. W.-Y. Chu, T.-Y. Zhang, and C.-M. Hsiao: Corrosion-Nace, 40, 4, April (1984) p. 197
6. W.-Y. Chu, C.-M. Haiao, and B.-J. Xu: Metallurgical Transactions, 17A, April (1986), p. 711
7. T. Okamura, Y. Murayama, K. Kawasaki, and E. Yamashita: Transactions of JSSR 45 (2000) p. 9
8. K. Tango, Y. Ono, M. Sugiyama, T. Akutsu and M. Ayada: Transactions of JSSR 47 (2002) p. 21
9. K. Tango, M. Sugiyama, T. Suzuki and M. Ayada: Transactions of JSSR 48 (2003) p. 15
10. T. Aoki, M. Wakita and T. Ozone: Transactions of JSSR 43 (1998) p. 43

2

Metallic Material for Springs

2.1 Steel Material

2.1.1 Manufacturing Method of Steel Material for Springs

(1) Introductions

The materials applied for springs are extended to metallic and nonmetallic types, in addition, among the metals there are many types, such as, spring steel, stainless steel, nickel alloy and so on. Their required properties vary accordingly. However, whatever the applications, it is certain that a high stress during cyclic loading and prolonged reliability should be required.

Steel material for springs, can be classified as, a) the materials which are heat treated during the spring manufacturing process, while the spring shape is formed by hot deformation, and b) the material where the required mechanical properties for the spring, particularly the tensile strength and hardness can be obtained before the spring forming.

In the case of a), the material size, for example, wire diameter for wire spring and thickness for plate spring, is comparatively large, and after hot forming, heat treatment is carried out (rarely formed by cold forming). This material can be applied to the springs, such as leaf spring, coil spring and torsion bar for the vehicles of railroads and/or automobile.

In the case of b), the size is comparatively small (wire diameter, thickness etc.) and assuming that the shape of the spring can be manufactured by cold forming, for example, the oil tempered wire where the quenching and tempering have been already done, the stainless steel wire which have been cold drawn to a predetermined tensile strength, and cold rolled steel strip for spring. The material, b) is described in Chap. 2.1.3.

As for the required characteristics of spring material, although there are many other factors which affect the dynamic strength of springs, besides modulus of elasticity, shape, dimensions, tolerance, surface roughness, corrosion resistance, and mechanical properties, it is essential to take the surface defects, decarburization, hardenability, non-metallic inclusion, segregation, etc, into consideration, for the requirements of high strength and improvement in reliability.

Recently, due to the progress of combined steel refining technology, reduction of harmful impurities and nonmetallic inclusions in industrial productions, and with low costs, becomes possible. The progress in rolling process, and non-destructive test technology in line has improved the accuracy of dimensions and the surface quality remarkably. By the combination between those technical progress and material development, it becomes possible for the design stress of the automotive suspension spring to increase from 980 MPa to 1079 MPa, furthermore, a stress up to 1275 MPa has been recently in practical use, to contribute the weight-saving of springs, as well as compact spacing in the suspension design.

(2) Steel grade

There are many steel materials for spring use, such as spring steel for hot forming, piano wire, hard drawn steel wire, stainless steel, and tool steel. Among these, the so-called spring steels used for hot forming are, carbon steel, Si steel, Mn steel, Si–Mn steel, Si–Cr steel, Mn–Cr steel, Cr–V steel, Si–Cr–V steel, Si–Ni–Cr steel, Ni–Cr–Mo steel, Cr–Mo steel, B steel, etc, which can be said as low alloy steel. The reason why most steel grade includes Si is that Si can raise the elastic limit of the steel and improve the resistance to permanent set of springs. The Ni–Cr–Mo steel, Cr–Mo steel, B steel, etc. with excellent hardenability were developed as steel for large-sized springs.

Table 2.1 shows the chemical compositions of spring steels designated or recommended by several countries. Oil quenching and tempering can be normal heat-treatments in any countries, and the carbon content is mostly ranged from 0.50 to 0.65%. except for the carbon steel. In Europe, water quenching has been partly applied for steels whose the carbon content is around 0.5%, considering the quenching crack. Although water quenching has been used in Japan, all the spring steel can be now oil quenched.

There is a Japanese standard, JIS G4801-1984 for hot formed spring steel, which designated the chemical compositions. The SUP3 is the eutectoid composition carbon steel and it has been widely used as spring steel material for many years. However, since its hardenability is not good enough to be used for large size spring, it has mainly been used for the leaf spring of railroad vehicles. SUP6 and SUP7 are high Si added Si–Mn steel type having high strength with good temper softening resistance, and their hardenability is also comparatively good. Such steel type acquires an excellent spring characteristic after quenching and tempering. Due to their high yield strength, these steels can be widely applicable and bear considerable severe operating conditions.

The SUP7 attains high permanent set resistance due to its higher Si content than SUP6 and can be widely applied to coil spring, leaf spring of automobiles. SUP9, Mn–Cr steel, shows good hot deformability and good hardenability to be applied for the relatively large sized stabilizers, torsion bars, and coil springs. SUP9A, which is equivalent to SAE5160 steel, has basically the same chemical composition as the SUP9 with a little bit higher carbon

Table 2.1. Chemical composition of spring steel of each country

(a) JIS (JIS G 4801-1984) Spring steel

Grade	Chemical compositions %							
	C	Si	Mn	P	S	Cr	Mo	V	B
SUP3	0.75–0.90	0.15–0.35	0.30–0.60	≦ 0.035	≦ 0.035	—	—	—	—
SUP6	0.56–0.64	1.50–1.80	0.70–1.00	≦ 0.035	≦ 0.035	—	—	—	—
SUP7	0.56–0.64	1.80–2.20	0.70–1.00	≦ 0.035	≦ 0.035	—	—	—	—
SUP9	0.52–0.60	0.15–0.35	0.65–0.95	≦ 0.035	≦ 0.035	0.65–0.95	—	—	—
SUP9A	0.56–0.64	0.15–0.35	0.70–1.00	≦ 0.035	≦ 0.035	0.70–1.00	—	—	—
SUP10	0.47–0.55	0.15–0.35	0.65–0.95	≦ 0.035	≦ 0.035	0.80–1.10	—	0.15–0.25	—
SUP11A	0.56–0.64	0.15–0.35	0.70–1.00	≦ 0.035	≦ 0.035	0.70–1.00	—	—	≧ 0.0005
SUP12	0.51–0.59	1.20–1.60	0.60–0.90	≦ 0.035	≦ 0.035	0.60–0.90	—	—	—
SUP13	0.56–0.64	0.15–0.35	0.70–1.00	≦ 0.035	≦ 0.035	0.70–0.90	0.25–0.35	—	—

(b) ISO 683-14 (1992-08-15) Spring steel (%)

No.	Steel Grade	ISO 683-14:1973 Grade	C	Si	Mn	P_{max}	S_{max}	Cr	Mo	
1	59 Si 7	5	0.55–0.63	1.60–2.00	0.60–1.00	0.030	0.030	—	—	
2	56 SiCr 7	—	0.52–0.59	1.60–2.00	0.70–1.00	0.030	0.030	0.20–0.40	—	
3	61 SiCr 7	7	0.57–0.65	1.60–2.00	0.70–1.00	0.030	0.030	0.20–0.40	—	
4	55 SiCr 63	—	0.51–0.59	1.20–1.60	0.50–0.80	0.030	0.030	0.55–0.85	—	
5	55 Cr 3	8	0.52–0.59	0.15–0.40	0.70–1.00	0.030	0.030	0.70–1.00	—	
6	60 CrMo 31	—	0.56–0.64	0.15–0.40	0.70–1.00	0.030	0.030	0.70–1.00	0.08–0.15	
7	60 CrB-3	10	0.56–0.64	0.15–0.40	0.70–1.00	0.030	0.030	0.60–0.90	—	B : 0.0008min
8	60 CrMo 33	12	0.56–0.64	0.15–0.40	0.70–1.00	0.030	0.030	0.70–1.00	0.25–0.35	—
9	51 CrV 4	13	0.47–0.55	0.10–0.40	0.60–1.00	0.030	0.030	0.80–1.10	—	V : 0.10–0.25
10	52 CrMoV 4	14	0.48–0.56	0.15–0.40	0.70–1.00	0.030	0.030	0.90–1.70	0.15–0.25	V : 0.07–0.15

(c) BS 970 : Part 2 (1988) Spring steel (%)

	Gr	C	Si	Mn	P_{max}	S_{max}	Cr	Mo	V	Ni	Cu	Sn
Si–Mn	251 A 58	0.55–0.60	1.80–2.10	0.80–1.00	0.035	0.035	0.15–0.30	0.10max	—	≦0.40	≦0.35	≦0.035
	251 A 60	0.57–0.62	1.80–2.10	0.80–1.00	0.035	0.035	0.25–0.40	0.12max	—	≦0.40	≦0.35	≦0.035
Alloy steel	525 A 58	0.55–0.60	0.20–0.35	0.80–0.95	0.035	0.035	0.70–0.85	0.10max	—	≦0.40	≦0.35	≦0.035
	525 A 60	0.57–0.62	0.20–0.35	0.85–1.00	0.035	0.035	0.80–0.95	0.06min	—	≦0.40	≦0.35	≦0.035
	525 A 61	0.57–0.63	0.20–0.35	0.85–1.00	0.035	0.035	0.85–1.00	0.08–0.15	—	≦0.40	≦0.35	≦0.035
	685 A 57	0.55–0.60	1.20–1.60	0.70–0.90	0.035	0.035	0.60–0.85	—	—	≦0.40	≦0.35	≦0.035
	704 A 60	0.57–0.62	0.20–0.35	0.85–1.0	0.035	0.035	0.80–0.95	0.15–0.25	—	≦0.40	≦0.35	≦0.035
	705 A 60	0.57–0.62	0.20–0.35	0.85–1.0	0.035	0.035	0.85–1.00	0.25–0.35	—	≦0.40	≦0.35	≦0.035
	735 A 51	0.48–0.54	0.20–0.35	0.70–1.00	0.035	0.035	0.90–1.20	—	0.10–0.20	≦0.40	≦0.35	≦0.035
	735 A 54	0.52–0.57	0.20–0.35	0.90–1.15	0.035	0.035	1.05–1.20	—	0.12–0.20	≦0.40	≦0.35	≦0.035
	925 A 60	0.55–0.65	1.70–2.10	0.70–1.00	0.035	0.035	0.20–0.40	0.20–0.30	—	≦0.40	≦0.35	≦0.035

Spring steel (%)

	Gr	C	Si	Mn	P_{max}	S_{max}	Cr	Mo	NI	V	Cu	Sn
Si–Mn	251 H 60	0.56–0.64	1.60–2.20	0.70–1.00	0.035	0.035	0.40_{max}	0.12max	≦0.40	—	≦0.35	≦0.035
Alloy steel	525 H 60	0.55–0.60	0.15–0.40	0.65–1.00	0.035	0.035	0.60–1.00	0.15max	≦0.40	—	≦0.35	≦0.035
	685 H 57	0.54–0.62	1.20–1.60	0.50–0.80	0.035	0.035	0.50–0.80	—	≦0.40	—	≦0.35	≦0.035
	704 H 60	0.55–0.64	0.15–0.40	0.65–1.10	0.035	0.035	0.40–1.00	0.15–0.25	≦0.40	—	≦0.35	≦0.035
	705 H 60	0.55–0.64	0.15–0.40	0.65–1.10	0.035	0.035	0.60–1.00	0.25–0.35	≦0.40	—	≦0.35	≦0.035
	735 H 51	0.47–0.51	0.15–0.40	0.70–1.10	0.035	0.035	0.90–1.20	—	≦0.40	0.10–0.25	≦0.35	≦0.035
	805 H 60	0.55–0.64	0.15–0.40	0.65–1.05	0.035	0.035	0.35–0.65	0.15–0.25	0.35–0.75	—	≦0.35	≦0.035

and higher range of Mn and Cr to improve its hardenability. SUP10 is Cr–V steel having good hardenability with high toughness, and is used for higher hardness application. SUP11A is the same material as boron treated SUP9A of Mn–Cr steel, and its hardenability is excellent, generally the mechanical property does not change from that of SUP9A. The Si–Cr steel, SUP12 is mostly used for oil tempered wire material of cold-formed springs. SUP13 is Cr–Mo steel with higher hardenability than the boron added SUP11A. The

(d) DIN 17221 (Dce. 1988) Spring steel (%)

Material									
Grade	Number	C	Si	Mn	P_{max}	S_{max}	Cr	Mo	V
38 Si 7	1.5023	0.35–0.42	1.50–1.00	0.50–0.80	0.030	0.030	—	—	—
54 SiCr 6	1.7102	0.51–0.59	1.20–1.60	0.50–0.80	0.030	0.030	0.50–0.80	—	—
60 SiCr 7	1.7108	0.57–0.65	1.50–1.80	0.70–1.00	0.030	0.030	0.20–0.40	—	—
55 Cr 3	1.7176	0.52–0.59	0.25–0.50	0.70–0.10	0.030	0.030	0.70–1.00	—	—
50 CrV 4	1.8159	0.47–0.55	0.15–0.40	0.70–1.10	0.030	0.030	0.90–1.20	—	0.10–0.20
51 CrMoV 4	1.7701	0.48–0.56	0.15–0.40	0.70–1.10	0.030	0.030	0.90–1.20	0.15–0.25	0.08–0.15

DIN 17224 (Feb. 1982) Stainless steel wire and steel strip for spring (%)

Material										
Grade	No.	C_{max}	Si_{max}	Mn_{max}	Al	Cr	Mo	Ni	P	S
X 12 CrNi 7	1.4310	0.12	1.5	2.0	—	16.0–18.0	≦ 0.8	5.0–9.0	≦ 0.045	≦ 0.030
X 5 CrNiMo 18 10	1.4401	0.07	1.0	2.0	—	16.5–18.5	2.0–2.5	10.5–13.5	≦ 0.045	≦ 0.030
X 7 CrNiAl 17 7	1.4568	0.09	1.0	1.0	0.75–1.50	16.0–18.0	—	6.5–7.75	≦ 0.045	≦ 0.030

(e) SAE J 403 (May 94) Carbon steel (%)

UNS No.	SAE/AISI No.	C	Mn	P_{max}	S_{max}
C 10050	1005	0.06 max	0.35 max	0.030	0.050
C 10060	1006	0.08 max	0.25–0.40	0.030	0.050
C 10080	1008	0.10 max	0.30–0.50	0.030	0.050
C 10100	1010	0.08–0.13	0.30–0.60	0.030	0.050
C 10120	1012	0.10–0.15	0.30–0.60	0.030	0.050
C 10150	1015	0.13–0.18	0.30–0.60	0.030	0.050
C 10160	1016	0.13–0.18	0.60–0.90	0.030	0.050
C 10170	1017	0.15–0.20	0.30–0.60	0.030	0.050
C 10180	1018	0.15–0.20	0.60–0.90	0.030	0.050
C 10200	1020	0.18–0.23	0.30–0.60	0.030	0.060
C 10210	1021	0.18–0.23	0.60–0.90	0.030	0.050
C 10220	1022	0.18–0.23	0.70–1.00	0.030	0.050
C 10230	1023	0.20–0.25	0.30–0.60	0.030	0.050
C 10250	1025	0.22–0.28	0.30–0.60	0.030	0.050
C 10260	1026	0.22–0.28	0.60–0.90	0.030	0.050
C 10290	1029	0.25–0.31	0.60–0.90	0.030	0.050
C 10300	1030	0.28–0.34	0.60–0.90	0.030	0.050
C 10350	1035	0.32–0.38	0.60–0.90	0.030	0.050
C 10380	1038	0.35–0.42	0.60–0.90	0.030	0.050
C 10390	1039	0.37–0.44	0.70–1.00	0.030	0.050
C 10400	1040	0.37–0.44	0.60–0.90	0.030	0.050
C 10420	1042	0.40–0.47	0.60–0.90	0.030	0.050
C 10430	1043	0.40–0.47	0.70–1.00	0.030	0.050
C 10440	1044	0.43–0.50	0.30–0.60	0.030	0.050
C 10450	1045	0.43–0.50	0.60–0.90	0.030	0.050
C 10460	1046	0.43–0.50	0.70–1.00	0.030	0.050
C 10490	1049	0.46–0.53	0.60–0.90	0.030	0.050
C 10500	1050	0.48–0.55	0.60–0.90	0.030	0.050
C 10530	1053	0.48–0.55	0.70–1.00	0.030	0.050
C 10550	1055	0.50–0.60	0.60–0.90	0.030	0.050
C 10600	1060	0.65–0.65	0.60–0.90	0.030	0.050
C 10650	1065	0.60–0.70	0.60–0.90	0.030	0.050
C 10700	1070	0.65–0.75	0.60–0.90	0.030	0.050
C 10780	1078	0.72–0.85	0.30–0.60	0.030	0.050
C 10800	1080	0.75–0.88	0.60–0.90	0.030	0.050
C 10860	1086	0.80–0.93	0.30–0.60	0.030	0.050
C 10900	1090	0.85–0.98	0.60–0.90	0.030	0.050
C 10950	1095	0.90–1.03	0.30–0.50	0.030	0.050

SAE J 404 (Apr 94) Alloy steel (%)

UNS No.	SAE No.	C	Mn	P_{max}	S_{max}	Si	Ni	Cr	Mo	V
C 13050	1335	0.33–0.38	1.60–1.90	0.030	0.040	0.15–0.35	—	—	—	—
C 13400	1340	0.38–0.43	1.60–1.90	0.030	0.040	0.15–0.35	—	—	—	—
C 40230	4023	0.20–0.25	0.70–0.90	0.030	0.040	0.15–0.35	—	—	0.20–0.30	—
C 40270	4027	0.25–0.30	0.70–0.90	0.030	0.040	0.15–0.35	—	—	0.20–0.30	—
C 40370	4037	0.35–0.40	0.70–0.90	0.030	0.040	0.15–0.35	—	—	0.20–0.30	—
C 40470	4047	0.45–0.50	0.70–0.90	0.030	0.040	0.15–0.35	—	—	0.20–0.30	—
C 41180	4118	0.18–0.23	0.70–0.90	0.030	0.040	0.15–0.35	—	0.40–0.60	0.08–0.15	—
C 41200	4120	0.18–0.23	0.90–1.20	0.030	0.040	0.15–0.35	—	0.40–0.60	0.13–0.20	—
C 41300	4130	0.28–0.33	0.40–0.60	0.030	0.040	0.15–0.35	—	0.80–1.10	0.15–0.25	—
C 41370	4137	0.35–0.40	0.70–0.90	0.030	0.040	0.15–0.35	—	0.80–1.10	0.15–0.25	—
C 41400	4140	0.38–0.43	0.75–1.00	0.030	0.040	0.15–0.35	—	0.80–1.10	0.15–0.25	—
C 41420	4142	0.40–0.45	0.75–1.00	0.030	0.040	0.15–0.35	—	0.80–1.10	0.15–0.25	—
C 41450	4145	0.43–0.48	0.75–1.00	0.030	0.040	0.15–0.35	—	0.80–1.10	0.15–0.25	—
C 41500	4150	0.48–0.53	0.75–1.00	0.030	0.040	0.15–0.35	—	0.80–1.10	0.15–0.25	—
C 43200	4320	0.17–0.22	0.45–0.65	0.030	0.040	0.15–0.35	1.65–2.00	0.40–0.60	0.20–0.30	—
C 43400	4340	0.38–0.43	0.60–0.80	0.030	0.040	0.15–0.35	1.65–2.00	0.70–0.90	0.20–0.30	—
C 43406	E 4340	0.38–0.43	0.65–0.85	0.025	0.025	0.15–0.35	1.65–2.00	0.70–0.90	0.20–0.30	—
C 46200	4620	0.17–0.22	0.45–0.65	0.030	0.040	0.15–0.35	1.65–2.00	—	0.20–0.30	—
C 48200	4820	0.18–0.23	0.50–0.70	0.030	0.040	0.15–0.35	3.25–2.75	—	0.20–0.30	—
C 50461	50 B 46	0.44–0.49	0.75–1.00	0.030	0.040	0.15–0.35	—	0.20–0.35	—	—
C 51200	5120	0.17–0.22	0.70–0.90	0.030	0.040	0.15–0.35	—	0.70–0.90	—	—
C 51300	5130	0.28–0.33	0.70–0.90	0.030	0.040	0.15–0.35	—	0.80–1.10	—	—
C 51320	5132	0.30–0.35	0.60–0.80	0.030	0.040	0.15–0.35	—	0.75–1.00	—	—
C 51400	5140	0.38–0.43	0.70–0.90	0.030	0.040	0.15–0.35	—	0.70–0.90	—	—
C 51500	5150	0.48–0.53	0.70–0.90	0.030	0.040	0.15–0.35	—	0.70–0.90	—	—
C 51600	5160	0.56–0.64	0.75–1.00	0.030	0.040	0.15–0.35	—	0.70–0.90	—	—
C 51601	51 B 60	0.56–0.64	0.75–1.00	0.030	0.040	0.15–0.35	—	0.70–0.90	—	—
C 52986	652100	0.98–1.10	0.25–0.45	0.025	0.025	0.15–0.35	—	1.30–1.60	—	—
C 61500	6150	0.48–0.53	0.70–0.90	0.030	0.040	0.15–0.35	—	0.80–1.10	—	0.15 min
C 86150	8615	0.13–0.18	0.70–0.90	0.030	0.040	0.15–0.35	0.40–070	0.40–0.60	0.15–0.25	—
C 86170	8617	0.15–0.20	0.70–0.90	0.030	0.040	0.15–0.35	0.40–070	0.40–0.60	0.15–0.25	—
C 86200	8620	0.18–0.23	0.70–0.90	0.030	0.040	0.15–0.35	0.40–070	0.40–0.60	0.15–0.25	—
C 86220	8622	0.20–0.25	0.70–0.90	0.030	0.040	0.15–0.35	0.40–070	0.40–0.60	0.15–0.25	—
C 86300	8630	0.28–0.33	0.70–0.90	0.030	0.040	0.15–0.35	0.40–070	0.40–0.60	0.15–0.25	—
C 86400	8640	0.38–0.43	0.75–1.00	0.030	0.040	0.15–0.35	0.40–070	0.40–0.60	0.15–0.25	—
C 86450	8645	0.43–0.48	0.75–1.00	0.030	0.040	0.15–0.35	0.40–070	0.40–0.60	0.15–0.25	—
C 87200	8720	0.18–0.23	0.70–0.90	0.030	0.040	0.15–0.35	0.40–070	0.40–0.60	0.20–0.30	—
C 8 B 220	8822	0.20–0.25	0.75–1.00	0.030	0.040	0.15–0.35	0.40–070	0.40–0.60	0.30–0.40	—
C 92590	9259	0.56–0.64	0.75–1.00	0.030	0.040	0.70–1.10	—	0.45–0.65	—	—
C 92600	9260	0.56–0.64	0.75–1.00	0.030	0.040	1.80–2.20	—	—	—	—

SUP13 can be applicable for extremely large coil spring with over 100 mm diameter.

(3) Steel making process

(i) Arc furnace steel making process

Although the open hearth-furnace was used to make spring steels using pig-iron and ferrous scrap in the past, the arc furnace steel making process, blast furnace and converter steel making process have been popular recently. In the arc furnace steel making process, the steel scrap as main material can be melted and refined by applying arc heat. The refining process can be roughly divided into the oxidizing period and reducing period. In the oxidization period, the oxygen gas is blown into the molten steel to make decarburization and dephosphorization. In order to prevent re-phosphorization from slag in the

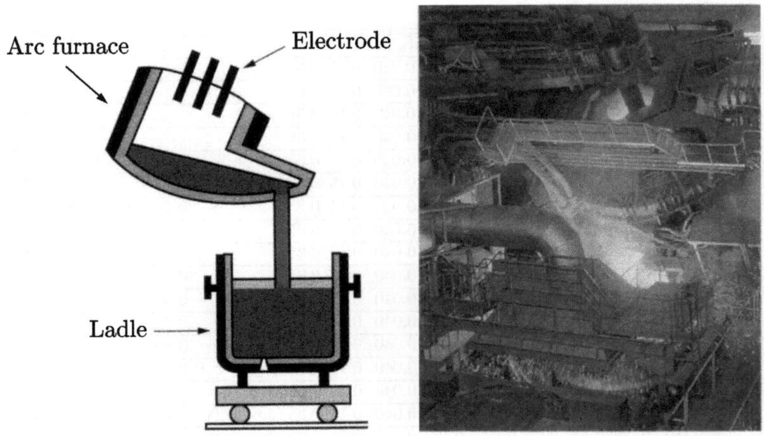

Fig. 2.1. Outline and Appearance of 140 t arc furnace

reducing period, the slag generated at the oxidization period is removed, and the new slag is remade to perform desulphurization, de-oxidization, composition adjustment, and temperature adjustment [1]. Due to the needs of high productivity and low cost the large size of furnace around 100 ton becomes popular. The appearance of a typical arc furnace can be shown in Fig. 2.1.

(ii) Converter steel making process

The converter usually uses molten pig iron as its main raw material for steel refining. When the converter was introduced, all refining processes were carried out with the converter. In recent years, however, due to the introduction of the molten pig-iron pretreatment process in pig-iron transportation ladle or in molten pig-iron pretreatment furnace and the ladle refining process following the converter process, the efficiency of refining and the quality of steel products have been improved significantly.

In the steel making process in converter, high pressure pure oxygen is normally blown from the upper part of the converter to the pretreated pig-iron melt surface, to oxidize and remove carbon, silicon and some other elements in the melt. Besides the top-blowning of oxygen gas described above, it is popular to blow the Argon gas into the melt from the bottom of the converter for the purpose of improving the refining efficiency. This process can be said upper-bottom blowing. The capacity of the converter is 80–300 Ton. The main features of the converter steel-making process can be summarized below.

1. High productivity with rapid chemical reactions.
2. Since dephosphorization and decarburization progress simultaneously at extremely high temperature with the help of slag formed in the early stage of blowing, steel with high carbon content such as spring steel can be produced efficiently.
3. Since the refining is made with the combustion reaction of C, Mn and Si in the melt by pure oxygen, and the inside of the converter above the melt

is filled with the reducing atmosphere of CO gas, low levels of P, H, O, and N in the refined steel become possible.

(iii) Ladle refining method

In the ladle refining process, the partly refined molten steel by the arc furnace or converter can be refined in the ladle. Thereby, the reducing refinement in the arc furnace or in the converter can be shortened. In the typical LF (Ladle Furnace), applying the additional heating to the molten steel by the arc, the molten metal can be simultaneously agitated by the gas blowing from the bottom of ladle to obtain the homogenization of molten steel. The adjustment of alloy composition can be easily made. The outline of LF is shown in Fig. 2.2.

Since stainless steel has high Cr content, compared with SUP6 and SUP9, the oxidization of Cr takes place preferentially to make the decarburization difficult in the decarburization refinement. In order to control the Cr oxidization to promote the decarburization, increasing the molten steel temperature and decreasing the CO partial pressure in the atmosphere can be required. However, due to the damage of refractory, there is a limit in refinement under high temperature. The decarburization by dropping the CO partial pressure falls into two methods-decarburization by the dilution with the inert gas represented by the AOD (Argon oxygen decarburization) method and the decarburization under the reduced pressure represented by VOD (Vacuum oxygen decarburization) method.

Melting, refining and casting the steel under the atmosphere, the steel can be normally polluted to some extent by the atmosphere such as oxygen, nitrogen, or moisture, to have the mechanical properties of steel deteriorated. Therefore, the practices to produce the steel with low gas content by preventing from such contaminations, have been made in various methods in the past. With the development of a large-sized steam ejector, the vacuum degassing method becomes possible under the commercial basis. At the beginning, the vacuum gas method was applied only to the white spot defect prevention of large-sized casting and forging steel or to the measure to the non-metallic

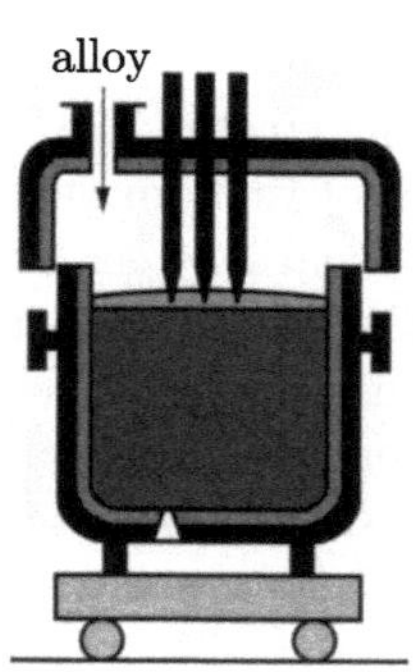

Fig. 2.2. Ladle Furnace (LF)

Fig. 2.3. RH method

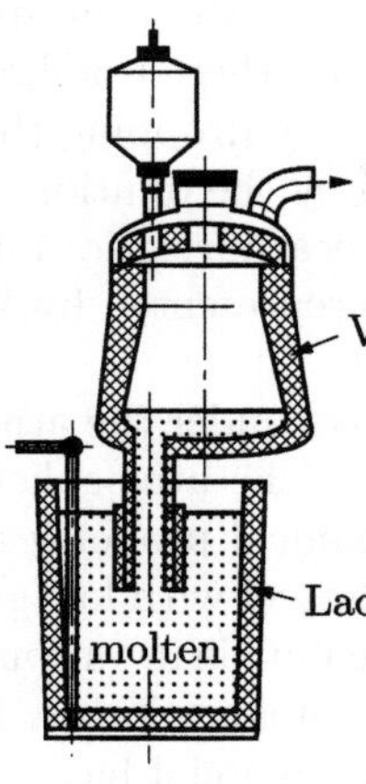

Fig. 2.4. DH method

inclusion of bearing steel. However, based on the customer's requirements, the vacuum degassing method can be widely applied to the mechanical and structural steel, including spring steel.

Although the various types of vacuum degassing method were developed, for the machine structual steel, bearing steel, and spring steel, the RH method, the DH method, and the LD-VAC method are mainly used for the main purpose of degassing, as shown in Fig. 2.3, 2.4 and 2.5.

Although the main gas composition emitted by the vacuum degassing method are H_2, N_2, O_2, CO, CO_2, CH_4, etc., it was experimentally found that the chemical composition of the extracted gas can be dependent upon the type of the degassing method. It may be difficult to compare the effect of various kinds of degassing methods at the same level. However, the following advantages can be obtained by the degassing process.

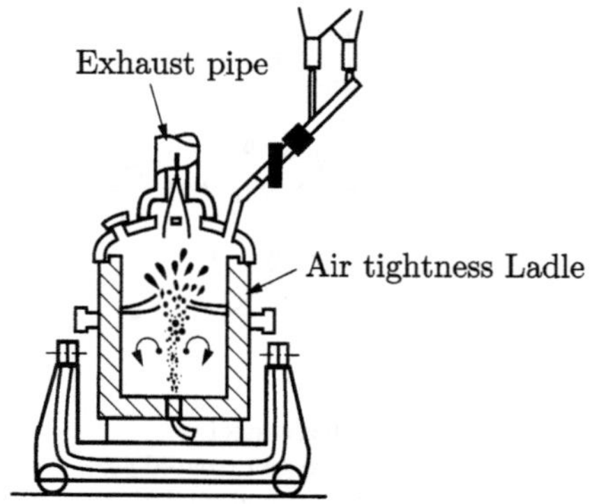

Fig. 2.5. LD-VAC method

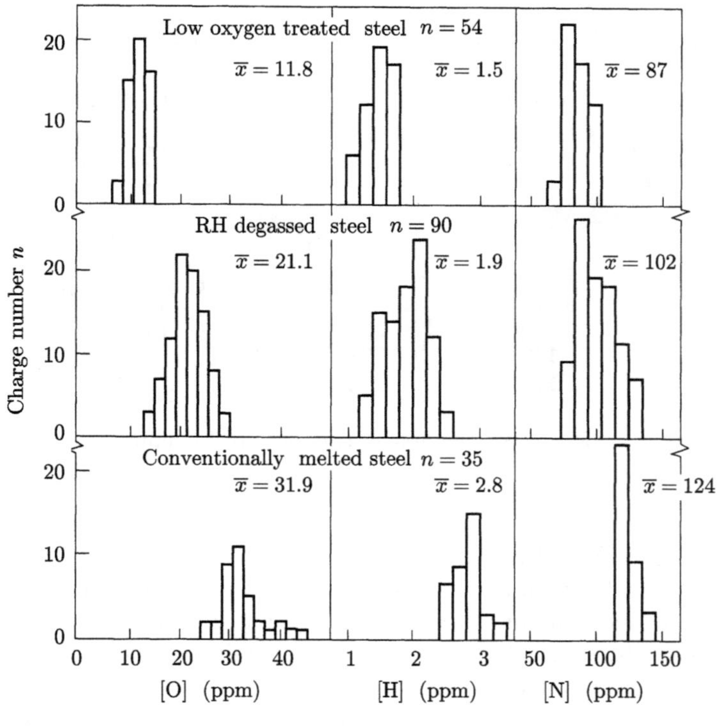

Fig. 2.6. Effect of degassing treatment in lowering Oxygen, Hydrogen, and Nitrogen [2]

1) Decrease the gas contents, such as, hydrogen, oxygen, and nitrogen
2) Decrease non-metallic inclusions
3) Improve the mechanical properties of steel

Figure 2.6 shows the difference of the gas content of the normal steel, RH degassing steel, and low oxygen processing steel. Decreasing the oxygen

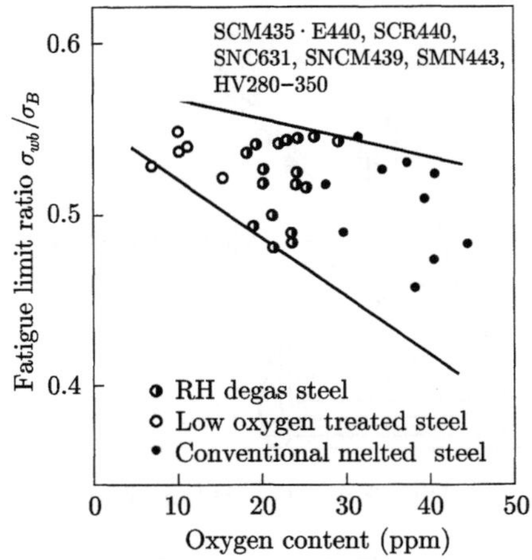

Fig. 2.7. Effect of degassing treatment in lowering oxygen, hydrogen and nitrogen [2]

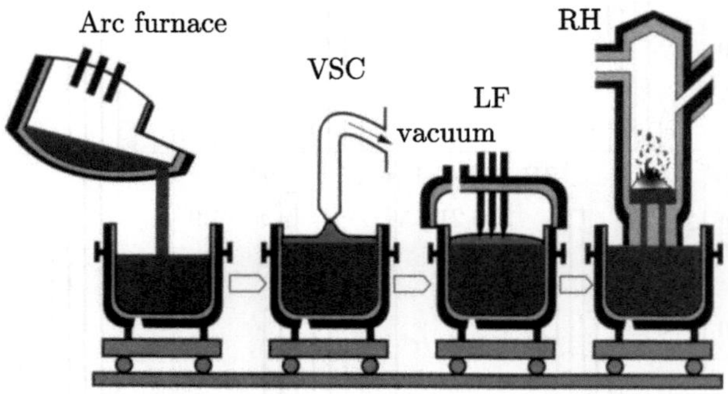

Fig. 2.8. Example of complex steel making

content can be expected to decrease the total amount of oxide and improve the fatigue limit ratio by decreasing the size of oxide type inclusions.

The relation between the fatigue limit ratio and oxygen content of various low alloy machine structural steels is shown in Fig. 2.7 [2]. In low alloy machine structural steels, 0.01–0.03% aluminum is contained. The oxygen normally exists in steel as alumina type inclusions. In such Al deoxidized steel, lowering the oxygen content can lead to the improvement of fatigue strength.

With the progress of Converter refining method, the segmentation of steel refining functions becomes popular. The functional segmentation means that in steel refining process the group of specialist at each stage and its combination can be responsible for the refining process, comparing with the conventional system where the many-skilled workers take charge of all the processes of refining. This can improve the productivity and quality remarkably.

An example of complex steel-manufacturing processes is shown in Fig. 2.8. The slag removal which was conventionally one of the function of an arc

inclusion type		Al_2O_3	$MgO{-}Al_2{-}O_3$	$CaO{-}Al_2O_3{-}SiO_2$	SiO_2
Nonmetallic Inclusions of Fracture surface 20 μm		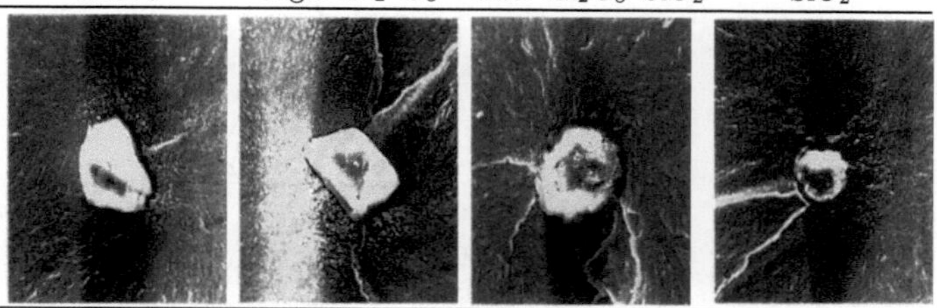			
Chemical Composition (%)	MgO	0	23.1	1.3	0
	CaO	0.2	0.2	20.4	0.2
	Al_2O_3	96.7	68.4	29.6	8.9
	SiO_2	2.9	5.3	44.8	86.5
	MnO	0.2	3.0	3.9	4.4

Fig. 2.9. Appearance of nonmetallic inclusions on fracture surface [3]

furnace, can be made at VSC (vacuum slag cleaner), the reduction refinement functions, such as desulfurization or the control of non-metallic inclusion can be made at LF, de-gassing and composition control function can be made at RH, and the melt oxidization refinement can be left only at the arc furnace.

The automotive engine valve spring which can be exposed at severe stress conditions for a long time duration, is well known that the fatigue fractures initiate from the nonmetallic inclusion of size 30–40 μm around the surface. At present, most steel materials for valve springs are manufactured by the converter steel making method. An example of nonmetallic inclusions, which appears in the fracture surface of a valve spring, is shown in Fig. 2.9 [3]. Oil tempered wire for valve springs was prepared from experimentally melted six kinds of SiCr steel, having six different kinds of oxide type non-metallic inclusions in the ternary phase diagram of $CaO{-}Al_2O_3{-}SiO_2$ in Fig. 2.10 (a), which marked by ○ from A to F respectively. Fig. 2.10 (b) shows the results of rotating bending fatigue test on those steels. It can be seen from the Fig. 2.10 (b), that the fatigue fracture by non-metallic inclusion can be prevented in the case of E, where the compositions of inclusions have low melting point.

The composition of non-metallic inclusions, which comes out in the fatigue fracture surface, can be said as alumina (Al_2O_3), spinal (MnO, MgO, or $CaO{-}Al_2O_3$), and silica (SiO_2). These non-metallic inclusions can be said to have high melting point with high hardness. Although the relation between the fatigue life and the size of non-metallic inclusion can be shown in Fig. 2.11, the size of nonmetallic inclusion and fatigue life show some correlations. It can be said that the decrease of absolute number of non-metallic inclusion is important from the viewpoint of reliability, and controlling the size of nonmetallic inclusions is essential for improving the fatigue life.

Under a constant pressure and temperature, the product of the amount of carbon and that of oxygen in equilibrium with the molten steel becomes constant. Therefore, as for the spring steel with high carbon, the amount of oxygen becomes lower, compared with mild steel. Under the reduced pressure

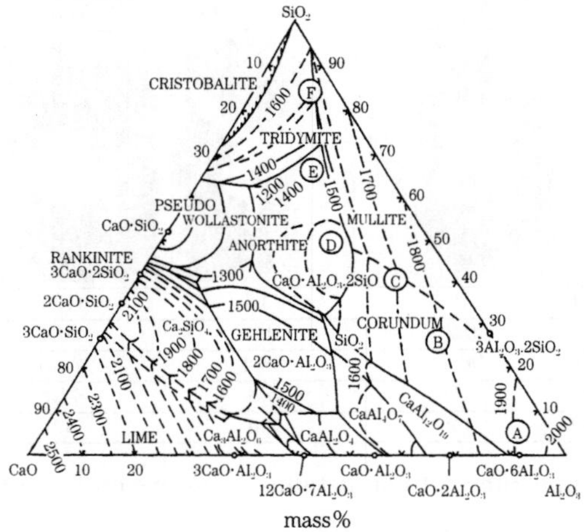

(a) Non metallic inclusion composition

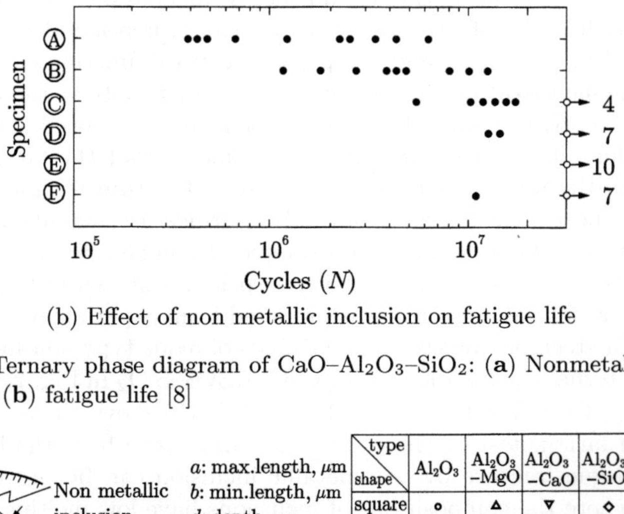

(b) Effect of non metallic inclusion on fatigue life

Fig. 2.10. Ternary phase diagram of CaO–Al_2O_3–SiO_2: (**a**) Nonmetallic inclusion composition (**b**) fatigue life [8]

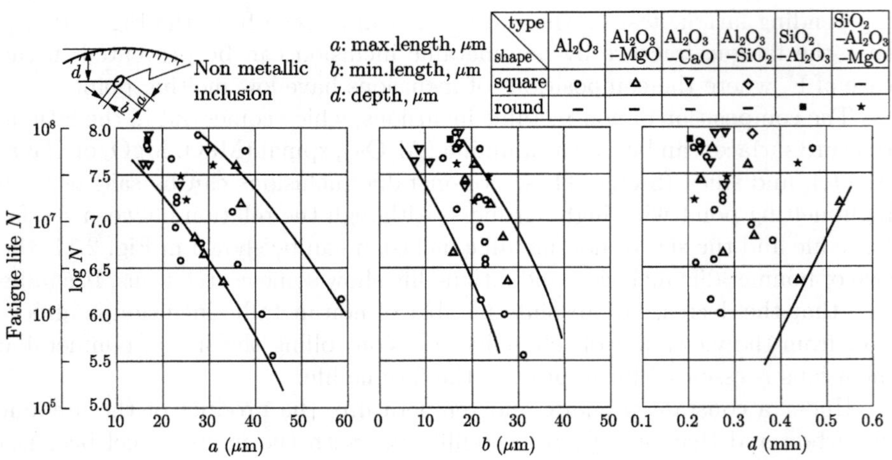

type / shape	Al_2O_3	Al_2O_3 -MgO	Al_2O_3 -CaO	Al_2O_3 -SiO_2	SiO_2 -Al_2O_3	SiO_2 -Al_2O_3 -MgO
square	○	△	▽	◇	–	–
round	–	–	–	–	■	★

Fig. 2.11. Appearance of nonmetallic inclusions on fracture surface

or in vacuum, the amount of oxygen in molten steel becomes lower compared with the normal pressure. In the solidified steel, since the solubility of oxygen in steel is very small, most oxygen becomes oxides. Therefore, the reduction of oxygen in steel is effective in reducing the absolute number of oxide type non-metallic inclusions. Besides the carbon, aluminum, silicon, manganese, calcium, can be used for deoxidizing. However, these metal-deoxidizing agents can form non-metallic inclusion (metal oxide) in steel. Since the specific gravity of the formed nonmetallic inclusions is smaller than the steel, they can gradually float up to be the slag while the part of them remain in steel.

Although in order to control non-metallic inclusions, the refining process out of furnace can be applied, typical measures where non-metallic inclusions can be made harmless, fall into methods – extreamly low oxygen method and control by Si deoxidizings as follows.

(v) Extremely low oxygen method

This is the method for reducing the absolute quantity of non-metallic inclusions while reducing the amount of oxygen in steel as much as possible. The steel can be classified into rimmed steel and killed steel. All spring steels, bearing steels and machine structural steels are killed steel. In killed steel, the oxygen in steel will prevent the formation of CO gas by reacting with the carbon during the solidification. This means that the oxygen can be fixed with the deoxidize agents of Al, Ca, Si, and Mn. The oxygen supply from slag, refractory, and the atmosphere is cut off after that, and the degassing processing can be carried out. The killed steel, depending upon their type of deoxidization, is also classified into Si killed steel, aluminum killed steel, calcium deoxidization type steel (an example, free cutting steel).

As for the material for valve springs, Cr–V steel (oil tempered wire for valve springs, material of JIS SWOCV-V), steel for bearing represented by SUJ2, and most steels for machine structures, the aluminum deoxidization method are applied to decrease the non-metallic inclusions. Although the low oxygen method by this aluminum deoxidization has the problems where hard non-metallic inclusions of Al_2O_3 type remain, and Al_2O_3 particles in the shape of cohesive clusters exist, the current steel making processes which can reduce the amount of oxygen, make it possible to control the amount of oxygen under 10 ppm. In steel for machine structure, around 0.02% of aluminum can be normally added. However, in these extremely low oxygen steel, the oxygen can mostly become Al_2O_3 without adding aluminum so much. Regarding the fatigue strength of spring steel with extremely low oxygen steel, there is a report by Saito [2]. This report also describes the influence of Ti type inclusions.

(vi) Control of non-metallic inclusions by Si deoxidization

The non-metallic inclusion composition controlling method by the Si deoxidization is that the non-metallic inclusions can be formed as multi-type low melting point oxides of which the main composition is SiO_2. The inclusions can be elongated or shortened by the forming processes such as rolling and cold drawing to be harmless [3–7]. The relation between the nonmetallic inclusion

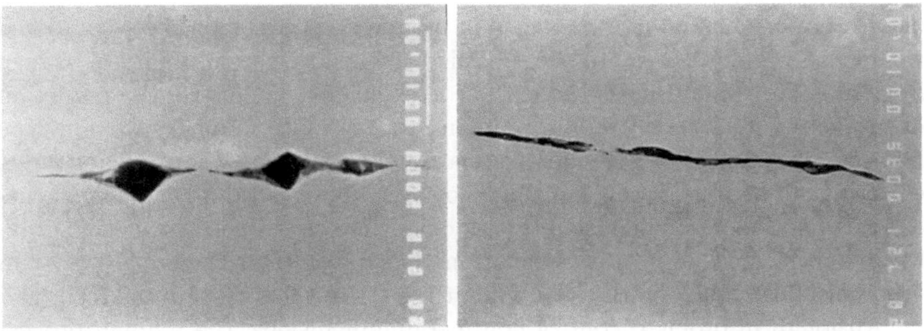

Chemical compositions (mass%)				
CaO	Al_2O_3	SiO_2	MnO	MgO
0	69.7	0	11.6	18.7

Chemical compositions (mass%)				
CaO	Al_2O_3	SiO_2	MnO	MgO
19.2	18.5	49.3	8.2	4.5

Fig. 2.12. Shape deference by chemical composition of non-metallic inclusion (billet) [3]

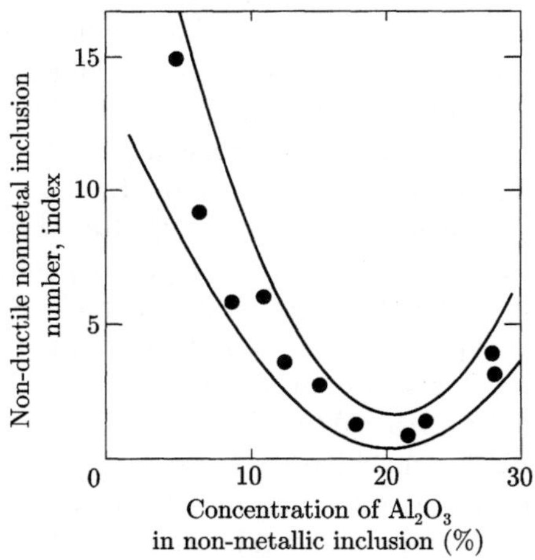

Fig. 2.13. Number of hard non-metallic inclusion and Al_2O_3 concentration in non-metallic inclusion [3]

compositions and shapes is shown in Fig. 2.12. It can be seen that when the compositions of nonmetallic inclusion are controlled, the shape of the inclusion becomes elongated. It was shown in the Fig. 2.10, that the low melting point nonmetallic inclusions whose melting point is around 1,200–1,400°C be not the origin of fatigue fracture. The relation between the number of inclusions with high hardness and the concentration of Al_2O_3 is shown in Fig. 2.13, in the case of CaO–Al_2O_3–SiO_2 type inclusions. It can be said from the Fig. 2.13, that the content of Al_2O_3 is around 20% to minimize the inclusions with high hardness. From this point, it can be said that the quantity of oxygen in steel, normally around 20 ppm, is required, compared with low oxygen method.

In the fatigue test result of the Si–Cr steel for valve springs manufactured by the conventional method, the extreme low oxygen method (Al: 0.033%, O: 0.0007%), and the non-metallic inclusion composition controlling method by Si deoxidization (Al: 0.001%, O: 0.0019%), there are reports that the inclusion distributions and the fatigue life show worse results with the above order [8]. In the sense of improvement in fuel consumption of engine and high power output with the valve spring, further high strengthening is required with further thinning of nonmetallic inclusion and reduction of their numbers.

References

1. Steel Manual 3rd edition, Maruzen, (1979) p. 551
2. Kiyoshi Morii, Makoto Saito: Electric Furnace Steel, 51 (1980) p. 4
3. Nishiyama commemoration technical lecture No. 127
4. Kiyoshi Shiwaku, Yoshiro Yamada, Jiro Koarai and Yasunobu Kawaguchi: (1986) SAE Technical Paper Series 850364 and Yoshiro Yamada, Jiro Koarai, Yasunobu Kawaguchi, Osamu Ishigami, Nobuhiko Ibaraki: Tran. JSSR No. 31, 1
5. Toshiaki Hagiwara, Akira Kawami, Akihiko Ueno, Atushi Kido, Michael J.Urbancic and Ralph D. Friedel: Wire J., International, April, 29 (1991)
6. Jiro Tominaga: SAE Tech. Paper Series 91044 (1991–2)
7. Jun Kawahara, Koji Tanabe, Toshio Banno and Masashi Yoshida: Wire J., International, 55 (1992)
8. Takehiko Ohshiro, Tatsuo Iked, Hiroyuki Matsuyama, Susumu Okushima, Yasuhiko Oki and Nobuhiko Ibaraki: Stahl und Eisen 109, 1011 (1989)

(4) Continuous casting method

In the past, the refined molten steel was poured into the mold to make ingots, which was supplied to the next rolling process. Most steel produced in large quantities by the advantage of yield ratio improvement has been produced recently by countinuous casting method. Spring steel such as SUP7 and SUP9 can be manufactured by this method. Fig. 2.14 shows the outline of a bloom continuous casting machine of special steel.

The rolling ratio (reduction of cross-section) can be an important point on spring steel rolled from the long and slender bloom manufactured by continuous casting. Although there are a lot of research works related to the rolling ratio, the conclusions can be said to be 6–10 for the normal steel, 8–12 for special steel and around 18 for high strength special steel. When the casting conditions in the continuous casting are appropriate, the quality should be the same as that of the conventional ingot casting – blooming processing.

The relation between the rolling ratio and mechanical properties in case of continuous cast billet is shown in Fig. 2.15.

Since the continuous cast steel is rapidly solidified in a water-cooled copper mold, it has been said that the surface separation of non-metallic inclusions becomes difficult compared with the steel made of the ingot. However, with the introduction of ladle refining furnace, developments of the improved reoxidization prevention technology in the continuous caster, and the electric

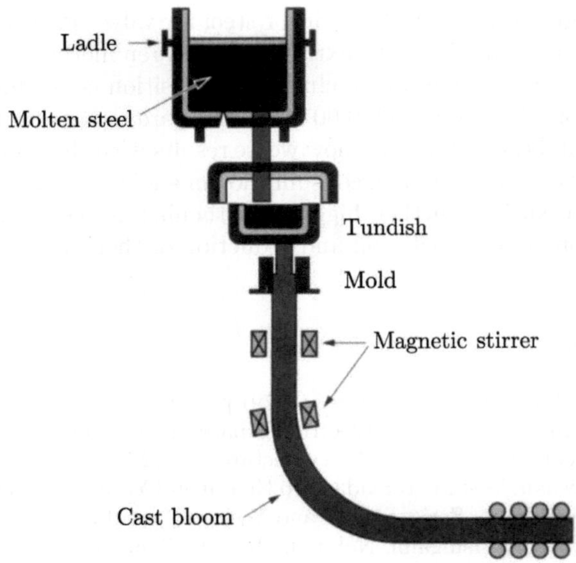

Fig. 2.14. Outline of bloom continuous casting

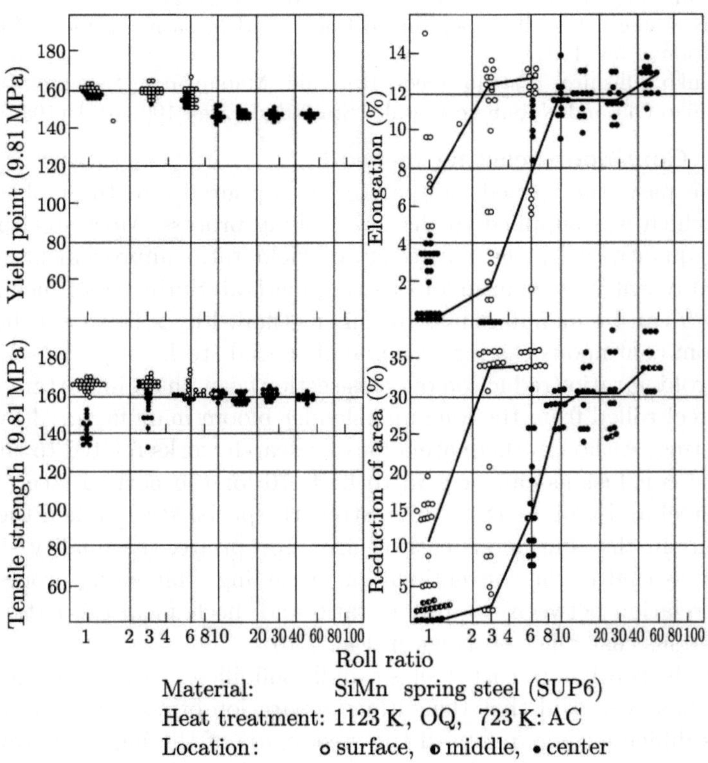

Fig. 2.15. Effect of roll ratio on the mechanical property of continuous cast billet

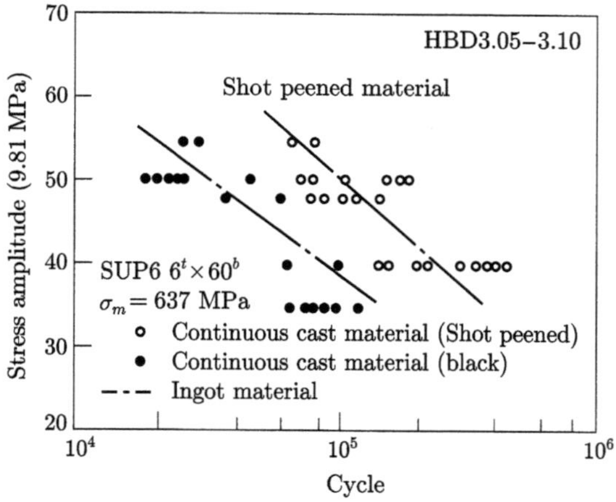

Fig. 2.16. Fatigue strength of continuous cast and ingot material

magnetic stirrer during the solidification process, the spring steel with sufficient degree of purity can be obtained. As for the fatigue strength of spring steel, Fig. 2.16 shows that the fatigue strength of the continuous cast steel is almost the same as that of the ingot cast steel.

(5) Rolling process

The bloom, the billet or the ingot can be hot-worked in the rolling process to have higher quality and the designated cross-section according to the customer's requirement. The main purpose of rolling can be the improvement of the cast structure and having a predetermined cross-section. By hot-rolling with the appropriate rolling ratio, the dendrite in as-cast structure can be improved to fine rolled microstructures, and the defects such as pipe can be also cured. However, if the rolling ratio is small, these defects can remain in the rolled product and possibly have the mechanical properties after heat treatment deteriorated. Therefore, for the selection of bloom size, it can be required to take enough rolling ratio into consideration.

Since the dimension of spring steel material can give a large impact to the spring rate, the commercial requirements for the accuracy of dimension can be more severe compared with the normal steel materials. Most of the rolled spring steel can be classified into flat steel (flat bar) and round steel (round bar and wire rod). In producing flat steel, rather thick square section bar is produced at the roughing train, which elongates the bar consecutively in the longitudinal direction to the required section finally, controlling both the thickness and width dimensions with flat rolls at the intermediate and finishing train. The change of the cross-section during the rolling processes can be shown in Fig. 2.17. The flat steel is mainly used for leaf springs. Since the cold secondary processing is not normally applied, the rolled size should be the same as the final product. Therefore, at rolling mills, the manufacturing

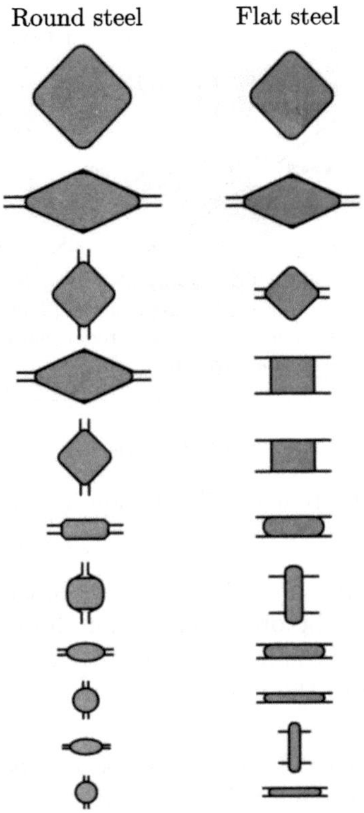

Fig. 2.17. Change of cross section during rolling processes of flat bar and round bar

line where the thickness and the width can meet with broad range size, is required.

In round steel, the method of being arranging the H-V (H: Horizontal, V: Vertical) tandem rolling mill from rough rolling to finish rolling, continuously can make it possible to produce the round steel with sufficient roundness.

In recent years, applying skin pass rolling at the final stage of finish rolling, the precision rolling material and ultra precise rolling material have been developed and used in production. As an example, the diameter of 20 mm with ±0.10 mm tolerance becomes possible, see Fig. 2.18 [1]. In order to have cost performance, these high precision material without the bar processing such as peeling or cold drawing, or with less cold drawing process, has been considered to apply to the commercial use.

In the rolling of spring steel, the surface defects and decarburization can be focal points. When these defects remain in the finished product, the fatigue property of springs can be affected to cause any early failure. As for the surface defects, severe attentions should be paid to the control of guide and rolls in

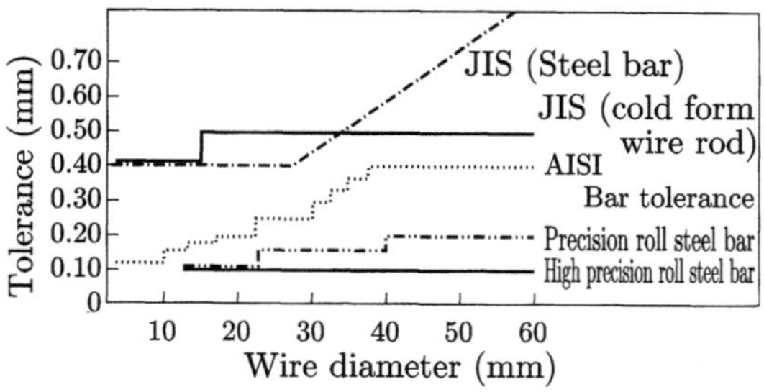

Fig. 2.18. example of size tolerance for high precision rolled material [1]

the rolling process of spring steel to prevent the surface defects such as scab, roll mark, lap, scratch, etc.

The surface decarburization occurs during the re-heating of the ingot, billet or bloom. Generally, the higher the heating temperature is and the longer the heating time is, the more the decarburization can be generated. Therefore, it is important that the heating temperature is kept as low as possible in the rolling range, and the heating time is kept as short as possible. By providing an induction heater around the exit of a re-heating furnace, the practice where the heating time at high temperature zone can be shortened to prevent the decarburization, is also reported [2]. In the case of wire rod productions, since the cooling speed between the wires becomes low, high Si steel like SUP7 can show partial ferrite decarburization. Therefore, increasing the cooling speed by applying the direct cooling by wind or making rapid cooling in the decarburization range between 600 and 900° in the conveyor type cooling bed, have been applied to the production. When the cooling speed in the cooling bed is too high, the hardness of rod becomes high to give some problem at the following processes such as cutting or machining. In this case, the annealing process is required for some material grade. When the cooling becomes unstable, it becomes difficult to keep the straightness of wire and the wire can show some bend. For the remarkable bend, the straightening process might be required. Special attention should be paid to the steel with good hardenability.

References

1. Ogura et al.: (1993) Tetsu and Hagane, 79, No. 3, p. 13
2. Matsue et al.: (1993) Aichi Gihou, 13

(6) Heat treatment and pickling process

The purpose of heat-treatment during the spring steel production can be to secure the sequent process such as secondary bar processing adequately. Most typical heat-treatments are low temperature annealing and spheroidizing annealing. Sometmes, full annealing and normalizing can be carried out [1] [3].

The low temperature annealing is keeping the materials under the temperature range between 450°C and A_1 transformation temperature and cooling gradually. The purposes are to remove the residual stress inside of the material, to recover the crystal distorted by the plastic deformation, and to make the material softened.

The spheroidizing annealing is keeping the materials around the A_{c1} transformation temperature (c means the heating status) for a certain time and cooling to the A_{r1} transformation temperature (r means the cooling status) by the cooling speed of 10–20°C/hr. The purpose is to have the cementite precipitates changed from the lamellar or net shape to spherical shape. Since this heat-treatment can soften the material by a great extent, this can be applied when the formability of materials is required.

Figure 2.19 shows the roller hearth type continuous annealing furnace. This furnace is used for the treatment of low temperature annealing or spheroidizing annealing. While moving at a constant speed inside the furnace divided into about 3–15 zones, the temperature described above can be obtained. Besides the temperature in each zone, the atmosphere can be automatically controlled. The purpose of controlling the atmosphere is to prevent the steel material from decarburization. Applying the indirect heating with radiant tubings and keeping the atmosphere under an appropriate concentration of hydrocarbon cracking gas (RX, DX) or other gas, depending upon the temperature and steel grade (carbon equivalent).

To improve the spring characteristics, the heat-treatment after secondary processing, is normally quenching and tempering. The low-temperature tempering for the purpose of improving the elastic limit, proof stress, and fatigue

Fig. 2.19. Roller hearth type continuous annealing furnace

limit, and the precipitation-hardening can be applied for stainless steel and super-alloy.

The quenching is a heat-treatment where the material is heated to the temperature about 30–50°C higher than the A_3 transformation temperature, and immersed into water or oil (sometimes air cooling in high alloy steel) and cooled rapidly, to have martensite structure, avoiding the primary ferrite and pearlite transformation. Although the martensite structure in the case of high-medium carbon steel and low alloy steel can be said to be the status of the alpha iron with carbon atoms in oversaturated solid solution, it has body centered tetragonal structure (B.C.T) with interstitial carbon atoms preferentially aligned so as to elongate one crystal direction (or C axis) only. Since the solution content of carbon is high, there are many lattice defects (dislocation) and the crystal structure is fine, the hardness of the martensite of medium-high carbon steel is as high as 60–65 HRC.

As for the sequent process, tempering with a comparatively high temperature below the transformation temperature A_1 point, and cooling it by a suitable speed, parts of the carbon, which have been in solid solution in alpha-iron, react with iron, and precipitate as fine cementite, and it is called tempered martensitic structure. By this tempering treatment, the hard and brittle martensite obtained by hardening can be softened in predetermined hardness, and the ductility is recovered. Moreover, if the material is left as hardened, a quench crack may be caused by internal stress, that is why tempering is performed immediately in many cases. The spring steel has a high yield point; besides suitable high ductility is required, therefore, it is tempered at about 723 K (450°C).

There are cases where rolled spring steel is acid pickled either to remove scale generated during rod rolling or heat treatment or to inspect surface defects. The optimal pickling condition depends not only on steel grade to be pickled, but also on the aspect of scale which is affected by the prior heat treatment condition such as the atmosphere, temperature and time. In acid pickling spring steels, sulfuric acid or hydrochloric acid solution is used. Sulfuric acid is being used widely since it is cheap although heating is necessary and smut is easily generated on the steel surface. When using hydrochloric acid, fume removal equipment such as roof exhaust is often necessary and enough water rinse is desired to avoid pitting corrosion caused by acid remaining on steel surface, although de-scaling is possible at relatively low temperature or at room temperature.

As examples of reaction between scale and acid solution, magnetite and sulfuric acid reaction equation, and wustite and hydrochloric acid reaction equation are shown below.

$$Fe_3O_4 + 4H_2SO_4 \rightarrow FeSO_4 + Fe_2(SO_4)_3 + 4H_2O \quad (2.1)$$

$$FeO + 2HCl \rightarrow FeCl_2 + H_2O \quad (2.2)$$

In fact, the de-scale effect by the reaction with base metal is also large, and various reactions advance simultaneously.

References

1. Iron and Steel Institute of Japan: Heat treatment of steel Latest trend of the 5th edition of revision: Maruzen, (1989)
2. Special edition spring material: Special Steel, 38, and (1989) p. 7
3. Special-edition wire drawing secondary processing: Special Steel, 39, (1990), p. 6

(7) Secondary bar processing

Although the flat shaped material for leaf springs (flat bar) can be produced by hot rolling, the wire rod for coil spring can be normally applied the secondary bar processing such as cold drawing or peeling. It can be said that the secondary processing is carried out, in order to remove the decarburization and surface defects of the hot rolled steel, and also to have an intermediate diameter between the diameters obtained by hot-rolling processes and the diameter with small tolerances. In hot working processes, since the surface is exposed to air, steam, and carbon dioxide, the oxidization and decarburization of steel surface can occur. The oxidization and decarburization can be normally generated simultaneously. The oxidized scale can be easily removed by the sequent shot blast, or pickling process. However, since the decarburization can be generated from the surface to inside, it can be required to have the decarburization layer shallow by carbon restoration, or to have the decarburization layer removed mechanically. While the available diameters of hot-rolled bars and wire rods are normally limited to the sizes at intervals of 0.5–1.0 mm, spring makers require the sizes at intervals of 0.1 mm in diameter. For this reason, secondary processing shown in Table 2.2 such as peeling, grinding or wire drawing is necessary. The peeling (turning) machine shown in Fig. 2.20, makes it possible to remove the suface layer of 0.2–0.7 mm in the radius by the cutter with four cutting edges. Since the surface as peeled is rough, the light reduction by polishing roll can be normally carried out to make the surface roughness less than 10μm $R_{\max}$. By using the peeling machine, it becomes possible to have the surface defects removed completely and have any preferable diameter. The tapered bar can be also manufactured by the numerical control. As for the shaving machine shown in Fig. 2.21, the application can be for the wire diameter less than 15 mm. This process is that the wire goes through the cutting die of cylindrical shape of which the inner diameter is the finish diameter. Although the cutting speed is higher than that of peeling machine, the cutting depth is around 0.1–0.15 mm in the radius.

Although a partial flaw removal machine is the method of not carrying out all circumference and full length cutting, pinpointing (above) the flaw position with an eddy current machine, and removing only the portion and is good for the surface flaw removal and improvement in size accuracy, but removal of a decarburization layer cannot be performed.

The grinding processing by the centerless grinder is shown in Fig. 2.22. Although the surface roughness less than 5 μm and the dimensional accuracy can be obtained, the grinding amount is less than 0.15 mm in the diameter for one grinding. Therefore, it is unsuitable to be applied for intermediate surface

Table 2.2. Secondary bar processing

Processing type	Machine	Application		
Cutting	Bar turning machine	Bar	→	Bar
	Peeling machine	Wire rod	→	Wire rod
			→	Steel bar
	Shaving Machine	Wire rod	→	Wire rod
	Partial shaving	Wire rod	→	Wire rod
Grinding	Centerless grinder	Steel bar	→	Steel bar
	Wire drawing	Wire rod	→	Wire rod
Drawing	Draw bench	bar	→	bar
	Combined machine	Wire rod	→	Steel bar

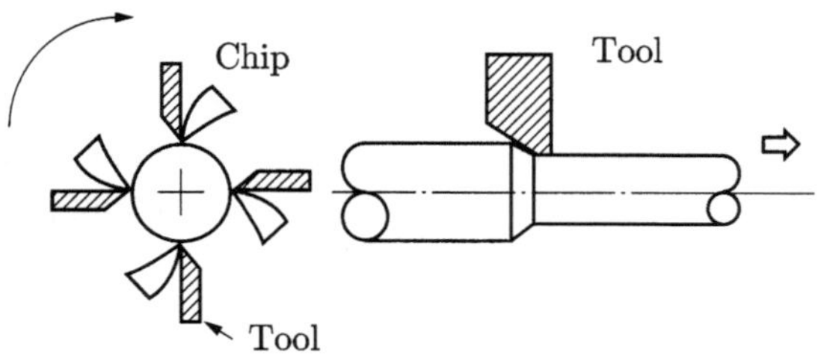

Fig. 2.20. Cutting by 4 piece tools

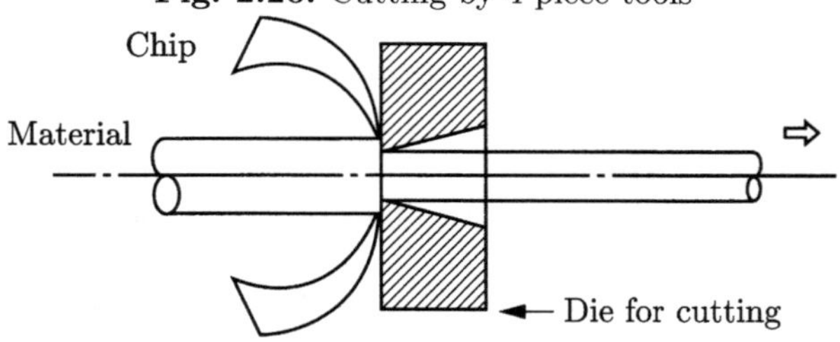

Fig. 2.21. Rod shaving

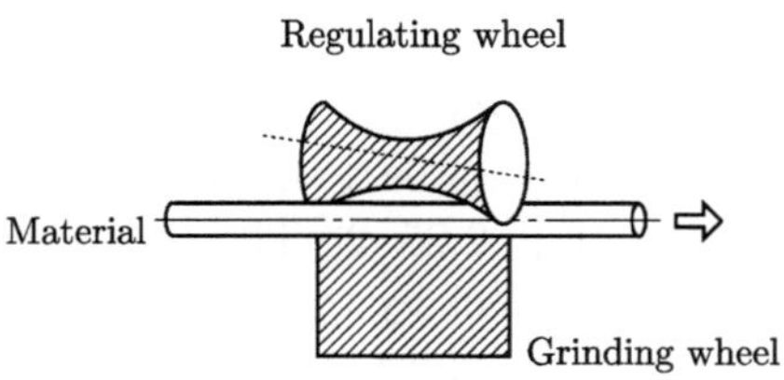

Fig. 2.22. Centerless Grinder

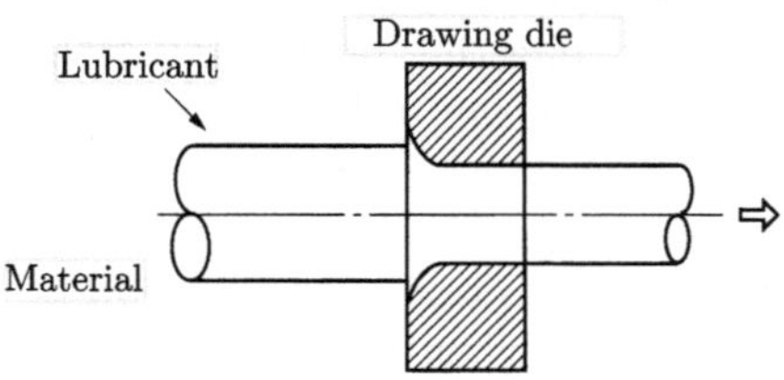

Fig. 2.23. Drawing process

removing. Although the application to hot rolled round bars can be seen, the application to peeled bar or cold drawn bar can be said to be normal. Since the cold drawing process shown in Fig. 2.23 is not the processing for removing the surface layer such as peeling or grinding, the removal of surface defects is not possible. However, due to the good productivity, yield rate and dimensional

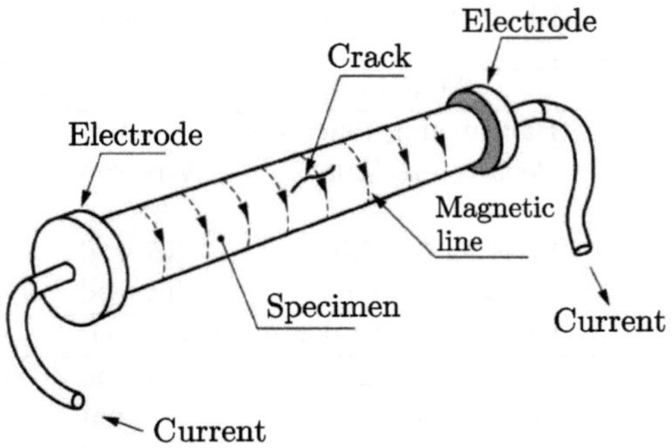

Fig. 2.24. Magnetic particle detector

accuracy, it is suitable to be applied for mass production. As shown in the Table 2.2, there are three kinds of processing machines. As shown in the Fig. 2.23, the wire rod or bar de-scaled by shot blasting or pickled, can be drawn to the predetermined diameter. The appropriate lubrication is required. Although the outline of secondary bar processing was described, in most case, the bar can be manufactured by each independent processing or combined process such as drawing plus centerless grinding. A good understanding of the manufacturing process of springs and its application can be required to decide the appropriate manufacturing process under the appropriate quality and reasonable cost.

(8) Inspection process

The main inspection items for steel bars are dimension, length, bend, surface defect and internal defect. The surface and internal defect should be inspected carefully for spring steel. Besides the visual inspection of surface defects, the non-destructive testing such as magnetic particle flaw detection, leakage magnetic flux detection, or eddy current flaw detection can be applied. The magnetic particle flaw detection is the method of magnetizing the product, and collecting magnetic particles around the defect to make the defects visible, as shown in Fig. 2.24 [1]. The leakage magnetic flux detection is the method of detecting the flux leaked from the defect directly, by a magnetic sensor instead of magnetic particles, as shown in Fig. 2.25 [2]. The eddy current flaw detection is the method of catching the defect by the eddy current disturbance when the alternating magnetic field is applied to the product. There is the through type method using circumferential through-type coil, where the bar can pass through the fixed coil and the rotating eddy probe coil method, where the detection coil rotates around the bar with high-speed, as shown in Fig. 2.26 and Fig. 2.27.

The internal defects can be represented as pipe, segregation and nonmetallic inclusion inside of steel bar. The inspection can be normally conducted with the ultrasonic flaw detector.

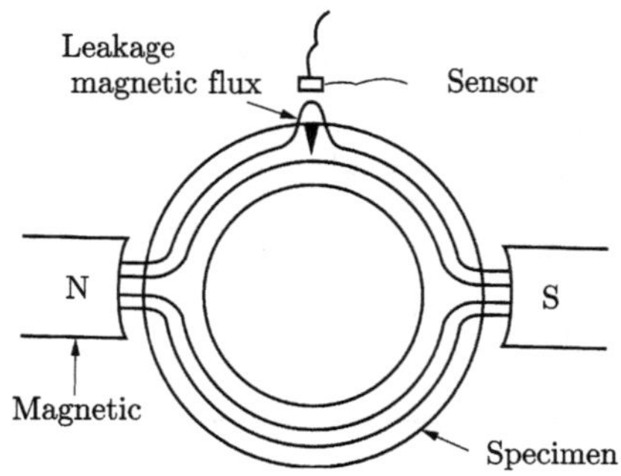

Fig. 2.25. Leakage magnetic flux detector

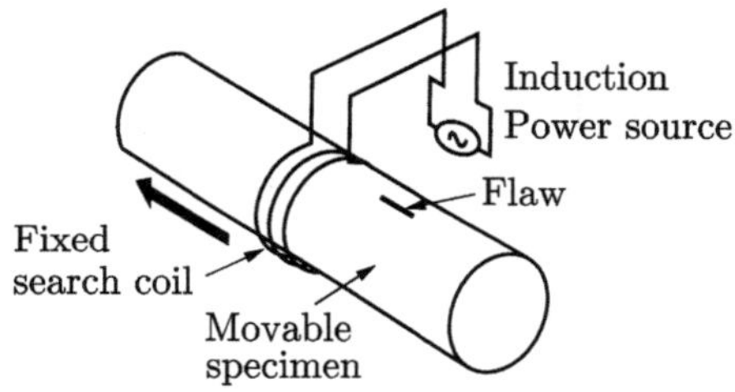

Fig. 2.26. Through type eddy current flaw detector

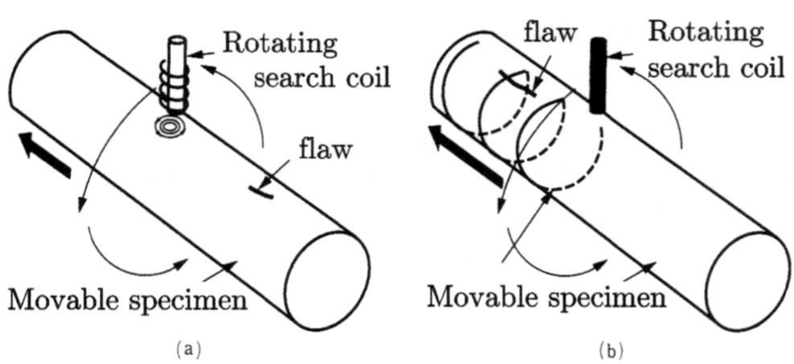

Fig. 2.27. Rotating coil type eddy current flaw detector

The ultrasonic flaw detection is the method of catching internal defects by the change of the ultrasonic wave from transmitting to receiving, when the ultrasonic is applied to the bar through the search unit, as shown in Fig. 2.28.

A wire rod product is coiled to a ring shape. The coiled wire rod can be inspected after cutting off each one sample from both ends. Although the main items of inspection are normally dimensions and surface defects, the cut sample at the coil end can be inspected the surface defects after shot blasting by visual method or magnetic-particles flaw detection. However, since this inspection cannot cover the full coil length, the hot eddy current flaw detection before coiling, has recently been applied.

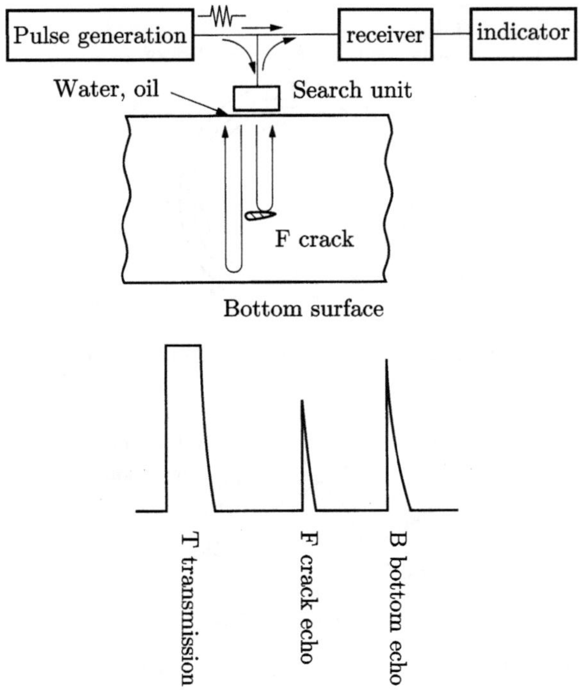

Fig. 2.28. Ultrasonic flaw detector

As for flat bar material, the inspection items can be also dimensions, length, bend, and surface defects. Although the inspection of surface defects can be carried out visually, the non-destructive test by the eddy current flaw detection has recently been in practical use.

References

1. Iron and Steel Institute of Japan: The magnetic-particle testing (1990) of a steel product, the disclosure magnetic-particle method (1990), P. 67
2. Iron and Steel Institute of Japan: Leakage magnetic flux flaw detection P. 44
3. Iron and Steel Institute of Japan: Ultrasonic magnetic-particle method (1993), P. 31

2.1.2 Various Characteristics Required for Hot-Formed Spring Steel

(1) Surface quality

As for the hot-formed spring, the surface of steel material can be normally used as the surface of springs. Therefore, the surface quality of steel material can directly give the impact on the fatigue life of spring. In case of hot rolled steel material, there would be uneven surface roughness around 10 to 100 μm because of high temperature heating and other factors.

In case of spring steel, since the surface roughness can be one of great concerns for steel mill, it has been controlled around 10 to 20 μm. As comparison, it can be noted that the surface roughness of grinded materials is less than 12 μm. In the manufacturing processes of hot rolled spring steel, there are many causes to make the surface quality deteriorated such as surface flaw, roll pass design and its regulation, impropriety nature of surface state, unsuitable handling and so on. Figure 2.29–2.37 shows examples of the main surface defects on hot-rolled steel material [1].

Since the overheat can cause not only rough surface but also austenite grain growth, oxidization and decarburization of the austenite grain boundary below the surface, careful attentions during the processes should be required. The scabs and dents can be seen partially, and shall be removed carefully. The burning, scratch and dent can mostly occur during the manufacturing processes of hot-formed spring. Considering the conditions of spring applications, the flaw perpendicular to the rolling direction, such as scab and dent can be said to be harmful, as shown in Fig. 2.38 [1]. Figure 2.39 is an example of the fatigue fracture that occurs from a surface defect of the spring steel [1]. Figure 2.40 shows the relation between fatigue life and the depth of seam of coil spring made of SUP6. It can be seen that even small defects affect the fatigue life remarkably [2].

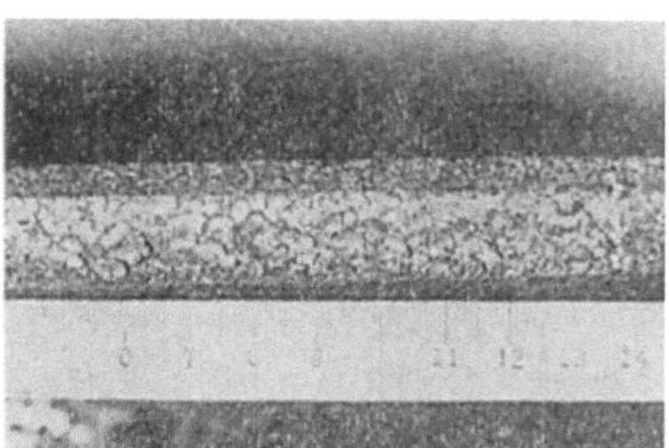

Fig. 2.29. Overheat (Burning)

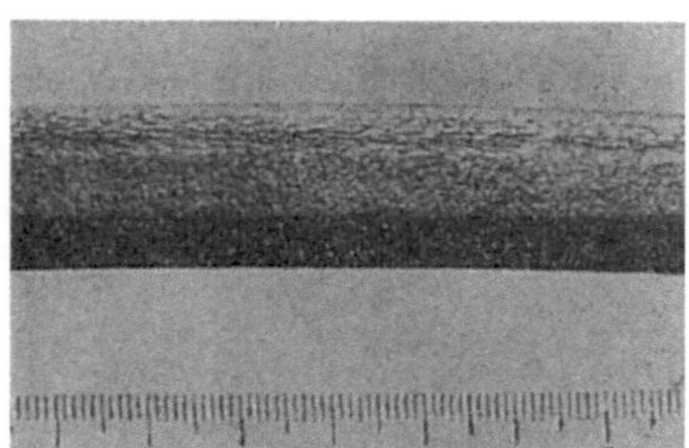

Fig. 2.30. Rough surface

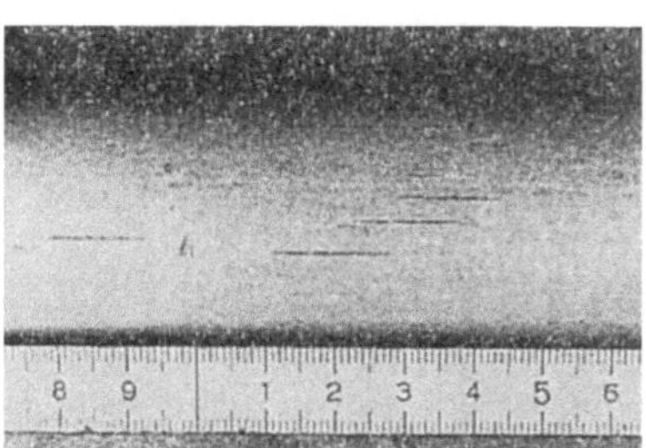

Fig. 2.31. Seam

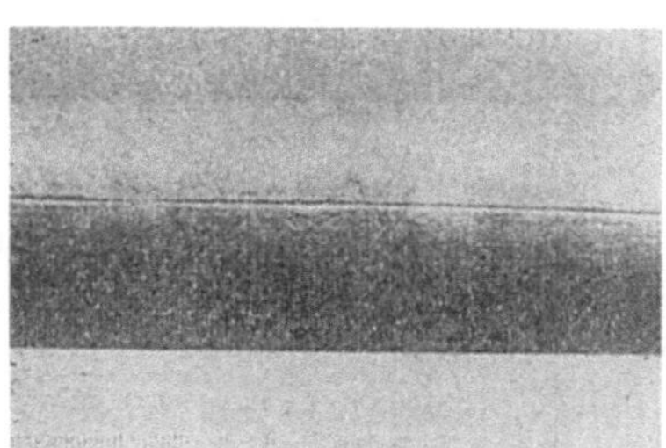

Fig. 2.32. Scratch

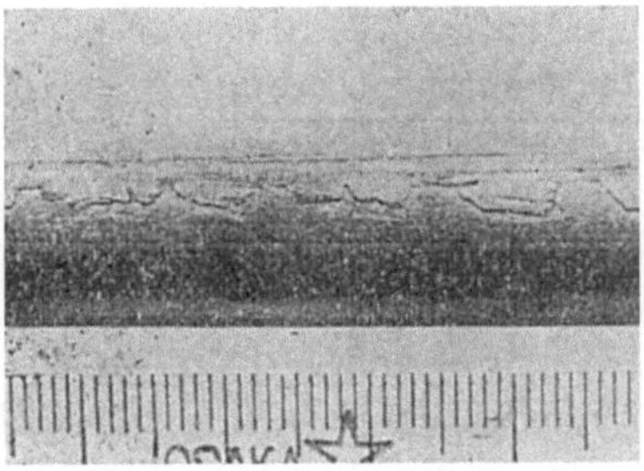

Fig. 2.33. Overlap

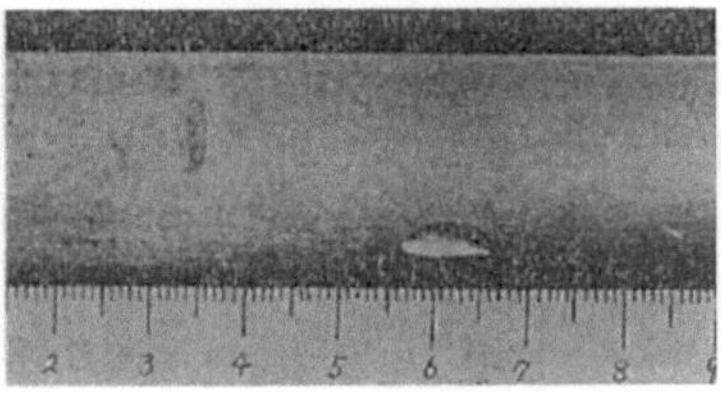

Fig. 2.34. Dent

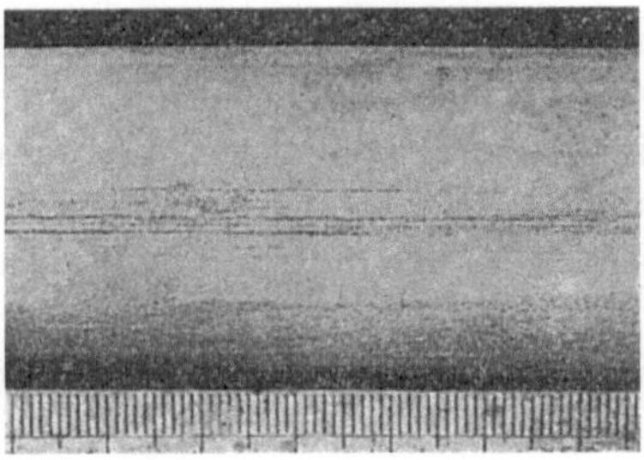

Fig. 2.35. Wrinkle

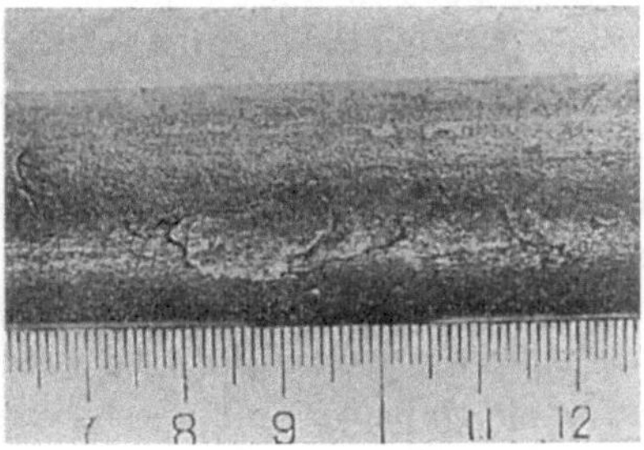

Fig. 2.36. Scab

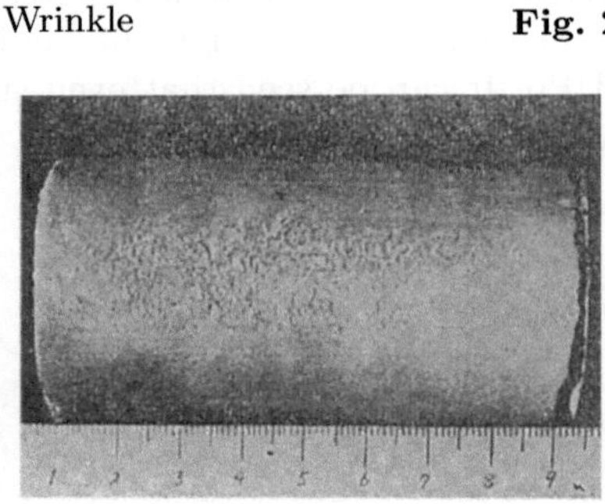

Fig. 2.37. Scale mark

References

1. The 3rd edition of Spring, Maruzen, (1982), pp. 33–35
2. K. Koyama and K. Ota: Transactions of JSSR, (1973), p. 64

(2) Decarburization

When steel is heated to high temperature, the surface layer will react with the oxygen in atmosphere, carbon dioxide, and steam, to cause oxidization and decarburization.

The oxidization is the phenomenon, where the base metal of steel reacts with the oxygen in atmosphere to form the oxide film. If the oxidization is progressed further, it will become the scale which can be peeled off from

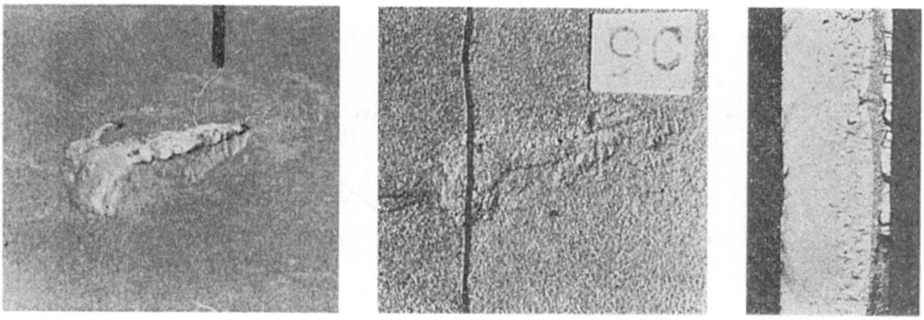

(a) As hot-rolled　(b) Relation between the defect and fracture position　(c) Fracture surface

Fig. 2.38. Example of fatigue fracture caused by scab formation (Fracture at 21.6×10^4 cycles against the average fatigue life, 200×10^4) [1]

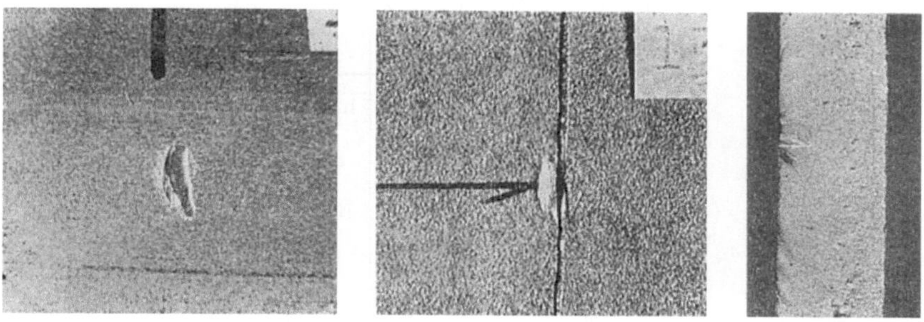

(a) As hot-rolled　(b) Relation between the defect and fracture position　(c) Fracture surface

Fig. 2.39. Example of fatigue fracture caused by dent (Fracture at 21.6×10^4 cycles against the average fatigue life, 200×10^4) [1]

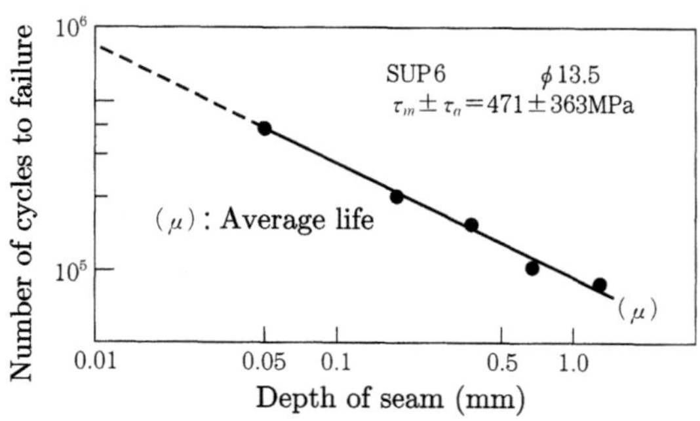

Fig. 2.40. Seam and fatigue life [2]

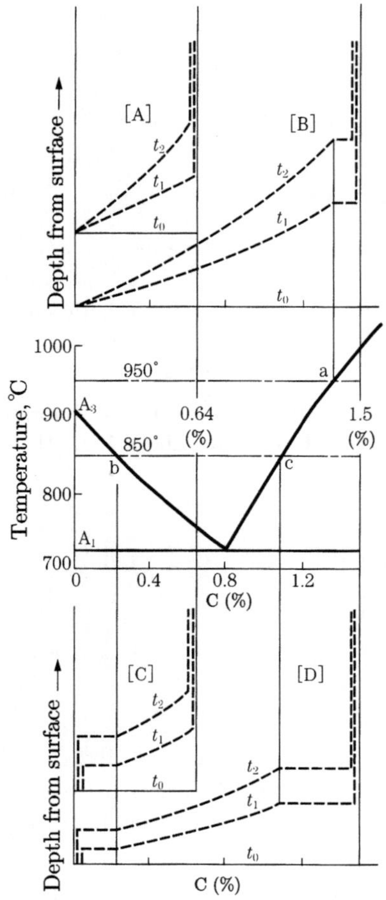

Fig. 2.41. Carbon content of decarburized steel [1]

the surface. The oxidization can occur drastically to form thick scale, as the heating temperature is high and/or heating time is long.

The decarburization is the phenomenon where carbon in steel combine with oxygen in atmosphere, carbon dioxide or hydrogen, and escapes as carbon monoxide or methane from the steel surface to reduce the carbon content of steel surface. The decarburization can be generated when the speed of carbon oxidization is higher than that of iron oxidization.

The oxidization and decarburization can occur simultaneously. However when the speed of oxidization is equal to that of decarburization or is larger than that of decarburization, the only scale can be formed on the surface of steel without any decarburization under the scale. However, since the decarburization normally progresses more quickly than the oxidization, the decarburized layer and the oxidized scale can be simultaneously formed. Figure 2.41

shows the decarburized state of 0.64% carbon steel, [A] and [C], and 1.5% carbon steel, [B] and [D], under heating at high temperature. The center figure shows the part of Fe–C phase diagram. The upper and lower Figure show the change of carbon content, in case of heating at 950°C (1223 K) higher than the A_3 transformation point and in case of heating at 850°C (1123 K) between the A_1 and A_3 transformation point, respectively. The t_1 and t_2 show the lapsed time after the decarburization starts. It can be seen from the Fig. 2.41, that when the decarburization occurs above the A_3 point the deep decarburized layer is formed with changing the carbon content slowly, and when the decarburization occurs below the A_3 point the ferrite layer is formed in the surface with changing the carbon content sharply. Since the diffusion of carbon occurs easily at high temperature, the fall of carbon content is slow and the decarburized layer is deeper.

It is said that Cr can decrease the carbon diffusion speed, and form the tight scale on the surface to prevent the penetration of oxygen, and Si, W, Mo can promote the decarburization. Since the SUP7, which has 2% Silicon for high permanent set resistance and has A_3 point shifted to high temperature side, easily generates the above-mentioned ferrite decarburization, it can be required to pay attentions to decarburization when rolling or heating.

Figure 2.42 [2] shows the relation between the decarburized depth and heating temperature in atmosphere. Since springs with decarburized layer show insufficient hardness in the surface area, to deteriorate the fatigue life critically, the generation of decarburized layer shall be avoided. Figure 2.43 shows the impact of decarburized layer to fatigue life [3]. The decarburization can make the compressive residual stress of shot-peened spring surface lower, to decrease the fatigue life, as shown in Fig. 2.44 [4].

In order to prevent the decarburization, heating in protective gas such as dry nitrogen or controlled carbon enriched gas, or adjusting the air to combustion gas ratio to form reducing atmosphere have been in practical use. As for the method for decreasing the decarburization of high silicon spring steel, there is a report where the rapid cooling of hot-rolled wire rod after passing through the finishing mill reduces the ferrite decarburization at the surface, as shown in Fig. 2.45 [5]. The histogram of decarburized layer depth of spring flat bars is shown in Fig. 2.46 [6]. The decarburization has been improved every year, to be less than 0.15 mm. Since it is difficult to make decarburization zero for hot rolled material, the materials peeled or grinded can be sometimes applied for higher quality requirements. In this case, since the materials can be improved in the surface roughness besides the decarburization, it can be said to be suitable material for springs. However, needless to say, an appropriate heating condition during spring manufacturing processes is required.

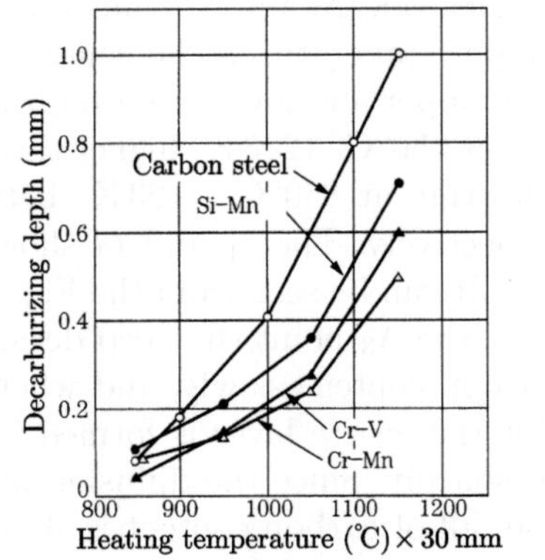

	C	Si	Mn	Cr	V
	0.80	0.24	0.86	–	–
Si–Mn	0.60	1.70	0.75	–	–
Cr–Mn	0.60	0.30	0.76	0.87	–
Cr–V	0.53	0.33	0.70	0.90	0.13

Fig. 2.42. Heating temperature and decarburization depth of spring steel [2]

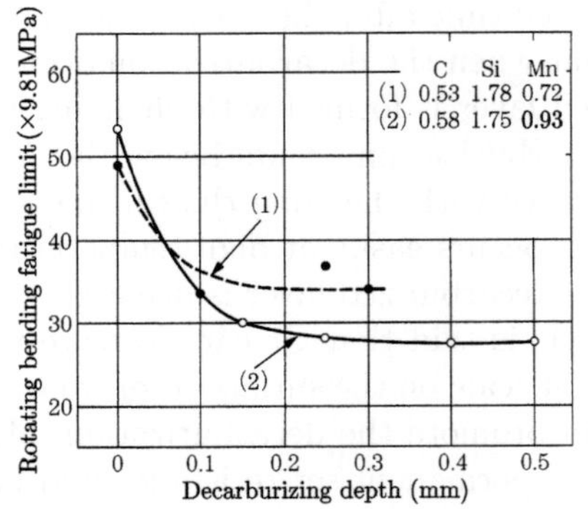

Fig. 2.43. Effect of decarburized layer on fatigue strength [3]

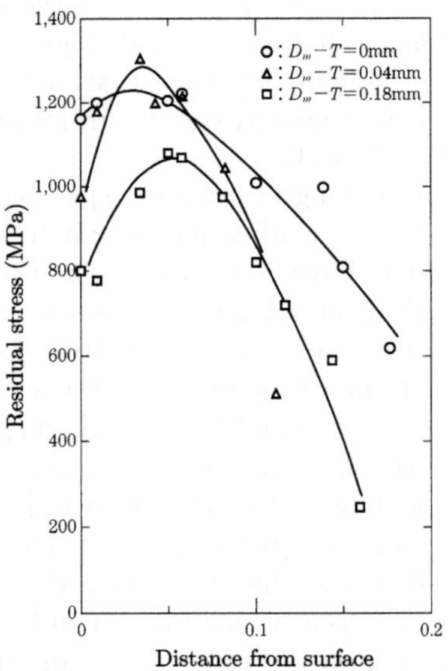

Fig. 2.44. Effect of decarburization on residual stress distributions [4]

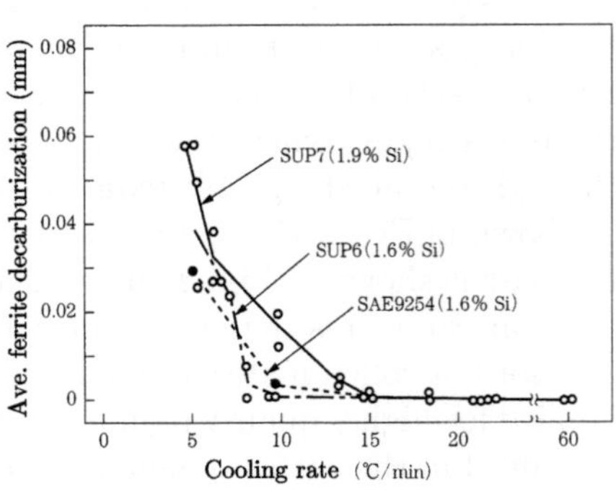

Fig. 2.45. Relationship between coil cooling rate and ferrite decarburization

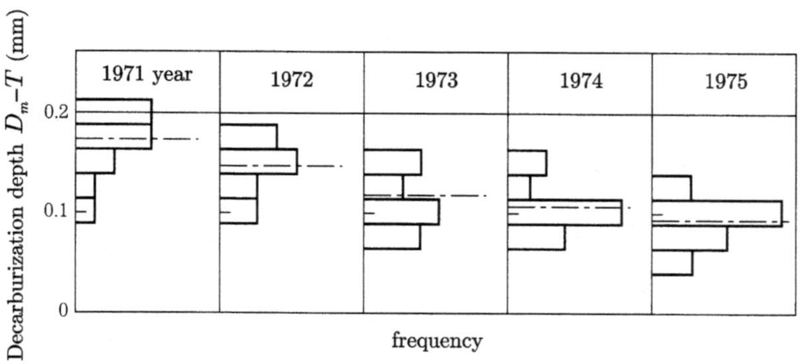

Fig. 2.46. Transition of decarburized layer depth of flat spring bar

References

1. M. Kawakami: Kinzoku-zairyou-rikougaku (Japanese), Phoenix Hall, and (1965) p. 285
2. Spring: Maruzen (Japanese), (1982) p. 36
3. A. Higuchi: Transaction of JSSR (Japanese), 1 (1952), p. 26
4. T. Osone: Proceeding of JSSR (Japan Society for Spring Research) (1991), p. 25
5. JSSR: New spring technology (1988), p. 84
6. A. Ohno and M. Shibata: Kikai-no-kennkyu (Japanese), Vol. 29 January (1977), p. 123

(3) Heat treatment characteristic

The steel materials for hot formed springs are always quenched and tempered. If the quenching is incomplete, the microstructure other than the full quenched martensite can be contained, and show lower hardness, compared with the complete martensite. Even if the same tensile strength is obtained by tempering, the elongation, reduction of area, impact value, ductility, toughness and proof stress, and fatigue strength become lower, compared with the full martensite. Therefore, in order to obtain the performance of steel fully, the full quenching should be carried out. Since the hardness of martensite of low alloy steel can be determined by carbon content, the completeness of quenching can be estimated by the hardness.

On the other hand, the depth where the steel can be hardened depends upon external factors, such as cooling ability of quenching media, and surface roughness of materials, and also depend upon the hardenability of steel itself. The hardenability of steel can be mainly determined by the chemical composition, and the austenite grain size.

For the comparison of hardenability, the Jominy end quench test is normally carried out. The hardenability band of the spring steel is shown in Fig. 2.47–2.54 [1]. Recent technologies make it possible to calculate a Jominy curve based on the chemical composition of molten steel, to detect the difference from the target Jominy curve, and to feedback to alloy addition, controlling the hardenability precisely with little variation. When selecting the spring

ISO/TC 17/4 N 868 E steel type J-2
C : 0.55~0.65% Si : 1.50~1.80% Mn : 0.70~1.00%

Distance from quenching end and its hardness																Temperature	
HRC \ mm	1.5	3	5	7	9	11	13	15	20	25	30	35	40	45	50	annealing	quenching
Upper limit	65	65	65	64	63	62	58	53	44	40	37	35	34	33	32	—	850±5
Lower limit	60	59	53	46	37	34	32	31	28	27	25	24	24	24	24		

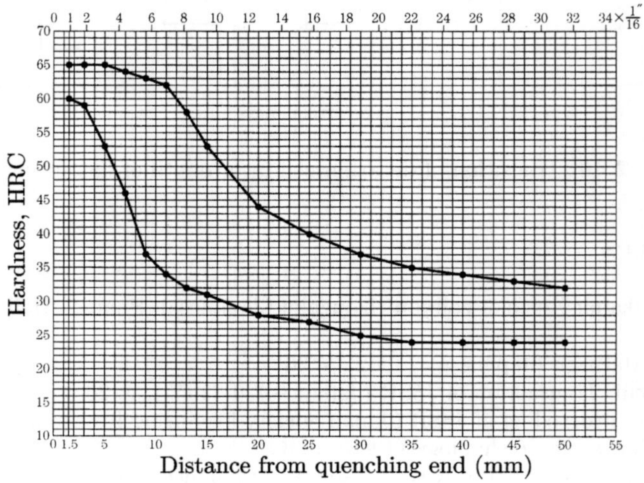

Fig. 2.47. Hardenability band for SUP 6

steel, it is necessary to choose the steel having appropriate hardenability to its cross-section. Although it had been considered the minimum quenching level as the 50% or more marteniste at the center of the spring cross-section, it has recently been required to be more martensite in the center, for the improvement of permanent set resistance and high fatigue strength. As examples, the maximum product dimensions recommended by ISO, where the central part becomes more than the 80% martensite is shown in Table 2.3.

The hardenability of steel varies with the austenite grain size. It becomes better as the austenite grain size becomes larger. However, as the steel quenched and tempered from the coarse austenite tends to show poor toughness, it can not be recommended to make the austenite grain size larger. In spring steel, the austenite grain size with usual heating temperature, can be said to be ASTM No. 5 or finer grain size. When the heating temperature is too high, the excess growth of austenite grain may occur. It can be recommended that the maximum heating temperature should be 950°C or less.

References

1. JIS G 4801-1984

〈ISO〉 ISO 683/XIV (1973) steel type No. 6
C : 0.57~0.64% Si : 1.70~2.20% Mn : 0.70~1.00%

HRC \ mm	Distance from quench end and hardness															Temperature, ℃	
	1.5	3	5	7	9	11	13	15	20	25	30	35	40	45	50	annealing	quenching
Upper limit	66	66	65	64	62	60	58	55	48	43	41	39	38	37	36	—	850±5
Lower limit	59	58	56	50	43	38	36	34	31	28	27	26	25	24	23		

〈SAE〉 SAE J 1268 (JUN 81) steel No. 9260 H
C : 0.55~0.65% Si : 1.70~2.20% Mn : 0.65~1.10%

HRC \ ×1/16 in	Distance from quench end and hardness																							Temperature, ℃	
	1	2	3	4	5	6	7	8	9	10	11	12	13	14	15	16	18	20	22	24	26	28	30	32	annealing	quenching
Upper limit	—	—	65	64	63	62	60	58	55	52	49	47	45	43	42	40	38	37	36	36	35	35	35	34	900±5	870±5
Lower limit	60	60	57	53	46	41	38	36	36	35	34	34	33	33	32	32	31	31	30	30	29	29	28	28		

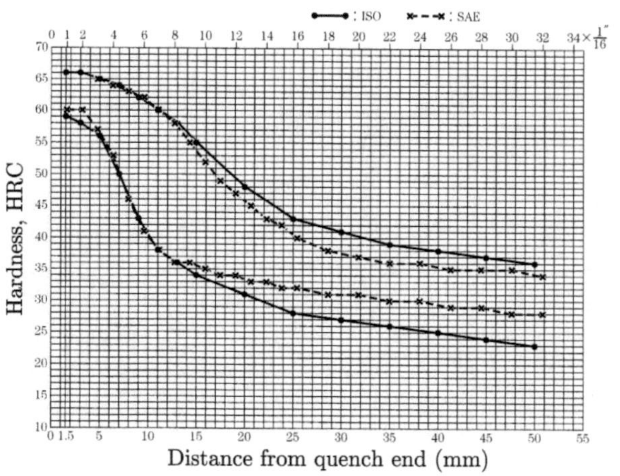

Fig. 2.48. Hardenability band for SUP 7 steel

Microstructure of spring steel

Optical microscope photo ×400

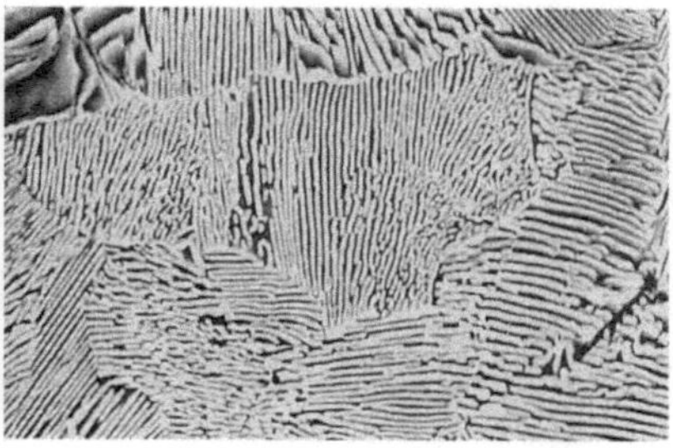

Scanning electron microscope photo ×4,000

Patenting microstructure of Si–Cr steel for valve spring
Mainly fine pearlite, including little proeutectiod ferrite (Mr. Y. Uematsu)

⟨ISO⟩ ISO 683/XIV (1973) steel type No. 8
C : 0.52~0.59% Mn : 0.70~1.00% Cr : 0.60~0.90%

	Distance from quench end and hardness															Temperature, ℃	
HRC \ mm	1.5	3	5	7	9	11	13	15	20	25	30	35	40	45	50	annealing	quench
Upper limit	65	65	64	63	63	62	61	60	57	52	48	45	42	40	39	—	850±5
Lower limit	57	56	55	54	52	48	43	39	33	30	28	27	26	25	24		

⟨SAE⟩ SAE J 1268 (JUN 81) steel No. 5155 H
C : 0.50~0.60% Mn : 0.60~1.00% Cr : 0.60~1.00%

	Distance from quench end and hardness																								Temperature, ℃	
HRC \ $\times\frac{1}{16}$ in	1	2	3	4	5	6	7	8	9	10	11	12	13	14	15	16	18	20	22	24	26	28	30	32	annealing	quench
Upper limit	—	65	64	64	63	63	62	62	61	60	59	57	55	52	51	49	47	45	44	43	42	41	41	40	870±5	854±5
Lower limit	60	59	58	57	55	52	47	41	37	36	35	34	34	33	33	32	31	31	30	29	28	27	26	25		

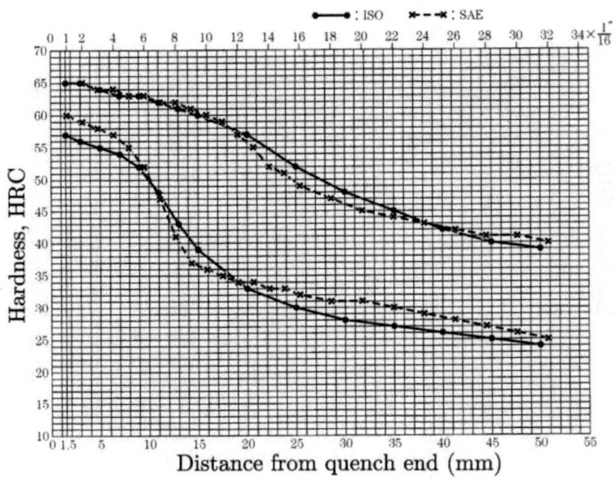

Fig. 2.49. Hardenability band for SUP 9 steel

Microstructure of spring steel

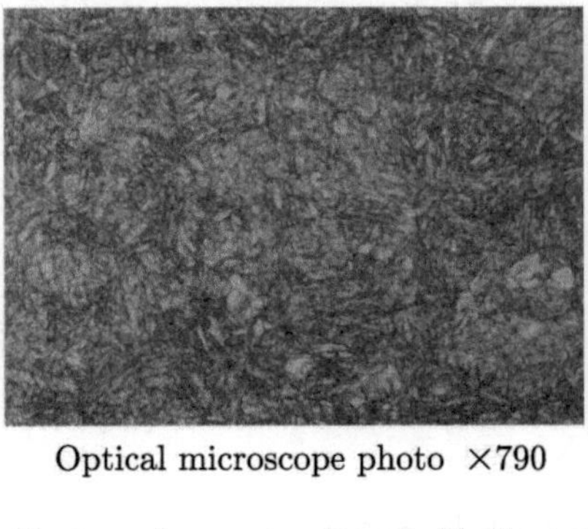

Optical microscope photo ×790

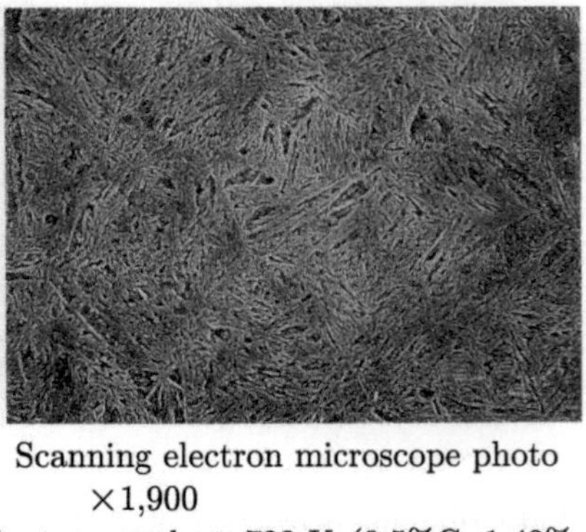

Scanning electron microscope photo ×1,900

Tempered martensite of Si–Cr steel, tempered at 723 K (0.5%C–1.42% Si–0.76% Mn–0.79%Cr) (Mr. Hayashi)

〈ISO〉 ISO 683/XIV (1973) steel type No. 9
C : 0.56~0.64% Mn : 0.70~1.00% Cr : 0.60~0.90%

	Distance from quench end and hardness															Temperature, ℃	
HRC \ mm	1.5	3	5	7	9	11	13	15	20	25	30	35	40	45	50	annealing	quencing
Upper limit	66	66	65	65	64	63	62	61	58	54	50	47	44	42	41	—	850±5
Lower limit	59	59	57	56	53	50	45	41	35	32	30	29	28	27	26		

〈SAE〉 SAE J 1268 (JUN 81) steel No. 5160 H
C : 0.55~0.65% Mn : 0.65~1.10% Cr : 0.60~1.00%

	Distance from quench end and hardness																								Temperature, ℃	
HRC \ $\times\frac{1}{16}$ in	1	2	3	4	5	6	7	8	9	10	11	12	13	14	15	16	18	20	22	24	26	28	30	32	annealing	quencing
Upper limit	—	—	—	65	65	64	64	63	62	61	60	59	58	56	54	52	48	47	46	45	44	43	43	42	870±5	845±5
Lower limit	60	60	60	59	58	56	52	47	42	39	37	36	35	35	34	34	33	32	31	30	29	28	28	27		

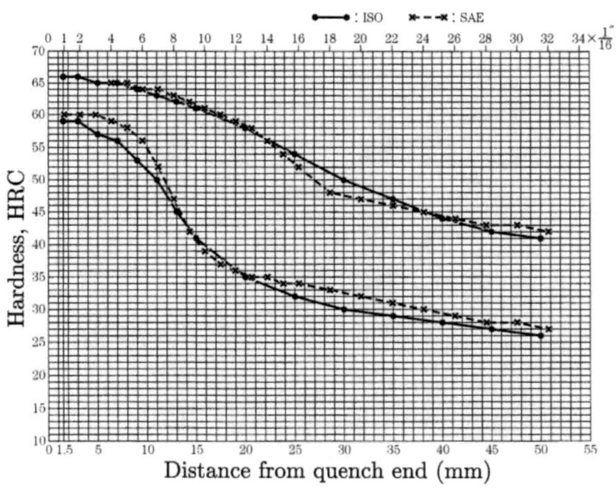

Fig. 2.50. Hardenability band for SUP 9A steel

Microstructure of spring steel

Optical microscope photo ×400

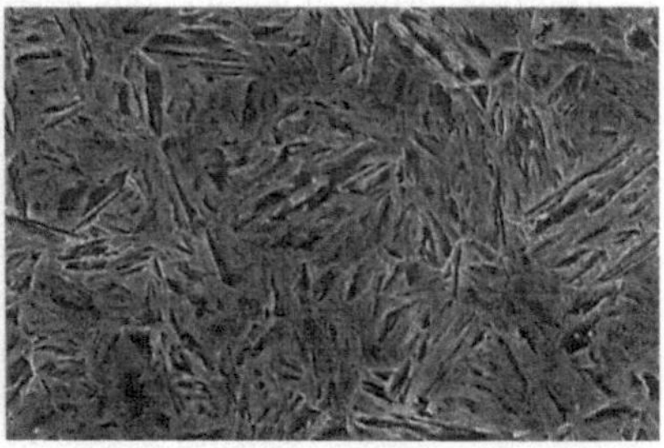

Scanning electron microscope photo ×2,000

Martensite of medium carbon spring steel as oil quenched (0.4%C-1.77% Si-0.18% Mn-0.22%Cu-0.52%Ni-1.05%Cr-0.17%V-0.07%Ti) (Mr. M. Sugıyama)

〈ISO〉 ISO 683/XIV(1973) steel type No.13
C : 0.48~0.55% Mn : 0.70~1.00% Cr : 0.9~1.20% V : 0.10~0.20%

Distance from quench end and hardness																Temperature, ℃	
HRC \ mm	1.5	3	5	7	9	11	13	15	20	25	30	35	40	45	50	annealing	quench
Upper limit	65	65	64	64	63	62	61	60	57	53	50	48	45	44	43	—	850±5
Lower limit	57	56	56	55	53	50	46	43	37	35	33	32	30	29	28		

〈SAE〉 SAE J 1268(JUN 81) steel No.6150 H
C : 0.47~0.54% Mn : 0.60~1.00% Cr : 0.75~1.20% V : 0.15%min

Distance from quench end and hardness																								Temperature, ℃	
HRC \ $\times\frac{1}{16}$ in	1	2	3	4	5	6	7	8	9	10	11	12	13	14	15	16	18	20	22	24	26	28	30	32	annealing	quench
Upper limit	65	65	64	64	63	63	62	61	61	60	59	58	57	55	54	52	50	48	47	46	45	44	43	42	870±5	845±5
Lower limit	59	58	57	56	55	53	50	47	43	41	39	38	37	36	35	35	34	32	31	30	29	27	26	25		

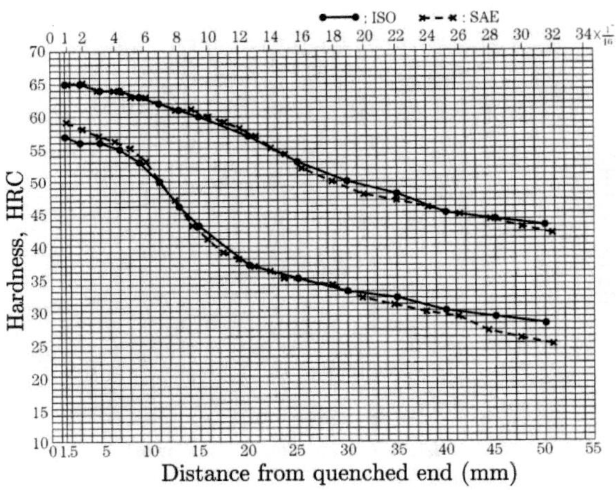

Fig. 2.51. Hardenability band for SUP 10 steel

Microstructure of spring steel

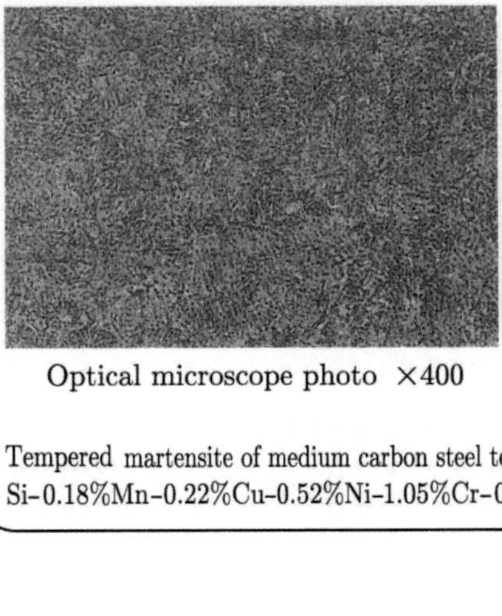

Optical microscope photo ×400

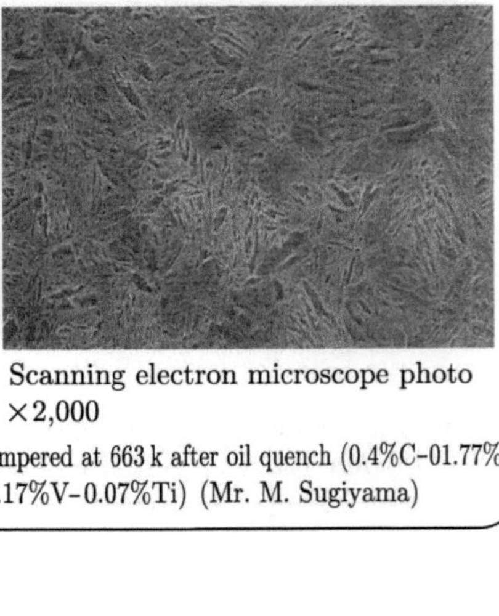

Scanning electron microscope photo ×2,000

Tempered martensite of medium carbon steel tempered at 663 k after oil quench (0.4%C-01.77% Si-0.18%Mn-0.22%Cu-0.52%Ni-1.05%Cr-0.17%V-0.07%Ti) (Mr. M. Sugiyama)

〈ISO〉 ISO 683/XIV(1973) steel type No.10
C : 0.56~0.64% Mn : 0.70~1.00% Cr : 0.60~0.90% B : 0.0005%min

Distance from quench end and hardness																Temperature, ℃	
HRC \ mm	1.5	3	5	7	9	11	13	15	20	25	30	35	40	45	50	annealing	quench
Upper limit	65	65	65	65	65	65	65	65	65	63	60	57	54	50	47	—	850±5
Lower limit	60	60	60	59	59	57	54	49	39	37	34	32	30	28	25		

〈SAE〉 SAE J 1268(JUN 81) steel No.51 B 60 H
C : 0.55~0.65% Mn : 0.65~1.10% Cr : 0.60~1.00% B : 0.0005%min

Distance from quench end and hardness																								Temperature, ℃	
HRC \ $\times\frac{1}{16}$ in	1	2	3	4	5	6	7	8	9	10	11	12	13	14	15	16	18	20	22	24	26	28	30	32	annealing	quench
Upper limit	—	—	—	—	—	—	—	—	—	—	—	65	65	64	64	63	61	59	57	55	53	51	49	47	870±5	845±5
Lower limit	60	60	60	60	60	59	58	57	54	50	44	41	40	39	38	37	36	34	33	31	30	28	27	25		

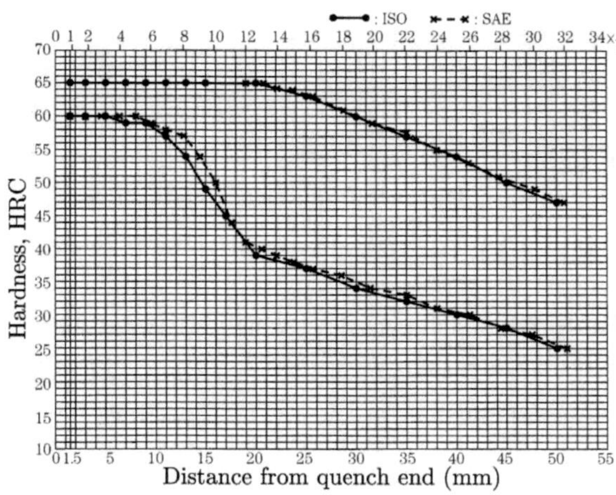

Fig. 2.52. Hardenability band for SUP 11A steel

Microstructure of spring steel

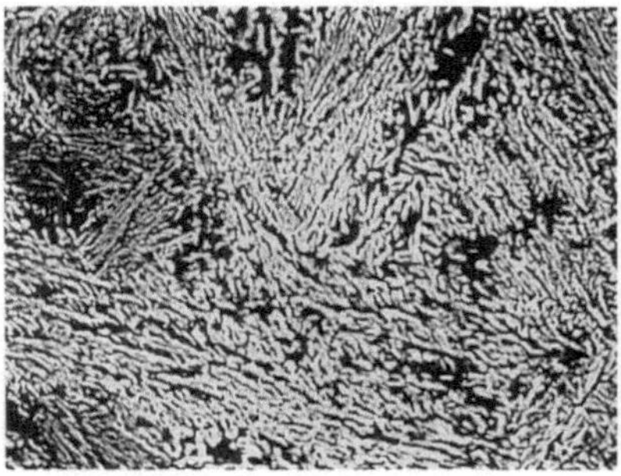

Upper bainite ×1,750

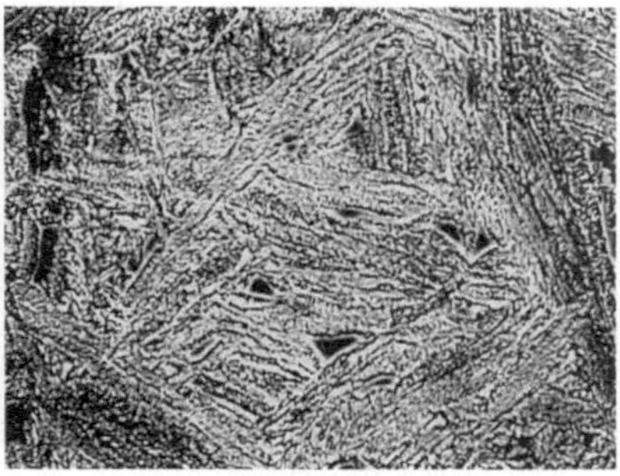

Lower bainite ×1,750

Microstructure observation by SEM of SUP11A spring steel. Tempered at 723 K (left) and isothermal transformation at 573 K (right) (Mr. N. Tsuji et al)

(%)

		C	Si	Mn	Cr
SAE 9254		0.51/0.59	1.20/1.60	0.60/0.80	0.60/0.80
$n = 24$	$\bar{x}$	0.561	1.373	0.724	0.715
	σ	0.022	0.091	0.058	0.034

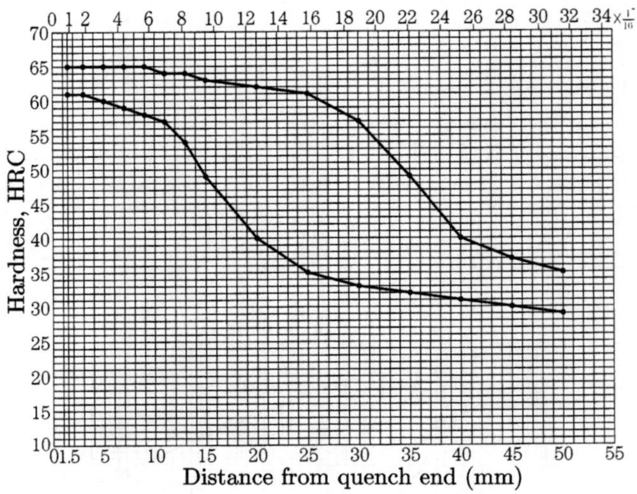

Fig. 2.53. Hardenability band for SUP 12 steel

Table 2.3. Examples of hardenability of spring steel

Materials		Size obtained the 80% of martensite at the center, by oil quenching mm	
ISO Steel type No.	JIS steel type	thickness	diameter
2	SUP 3	8	12
5	SUP 6	8	12
6	SUP 7	14	20
8	SUP 9	18	28
9	SUP 9A	22	33
13	SUP 10	27	40
10	SUP 11A	24	35
12	SUP 13	47	70

(: ISO 683/XIV 1973)

〈ISO〉 ISO 683/XIV(1973) steel type No.12
C : 0.56~0.64% Mn : 0.70~1.00% Cr : 0.70~0.90% Mo : 0.25~0.35%

	Distance from quench end and hardness															Temperature, ℃	
HRC \ mm	1.5	3	5	7	9	11	13	15	20	25	30	35	40	45	50	annealing	quench
Upper limit	65	65	65	65	65	65	65	65	64	64	63	63	63	63	63	—	850±5
Lower limit	60	60	60	60	60	60	60	59	58	56	54	50	46	43	41		

〈SAE〉 SAE J 1268(JUN 81) steel No.416 H
C : 0.55~0.65% Mn : 0.65~1.10% Cr : 0.65~0.95% Mo : 0.25~0.35%

	Distance from quench end and hardness																								Temperature, ℃	
HRC \ $\times\frac{1}{16}$ in	1	2	3	4	5	6	7	8	9	10	11	12	13	14	15	16	18	20	22	24	26	28	30	32	annealing	quench
Upper limit	65	65	65	65	65	65	65	65	65	65	65	64	64	64	64	64	64	63	63	63	63	63	63	63	871	843
Lower limit	60	60	60	60	60	60	60	60	59	59	59	59	58	58	57	56	55	53	50	48	45	43	42	41		

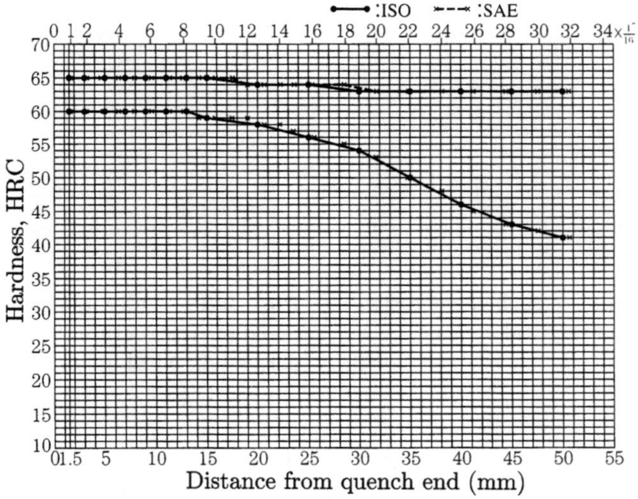

Fig. 2.54. Hardenability band for SUP 13 steel

(4) Microstructure and grain size

Since most spring steels are hypo-eutectoid steels, the microstructure as hot rolled is usually mixed structures of a little pro-eutectoid ferrite and mostly pearlite. Since some carbon spring steels are hyper-eutectoid steels, the microstructure as hot-rolled becomes mixed pearlite and a little pro-eutectoid cementite. The microstructure of typical spring steels as hot rolled, is shown in the Fig. 2.55–2.56. The microstructure and hardness of the hot-rolled material can be affected by the finishing rolling temperature and cooling speed. Since the material hardness may affect the ability of cutting and drilling process, it is recommended by JIS G 4801-1984 that the hardness of flat steel bar should be less than 302 HB. In case of thin flat bar and round bar with small diameter, depending upon the cooling speed after rolling, the bainite microstructure can appear as shown in the Fig. 2.56. In this case, it is recommended to apply

Fig. 2.55. Microstructure of hot rolled hypo-eutectoid spring steel (SUP6), [×600, Picral etch.]

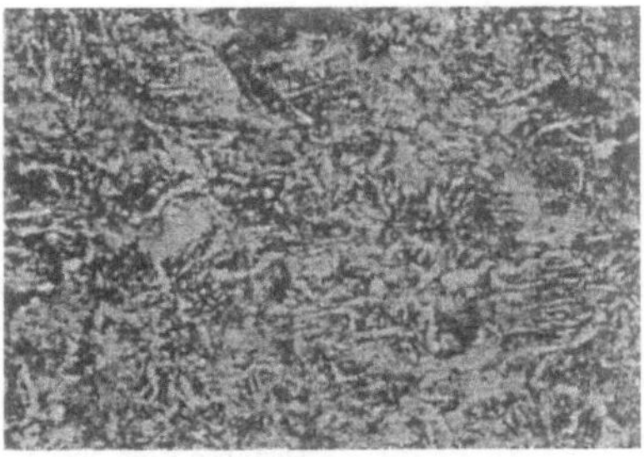

Fig. 2.56. Rolled microstructure including bainite [×600, Nital etch.]

Fig. 2.57. Fully quenched microstructure (SUP6, ×300, Nital etch.)

Fig. 2.58. Fully quenched and tempered microstructure (SUP6, 807 K temper, ×300, Nital etch.)

the annealing for softening the material. Prior to machining wire rod and bar materials by peeling machine, the spheroidizing annealing under the transformation point A_1, can keep the material hardness below 255 HB, and the full annealing above the transformation point A_3, can keep that below 299 HB.

The hot formed spring is normally tempered around 400–500°C after quenching, and its microstructure shows tempered martensite. If the quenching is incomplete, a mixed martensitic structure with ferrite and pearlite or bainite can be observed. Figure 2.57–2.60 shows typical examples of microstructure observed by the optical microscope.

As shown in Fig. 2.61 the austenite grain size, heated at higher temperature over the transformation point A_3, becomes larger. It can be noted that the microstructure transformed from coarse austenite grains is also coarse. Compared with the spring with fine grain, the spring with coarse grain can show lower toughness and lower fatigue strength. In addition, high temperature caused large grain, can be said to accompany with the oxidization and decarburization. Therefore, it can be recommended to avoid the overheat as much as possible. The steel with large grains by overheating can be recovered by normalizing, to have fine grains.

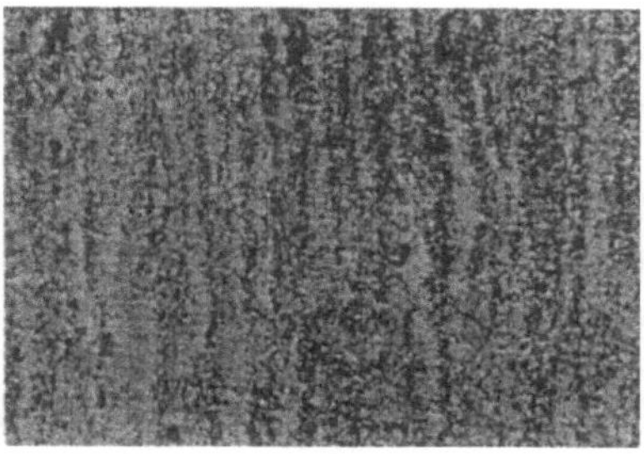

Fig. 2.59. Low temperature tempering microstructure (SUP6, 673 K temper, ×300, Nital etch.)

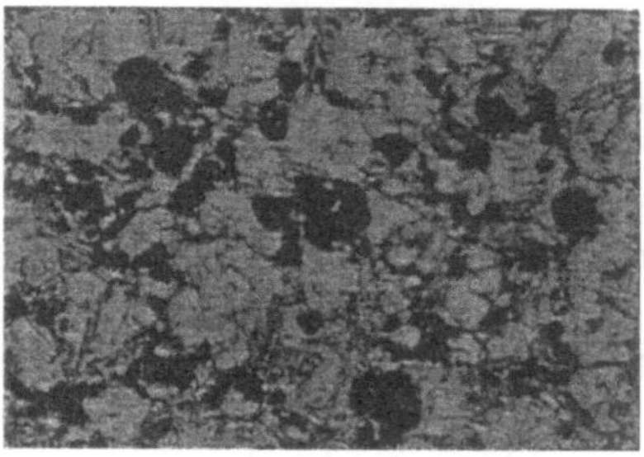

Fig. 2.60. Incomplete quenching microstructure (SUP6, ×300, Nital etch.)

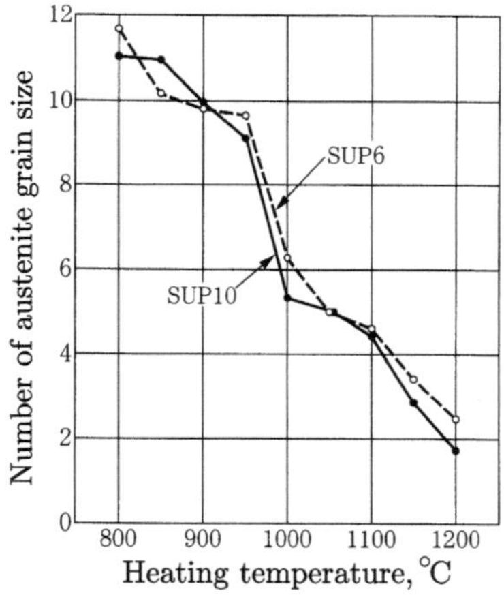

	Chemical composition (%)								
	C	Si	Mn	P	S	Cu	Ni	Cr	V
SUP6	0.61	1.64	0.64	0.017	0.020	0.15	0.07	0.09	–
SUP10	0.54	0.22	0.96	0.017	0.010	0.16	0.07	1.08	0.08

Fig. 2.61. Relationship of heating temperature and grain size

(5) Mechanical properties

As for the spring hardness, in the United States and Great Britain, it is recommended to use carbon steel and alloy steel with hardness of 352–444 HB and 375–461 HB, respectively. German, Russia and ISO designate the tensile strength value of spring steel. In France, the spring with Re-375 (for water quench) and Re-388 (for oil quench) has specified hardness over 375 HB and 388 HB respectively, for the other special use spring steel, the tensile strength is designated.

In Japan, Table 2.4 shows examples of the main mechanical properties of spring steels, based on JIS standards [1]. For alloy steel the hardness is 363–429 HB, and as for carbon steel slightly lower hardness is used. In order to satisfy this hardness, an appropriate tempering temperature is selected and the tensile strength can be determined. The hardness as springs can be chosen, apart from the material standard value. Although the hardness suitable for springs can vary according to the type of springs, the spring hardness tends to become harder gradually. The elongation value as a rough indication of ductility can be seen to be 5% of minimum values in Russia and a part of ISO standard. There are many examples where the elongation is specified as 6% or more, besides the German standard which specifies 6% or more of elongation uniformly. In Japan, the elongation is specified higher compared with the other countries. However, since the gauge length of tensile test specimen is 5 times the diameter in the other countries and the short gauge length is 3.54 times the diameter in JIS G 4801 for No. 4 type specimen, the substantial difference of both can be small.

Since the spring hardness can depend upon its application, it is said that for the spring expected to have a large impact load the hardness should be lower to keep the spring ductile, and for the spring not subjected to the impact load, the hardness can be higher.

As for the automotive suspension coil springs which have been demanded for the weight saving by high strength, the hardness has been specified to a higher range of 415–495 HB [2], and this trend has been accelerating to have the spring over 500 HB recently. However, there is a report [3] that the most sufficient hardness range with the 90% residual life can be 415–444 HB, and also another report [4] where if the hardness exceeds 477 HB the scattering of fatigue life becomes large. In case of high hardness applications, the appropriate considerations including manufacturing processes is required.

The spring steel can be tempered at lower temperature comparing with the temperature for the machine structural steel, 550–650°C. Although increasing the hardness (strength) can sacrifice the toughness such as elongation,

Table 2.4. Mechanical properties of spring steels

	Heat treatment		Mechanical properties				
					El. %	RA. %	
Steel grade	Quench (C°)	Temper (C°)	Yield strength (MPa)	Tensile strength (MPa)	JIS No. 4 or No. 7	JIS No. 4	Hardness HB
SUP3	830–860	450–500	≧ 834	≧ 1079	≧ 8	–	341–401
SUP6	830–860	480–540	≧ 1070	≧ 1226	≧ 9	≧ 20	363–429
SUP7	830–860	480–540	≧ 1079	≧ 1226	≧ 9	≧ 20	363–429
SUP9	830–860	460–510	≧ 1079	≧ 1226	≧ 9	≧ 20	363–429
SUP9A	830–860	460–520	≧ 1079	≧ 1226	≧ 9	≧ 20	363–429
SUP10	840–870	470–640	≧ 1079	≧ 1226	≧ 10	≧ 30	363–429
SUP11A	830–860	460–520	≧ 1079	≧ 1226	≧ 9	≧ 20	363–429
SUP12	830–860	510–570	≧ 1079	≧ 1226	≧ 9	≧ 20	363–429
SUP13	830–860	510–570	≧ 1079	≧ 1226	≧ 10	≧ 30	363–429

reduction of area, and impact value, it is needless to say that higher toughness with the same strength is better. The relations between the mechanical properties and tempering temperature for several spring steels are shown in Fig. 2.62–2.67.

The SUP10 has high tempering softening resistance, where the hardness does not decrease much, with tempering at high temperature. Therefore, the SUP10 can be said to be high strength and high toughness material. Since the Si–Mn steel such as SUP6 and SUP7, has high carbon and has high tempering softening resistance due to the content of silicon, they can be tempered at higher temperature, compared with the SUP9. The Mn–Cr steel such as SUP9, SUP9A, and SUP11A are required to be tempered at lower temperature than Si–Mn steel, to obtain the same spring hardness. Carbon steels are required to be tempered at lower temperature, compared with alloy spring steel. It can be said that the toughness of carbon steel is lower than that of alloy spring steel.

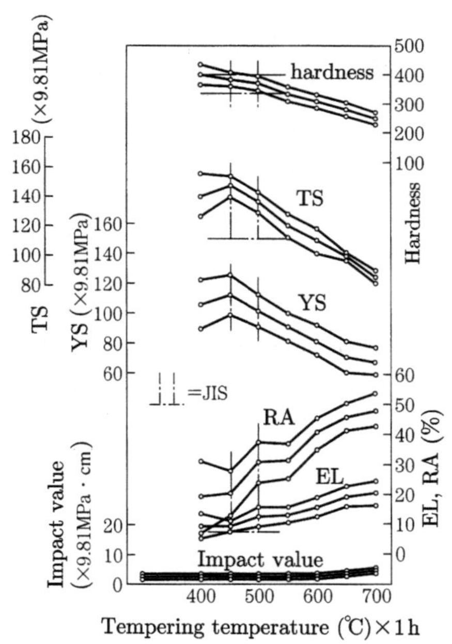

	Chemical compositions (%)							
	C	Si	Mn	P	S	Cu	Ni	Cr
$\overline{X}$	0.85	0.23	0.41	0.020	0.013	0.21	0.07	0.11
$\overline{X}+2s$	0.89	0.28	0.47	0.025	0.018	0.25	0.11	0.18
$\overline{X}-2s$	0.81	0.17	0.36	0.014	0.007	0.17	0.03	0.04
	0.75–0.90	0.15–0.35	0.30–0.60	$\leqq 0.035$	$\leqq 0.035$	$\leqq 0.35$	–	–

Specimen $n = 10$ annealing 745°C × 2h FC quench 845°C × 30 min OQ

Fig. 2.62. Tempering curve of Carbon steel (SUP3)

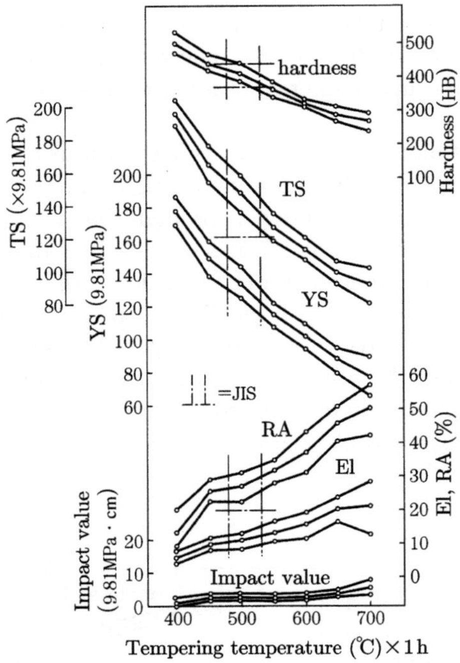

	Chemical compositions							
	C	Si	Mn	P	S	Cu	Ni	Cr
$\overline{X}$	0.58	1.61	0.83	0.020	0.012	0.18	0.11	0.11
$\overline{X}+2s$	0.61	1.73	0.89	0.023	0.015	0.21	0.14	0.14
$\overline{X}-2s$	0.55	1.49	0.77	0.017	0.010	0.16	0.09	0.08
	0.55–0.65	1.50–1.80	0.70–1.00	$\leqq 0.035$	$\leqq 0.035$	$\leqq 0.35$	–	–

Specimen $n = 10$ annealing 750°C × 2h FC quench 845°C × 30 min OQ

Fig. 2.63. Tempering curve of Si–Mn steel (SUP6)

References

1. JIS G 4801-1984
2. JASO 7211
3. T. Ozone and K. Ito: Trans. of JSSR, 22 (1977), p. 29
4. A. Ohno and M. Shibata: Kikai no kenkyu (Japanese), 29, 1 (1977), p. 123

(6) Permanent set resistance

The permanent set of spring indicates that the plastic deformation occurs with time during its use. Especially, it becomes an important issue for the automotive suspension coil spring. The permanent set of spring can make the height of vehicle lower to give bad appearances and also to cause the problem on safety due to the lower bumper height. It cannot be too much to say that the design stress of coil springs has been decided by the permissible

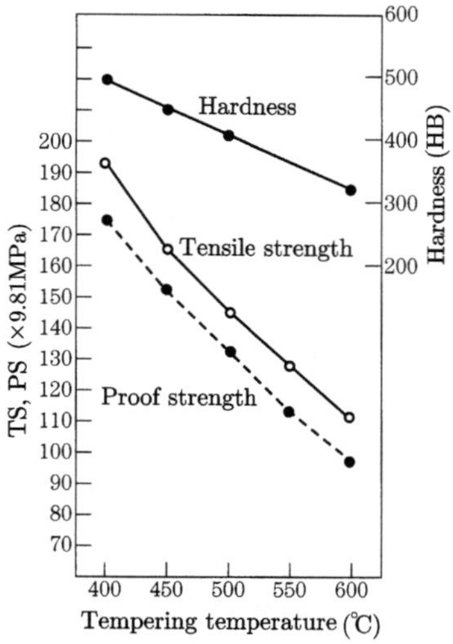

Chemical compositions						Heat treatment	specimen
C	Si	Mn	P	S	Cu		
0.58	1.85	0.86	0.016	0.008	0.16	850°C × 30 min OQ, tem. × 1 h	JIS4

Fig. 2.64. Tempering curve of Si–Mn steel (SUP7)

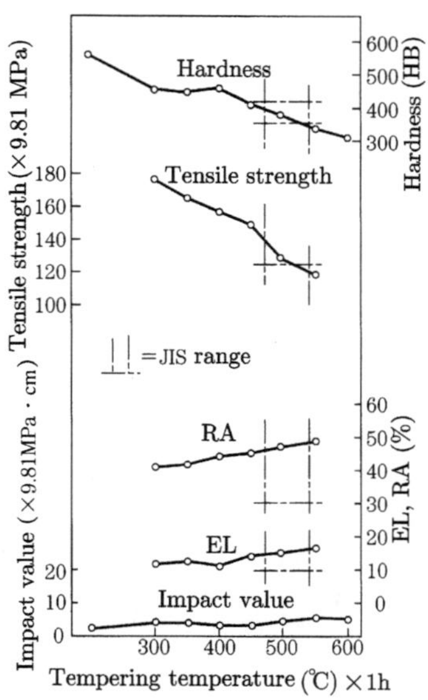

Chemical compositions					Heat treatment
C	Si	Mn	Cr	V	
0.48	0.21	0.74	1.10	0.22	850°C OQ

Fig. 2.65. Tempering curve of Cr–V steel (SUP10)

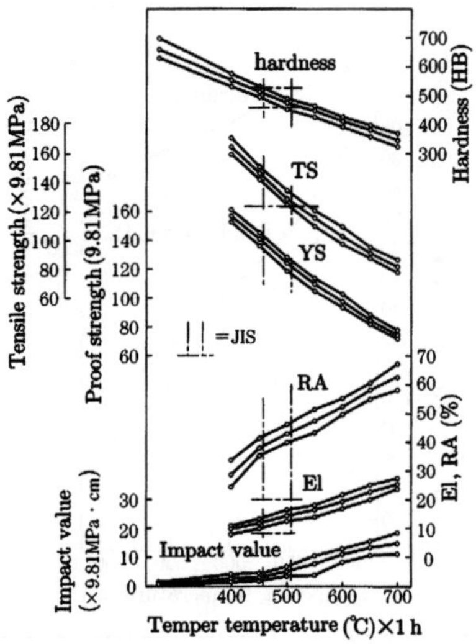

	Chemical compositions (%)							
	C	Si	Mn	P	S	Cu	Ni	Cr
$\overline{X}$	0.56	0.28	0.78	0.017	0.010	0.19	0.09	0.77
$\overline{X}+2s$	0.60	0.33	0.85	0.024	0.020	0.25	0.16	0.82
$\overline{X}-2s$	0.52	0.22	0.71	0.011	0.004	0.12	0.02	0.71
	0.56–0.60	0.15–0.35	0.65–0.95	≦ 0.035	≦ 0.035	≦ 0.35	-	0.65–0.95

Specimen $n = 10$ Annealing 780°C × 2h FC Quench 850°C × 30 min OQ

Fig. 2.66. Tempering curve of Mn–Cr steel (SUP9)

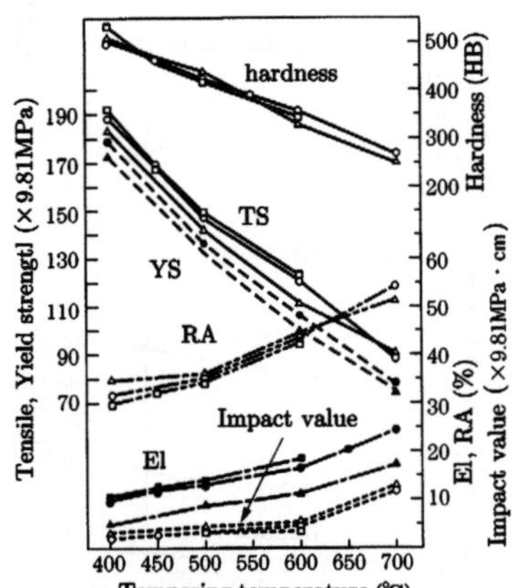

Symbol	Material	Compositions (%)							Heat treatment		specimen
		C	Si	Mn	P	S	Cu	Cr	quench	temper	
(1) ○●	JIS SUP8	0.57	1.08	0.46	0.015	0.011	0.14	0.92	850°C × 30 min OQ	Temp. × 60 min AC	25 mm ϕ
(2) △▲	SAE9254	0.51	1.43	0.66	0.014	0.016	1.14	0.67			JIS 2
(3) □■	SAE9254	0.55	1.64	0.86	0.031	0.009	0.17	0.75			JIS 4

Fig. 2.67. Tempering curve of Si–Cr steel

range of spring permanent set. The permanent set can be considered to be a deforming phenomenon by the movement of the active dislocations under the quenched and tempered metallurgical structure. Therefore, as for measures to improve the permanent set resistance, in order to increase the elastic limit (proof stress) of steel, high hardness (Fig. 2.68), Si contents (Fig. 2.69) and V and Nb contents (Fig. 2.70) have been studied.

(7) Fatigue strength

The relations between fatigue limit and hardness for various quenched and tempered steels are shown in Fig. 2.71. In the case of low hardness region, regardless of steel grade, the fatigue limit increases linearly with the increase of hardness. However, if the hardness becomes too high to some extent, the fatigue limit decreases rapidly. As for spring steel containing about 0.45 to 0.60% carbon, it can be seen in the Fig. 2.71, that if the hardness is less than 40HRC at least, there is no difference of the fatigue strength among the steel grades tested. When the hardness is not high enough due to the incomplete quenching it can be adjusted by lowering the tempering temperature to the predetermined spring hardness. However, the steel including such incomplete quenching structures can show lower fatigue strength, compared with complete quenching structures with the same hardness. Figure 2.72 shows the relation between the fatigue limit and the amount of martensite content as quenched. It can be seen from the Fig. 2.72, that comparing with the same

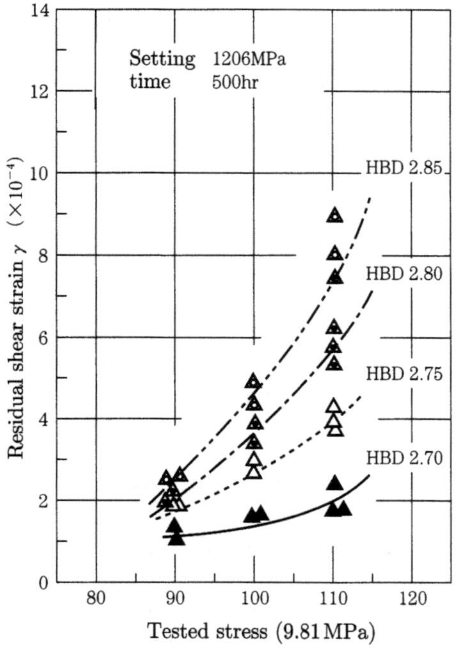

Residual shear strain, γ
$\gamma = \Delta\tau/G$.
where, $\Delta\tau$: shear stress relaxation
G: Modulus of rigidity

Fig. 2.68. Effect of hardness on permanent set resistance of coil spring [1]

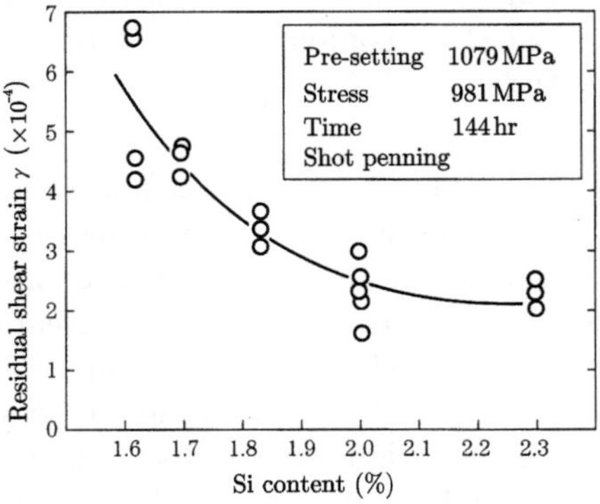

Fig. 2.69. Effect of Si content on permanent set of coil spring [2]

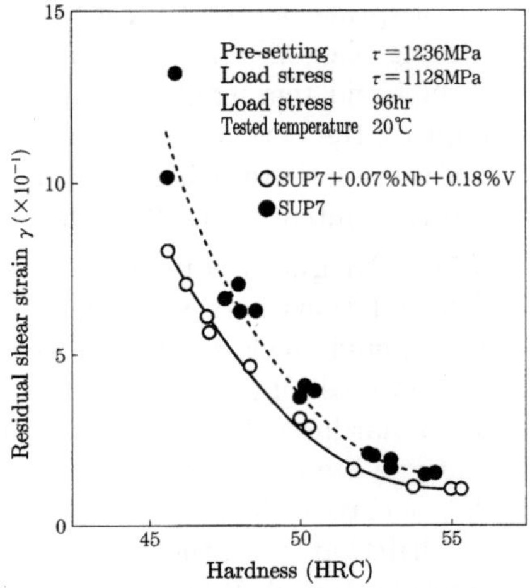

Fig. 2.70. Improvement of permanent set resistance by adding V and Nb [3]

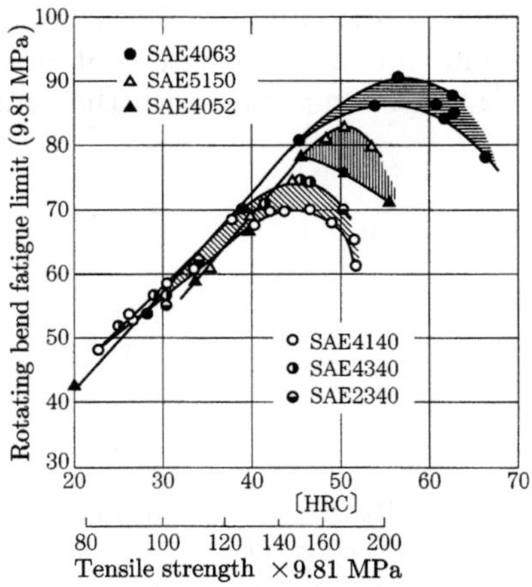

Fig. 2.71. Relations between fatigue limit and hardness of various materials [4]

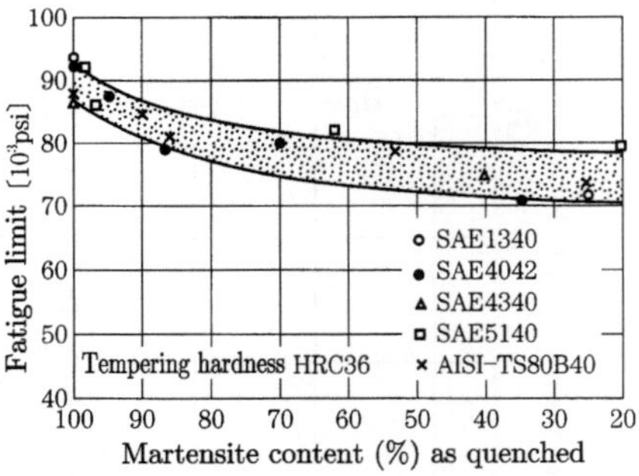

Fig. 2.72. Effect of incomplete quench on fatigue limit [5]

hardness, 36HRC, the mixture of a little incomplete quenching structures can reduce the fatigue limit remarkably.

In the processes of hot-formed springs, it is unavoidable to have rough surface while heating and forming. Figure 2.73 shows experimental results concerning with the effect of surface roughness and decarburization during the manufacturing processes of hot formed spring. The results of the Fig. 2.73 is the reversed torsional fatigue limits by using the spring material finished

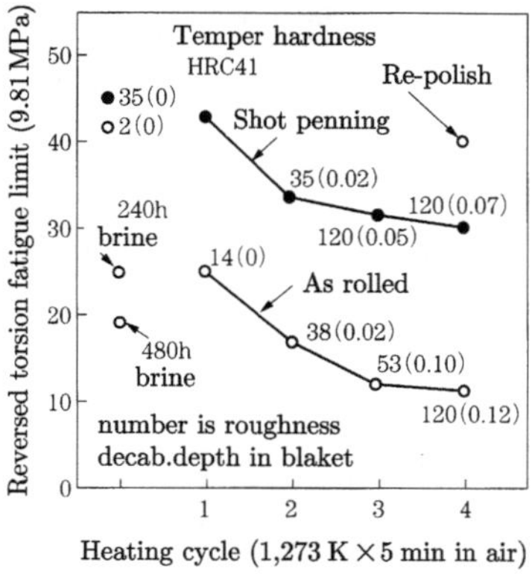

Fig. 2.73. Change of fatigue strength by repeated heating

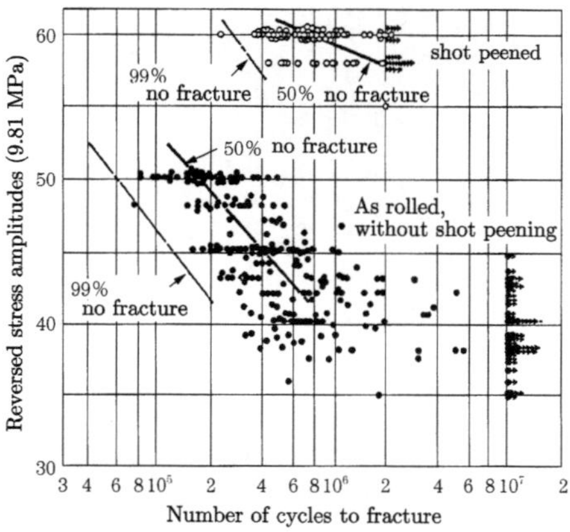

Fig. 2.74. Completely reversed bending fatigue test of heat treated SUP6 flat bar after shot peening

by polishing, applying repeated heating at 1000°C × 5 mins and keeping the hardness of specimens as 41HRC. It can be realized from the Fig. 2.73, that the fatigue limit of re-polished specimens after repeated heating, shows no more than the 7% decrease in comparison with the initial polished specimens. On the other hand, the decrease of fatigue limit can be seen to be remarkably large for the specimen immersed in a solution of salt. It can be also seen in the Fig. 2.73 that the influence of decarburized layer to fatigue strength can be recognized from comparatively small layer of 0.1 mm or less depth. Based on the above results, it can be said that the influence of the surface roughness to fatigue strength is extremely large. The rough surface and decarburization caused by heating can be recoverable to some extent by shot peening. Figure 2.74 shows examples of completely reversed bending fatigue strength of heat-treated Si–Mn steel as hot rolled and as shot peened.

(8) Corrosion fatigue resistance

The salt scattered on the road to melt the snow and ice in cold districts, can become a cause of corrosion fatigue for automotive suspension coil springs. Figure 2.75 shows the corrosion pits near the failure area of a spring which was failed in actual use [7]. The initiation of crack from U type, V type corrosion pit can be seen. The fatigue strength decreases as the corrosion pit becomes deeper, see Fig. 2.76, [8]. In recent years, for the purpose of weight saving of automotive suspension coil spring, the material strength has been increased to achieve higher design stress. Although the high strength can increase the fatigue strength, it can also increase fatigue notch sensitivity caused by corrosion. In order to improve the corrosion fatigue, the measures concerning with the spring design, processing such as shot peening and painting, and material have been made.

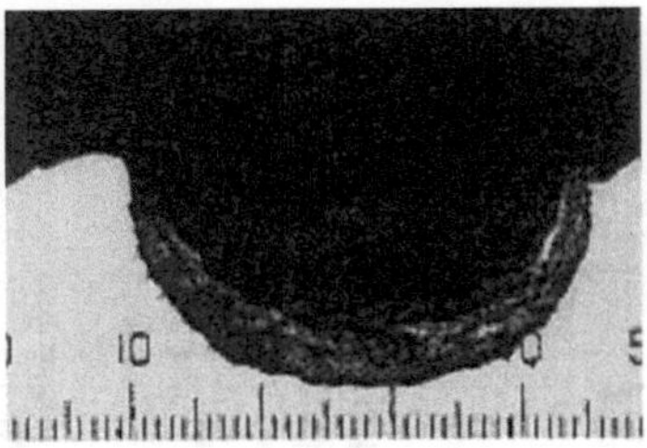

(a) Corrosion pit condition (No. S-33) × 100

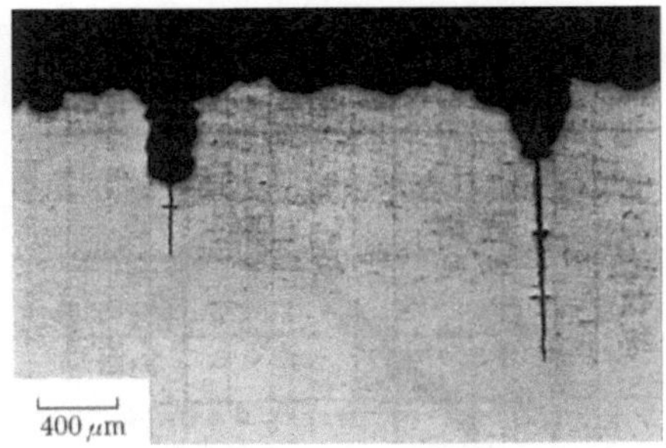

(b) Corrosion pit condition (No. S-19)

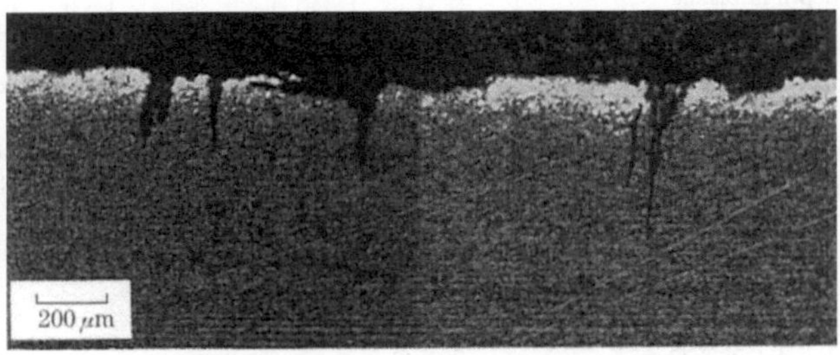

(c) Corrosion pit condition (No. S-4)

Fig. 2.75. Profile of corrosion pit of leaf spring collected from the commercial use

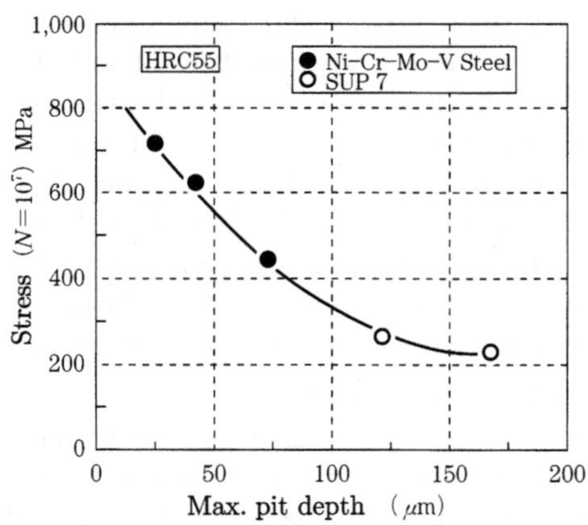

Fig. 2.76. Relations between fatigue strength and corrosion pit depth of high strength steel

From materials side, it was found that adding a small amount of Ni and Cu can make the configuration of corrosion pits smooth and shallow, to ease the harmful nature as a notch of corrosion pit. Additionally, new corrosion spring steel with lower carbon content for improving the toughness, reduction of impurities such as sulfur and adding a small amount of V and Ti for austenite grain refining, has been introduced to be used commercially. This new steel

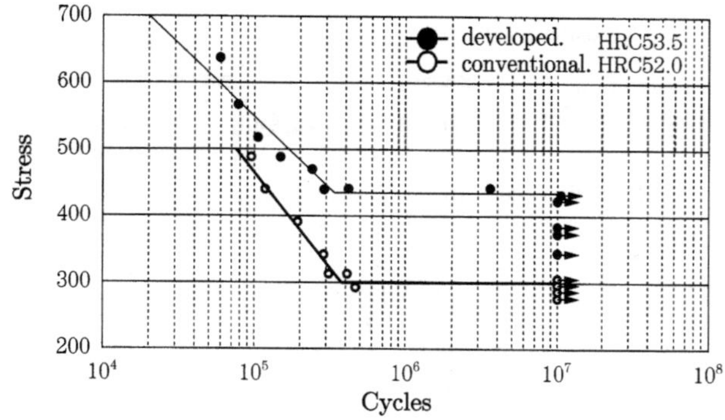

Fig. 2.77. S-N curves of new corrosion spring steels under SST corrosion, by Ono type rotary bending fatigue tests

shows excellent corrosion fatigue strengths with high strength spring design, as shown in Fig. 2.77 [9].

(9) Environmental brittleness resistance

As for the automotive suspension springs, due to the higher stress design for weight saving, the hydrogen embrittlement caused by very small quantity of hydrogen in steel and delayed fracture have been taken into considerations. In corrosion fatigue, the hydrolysis reaction of dissolved metal ion in the tip of corrosion pit or hydrogen invaded into steel due to pH fall by rust generation, can decrease the corrosion fatigue strength, see Fig. 2.78 [10]. Therefore, hydrogen embrittlement becomes an important issue for the high strength spring steel.

The evaluation methods for hydrogen embrittlement and delayed fracture of spring steel is under studies, including with new material developments.

References

1. JSMA, JSSR: Recent Spring Technologies (1988) p. 82
2. JSMA, JSSR: Recent Spring Technologies (1988) p. 18
3. JSMA, JSSR: Recent Spring Technologies (1988) p. 18
4. M. F. Garwood et al.: Trans. ASM 43 (1951) p. 1
5. T. J. Dolan and C. S. Yen: Proc. ASTM, 48 (1948) p. 664
6. K. Tomita: Gakushin 129 committee the 4th subcommit. (Japanese)
7. JSSR: Trans. of JSSR 40 (1995) p. 110
8. T. Nakayama et al.: Kobe Steel Engineering Reports 47 (1997) p. 50
9. J. Sugimoto: Tokushukou (Japanese) 48 (1999) 3 p. 12
10. S. Misawa: Okurehakai-kaimei (Japanese) ISIJ (1997) p. 82

(10) Measures to improve spring properties

(i) Improved performance through weight saving

Steel for forming hot-coiled springs is used for comparatively large springs that are difficult to form into springs through cold-forming, and for springs

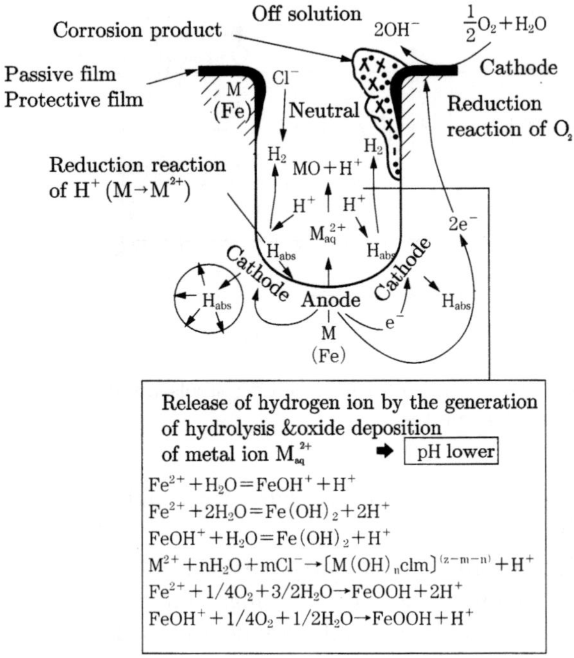

Fig. 2.78. schematic diagram of the hydrogen invasion caused by the hydrolysis in partial corrosion of steel and pH fall due to rust generation

Table 2.5. Spring steels defined by JIS G 4801 (mass%)

Grade	C	Si	Mn	Cr	Mo	V	B	Composition system
SUP3	0.82	0.25	0.45	–	–	–	–	High carbon steel
SUP6	0.60	1.65	0.85	–	–	–	–	Si–Mn steel
SUP7	0.60	2.00	0.85	–	–	–	–	
SUP9	0.56	0.25	0.80	0.80	–	–	–	Mn–Cr steel
SUP9A	0.60	0.25	0.85	0.85	–	–	–	
SUP10	0.51	0.25	0.80	0.95	–	0.20	–	Cr–V steel
SUP11A	0.60	0.25	0.85	0.85	–	–	0.0005	Mn–Cr–B steel
SUP12	0.55	1.40	0.75	0.75	–	–	–	Si–Cr steel
SUP13	0.60	0.25	0.85	0.80	0.30	–	–	Cr–Mo steel

with complex shapes. After the steel is formed into a spring through hot-coiling (in which processing is comparatively easy), the spring is quenched and tempered to give it the proper mechanical properties. Table 2.5 shows the chemical composition of the steel used for hot-formed springs as designated in JIS G 4801 (Spring steels).

With the exception of SUP3 which is carbon steels, most spring steels have a carbon content of 0.50–0.65%. As shown in Fig. 2.79, this is because the maximum hardness of the steel is determined by the carbon content, re-

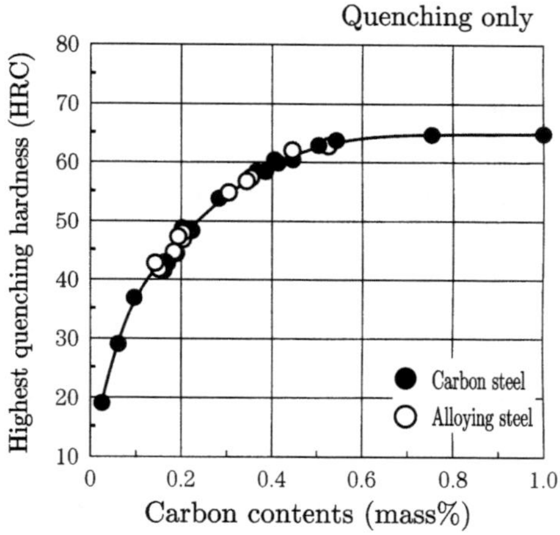

Fig. 2.79. Effect of carbon content on maximum hardness

gardless of whether or not alloy elements are present, and because even if the carbon content is increased above 0.6%, the maximum hardness will be saturated [1]. Accordingly, the carbon content of hot-formed spring steel specified by JIS is 0.5–0.6%. Table 2.6 shows the effect of chemical compositions on spring properties. Silicon, chromium, and manganese are added to improve the permanent set resistance and adjust the hardenability of the steel. Many JIS spring steels are used with a hardness of around 48HRC. However, SUP7 and SUP12 are used with a hardness of around 52HRC (tensile strength approximately 1,800 MPa).

With the aim of improving fatigue strength and permanent set resistance, high strength springs made of spring steel with increased hardness are now being used. Methods used to increase the hardness of the spring steel include the following:

(1) Increasing the strength by adding alloy elements
(2) Increasing the strength by changing the heat treatment conditions
(3) Improving the material properties through thermo-mechanical treatment such as ausforming

The properties required for springs include fatigue strength and permanent set resistance. However, in order to increase the fatigue strength by increasing the strength of the spring material, the followings are required:

(1) Reducing the number of defects such as non-metallic inclusions and surface flaws
(2) Reducing the austenitic grain size

Moreover, to improve the permanent set resistance, the followings are required:

Table 2.6. Effect of elements on steel and spring properties

Elements	Effect on steel properties	Effect on spring properties
Si	raises tempering resistance at 300 ℃ and below	Improvement in permanent set resistance
Mn	raises hardenability and raise strength	
Ni	raises hardenability, prevent low temperature brittleness	Improvement in corrosion resistance
Cr	raises hardenability and tempering resistance	Imorovement in permanent set resistance
Mo	raises tempering resistance	Improvement in permanent set registance
V	refines crystal grain and raises tempering resistance	Improvement in permanent set resistance
B	raises hardenability and prevent low temperature brittleness	
Nb	refines crystal grain	Improvement in permanent set resistance
Al	refines crystal grain causes remarkable surface hardening when nitriding	Improvement in fatigue strength

(1) Increasing the strength (hardness) of the steel material
(2) Increasing the softening resistance of the steel during tempering

Figure 2.80 shows the effect of adding elements on tensile strength. The addition of such elements as carbon, silicon and chromium is effective in increasing strength [2].

The simplest method of increasing the hardness of spring steel is to decrease the tempering temperature. It has been reported [3] that, for steels adjusted to the same hardness, the steel tempered at a higher temperature tends to show a higher fatigue limit, and studies are now being conducted into methods for obtaining high strength not simply by decreasing the tempering temperature but by adding chemical components as well. The high strength spring steels currently being used have an increased carbon content to provide high strength with the same tempering temperature as conventional steels, and with vanadium and niobium added to increase toughness and the yield ratio through reduction of the grain size. Figure 2.81 shows the martensitic structure of spring steel after tempering when vanadium is

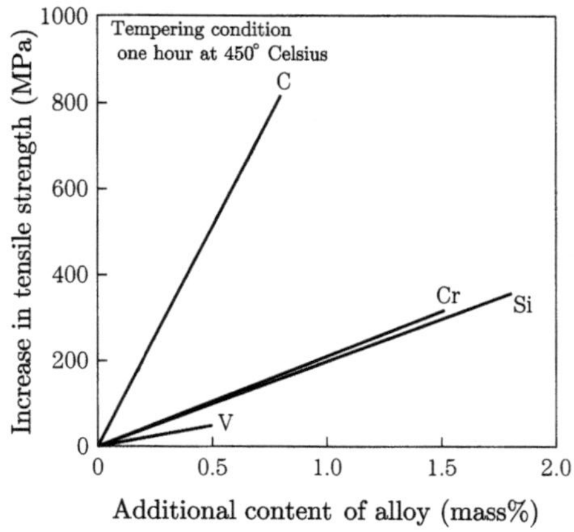

Fig. 2.80. Effect of alloy elements on tensile strength

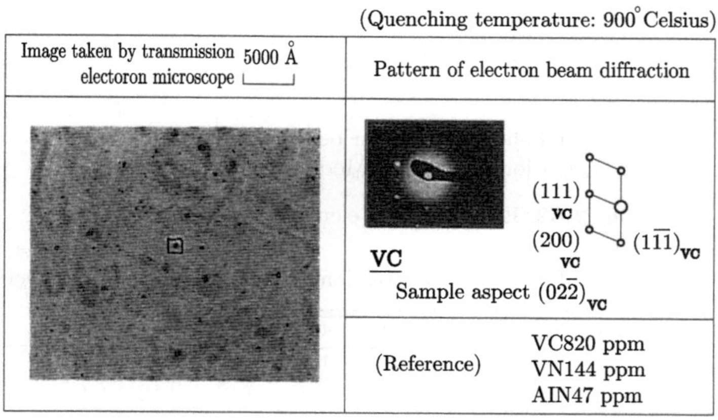

Fig. 2.81. Fine carbide in martensitic structure

added [4]. Fine vanadium cabonitrides precipitated in martensitic structures are finely dispersed and control austenitic grain growth.

When steel containing vanadium is tempered at a temperature of 450°C or above, secondary precipitation hardening occurs, increasing the softening resistance of the steel during tempering. This is also effective in improving permanent set resistance, making vanadium an element that is effective in increasing steel strength. The element niobium is also effective in reducing the austenitic grain size [5], and it is added to high strength spring steel in the same manner as vanadium. Figure 2.82 [6] shows the effect of alloy elements on permanent set resistance. Carbon, chromium and molybdenum increase the softening resistance of steel during tempering, and these alloy elements are effective in improving permanent set resistance.

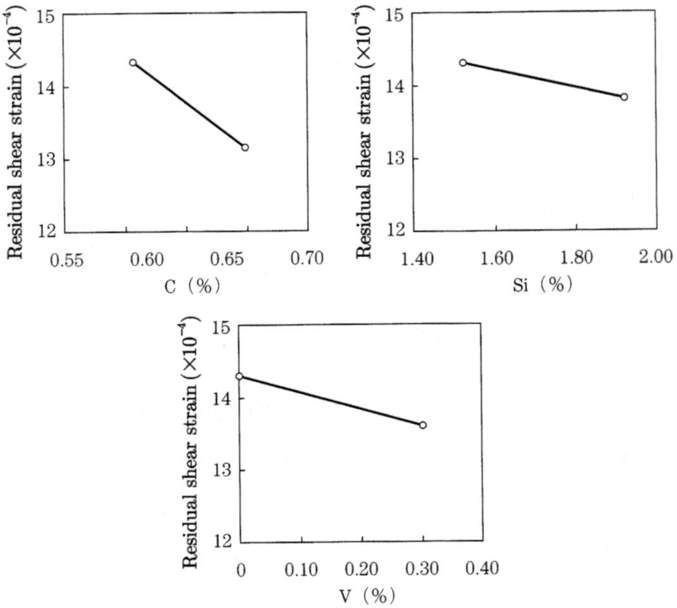

*) The residual shear strain can be obtained from the shear stress relaxation divided by Modulus of rigidity, G.

Fig. 2.82. Effect of alloy elements on sag resistance

Table 2.7. Example of chemical composition of high strength spring steel (mass%)

Grade	C	Si	Mn	Cr	Mo	V	Nb
A	0.60	1.45	0.5	0.55	–	0.18	–
B	0.55	1.60	0.50	0.80	0.10	0.18	0.04

Table 2.7 shows examples of typical chemical compositions of high strength spring steel to which these alloy elements are added, and which is used at a maximum spring shear stress of 1,100–1,200 MPa.

With springs that are used in corrosive environments, as typified by automotive suspension springs, the degradation of corrosion fatigue that occurs as strength increases is a major problem. Figure 2.83 shows the fracture surface for a corrosion fatigue test conducted after quenching and tempering of JIS SUP7 steel to HRC53. Figure 2.84 shows the results of fatigue tests conducted in air and in a corrosive environment.

As shown in the Fig. 2.84, fatigue life in air increases as hardness increases. In corrosive environments, however, fatigue life is known to decrease as hardness increases. The difference between the fracture surface produced in a corrosive environment and the fracture surface produced in air is that, in the case of the former,

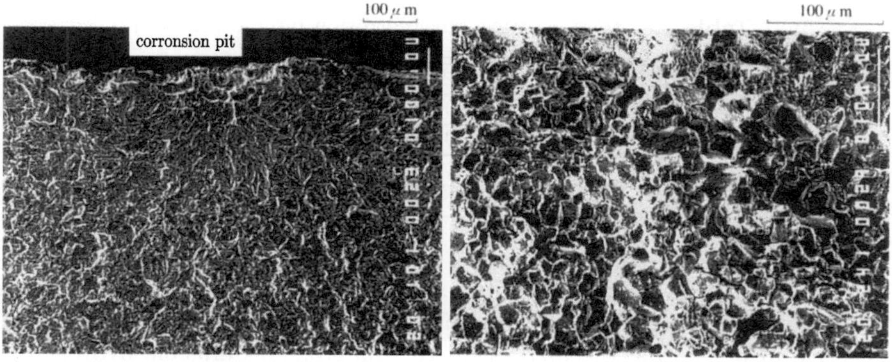

Fig. 2.83. Corrosion fatigue fracture in SUP7

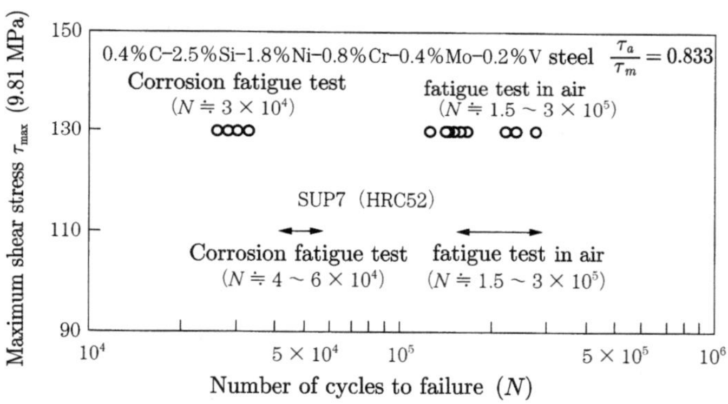

Fig. 2.84. Fatigue life in air and corrosive environments

(1) cracks are initiated from the corrosion pits, and
(2) cracks propagate along the prior austenite grain boundaries.

The development of cracks due to the concentration of stress at the bottom of the corrosion pits, and the accelerated crack propagation resulting from intergranular embrittlement which is caused by the hydrogen produced during the corrosion reaction, have a major impact on corrosion fatigue. For this reason, the design of alloy compositions for high strength springs that emphasize corrosion fatigue resistance focuses on two perspectives: improving corrosion resistance and reducing the susceptibility to hydrogen embrittlement. The elements copper, nickel and chromium are known to be effective in increasing corrosion resistance. In addition, nickel, chromium and molybdenum promote the precipitation of amorphous rust and prevent the intrusion of Cl^- and other elements that promote corrosion [8].

To reduce the susceptibility to hydrogen embrittlement, it is important to increase the toughness of the steel itself. For this purpose, the carbon content of the high strength spring steel that emphasizes corrosion fatigue resistance

Table 2.8. High strength steel corrosion fatigue (mass%)

Grade	C	Si	Mn	Cu	Ni	Cr	Mo	V	Nb	Ti
C	0.48	2.00	0.70	0.25	0.60	0.55	–	0.18	–	–
D	0.40	1.85	0.20	0.25	0.50	1.00	–	0.18	0.04	0.07
E	0.45	2.50	1.30	–	1.00	0.25	0.20	0.40	–	–
F	0.40	2.50	0.75	–	1.80	0.80	0.10	0.18	–	–

is kept low at 0.35–0.45%. In addition, titanium, vanadium and molybdenum precipitates are finely dispersed and used as hydrogen trap sites, as they are effective in preventing the increase in the quantity of diffusible hydrogen in the steel.

Table 2.8 shows examples of the chemical composition of spring steels that emphasize corrosion fatigue resistance and are currently used at a maximum spring shear stress of 1,200 MPa or greater [9], [10]. Spring steel with even greater strength will be needed in the future, and so improving corrosion fatigue properties will continue to be a major technical issue.

References

1. Metals, 26 (1938), p. 1
2. K. Tarui and N. Ibaraki: Japan society for technology of plasticity 39, 447 (1998) p. 297
3. S. Konuma and T. Furukawa: Zairyou 38, 433 (1989) p. 6
4. Y. Kawaguchi et al.: Technical reports of JSSR spring meeting (1991) p. 27
5. M. Minoguchi et al.: Technical reports of JSSR spring meeting (1993) p. 29
6. Y. Oki et al.: Transactions of JSSR 34 (1989) p. 26
7. Y. Yamada et al.: J. Japan Inst. Metals 23 16 (1984) p. 526
8. J. Inada et al.: Technical reports of JSSR autumn meeting (1996) p. 1
9. T. Akutsu et al.: Technical reports of JSSR spring meeting (1987) p. 1
10. J. Inada et al.: Transactions of JSSR 42 (1997) p. 15

(ii) Improvement in performance through thermo-mechanical treatment

In recent years, a thermo-mechanical technology known as the modified ausforming method (MAF) [1], [2], has been used for the leaf springs used in automobiles.

(1) MAF process

The modified ausforming (MAF) method is a modified version of the ausforming process designed to make it more suitable for practical application. Ausforming was first studied in 1951 by R. A. Harvey [3]. In 1968, the Ford Motor Company conducted research into the application to leaf springs [4].

Figure 2.85 is a simple explanation of the differences between the conventional heat-treatment, ausforming and MAF. The Fig. 2.85 shows a conceptual

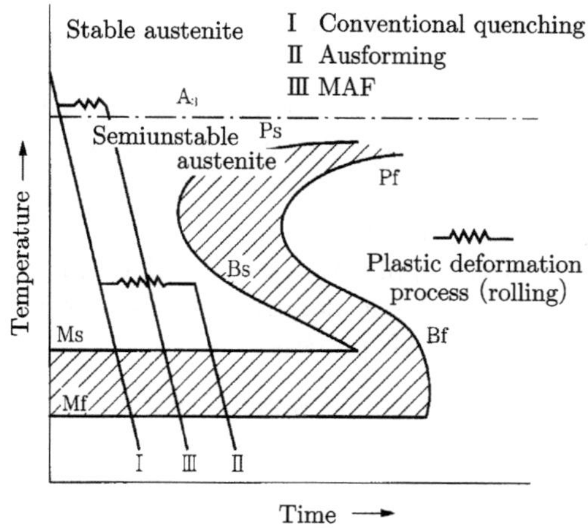

Fig. 2.85. TTT curve when quenching spring

depiction of the TTT curve during steel quenching. Line I in the Fig. 2.85 depicts the conventional heat treatment. The steel is heated to a temperature higher than the Ac_3 transformation temperature and then quickly cooled, resulting in a martensitic structure. Line II in the figure shows the ausforming process. The steel is heated to a temperature higher than the Ac_3 transformation temperature and then quickly cooled to the metastable austenitic range (approximately 500°C) applied plastic deformation of several tens of percentage, and quickly cooled to cause a martensite transformation. The principal feature of this method is that it increases strength without decreasing ductility or toughness. Line III depicts the MAF method. In this process, plastic deformation occurs in the region between the stable austenitic region and the metastable austenitic region, after which the steel is cooled quickly before pearlite transformation begins, resulting in a martensitic transformation. With this method, strength, ductility and toughness are all lower than with the ausforming process, but the steel has much greater toughness than that made with the conventional quenching process.

Despite the fact that the ausforming and modified ausforming processes are excellent thermo-mechanical methods, their application to leaf springs has lagged [5], [6]. The reasons are as follows:

(a) At the time that these methods were developed, in the 1960s, there was less need for lightweight springs than there is currently.
(b) The manufacture of leaf springs requires such processes as eye forming, and it could not be conducted prior to transformation with the conventional production equipment of the time.

(2) MAF processing conditions

(i) Material

The type of steel used in the MAF process is generally SUP10 (Cr–V steel), out of consideration for martensite lath and grain size refinement, carbides precipitation, etc [7].

(ii) Production equipment

The key to the practical application of the MAF process is to conduct quenching quickly following plastic deformation. Accordingly, the following equipment is needed for the MAF process:

(ii-1) Equipment for the taper rolling process that can do both one-side tapering and full tapering
(ii-2) Equipment that can perform eye rolling and other processes quickly following plastic deformation
(ii-3) High-speed transport of partially fabricated products during the heat treatment process
(ii-4) Integrated automated line that minimizes the involvement of human beings

(3) Mechanical properties

Figure 2.86 shows the hardness, proof strength, elongation and impact values when the tempering temperature is varied for steel processed by the MAF and the conventional method. For steel of the same hardness, the proof strength, elongation and impact values are greater for MAF steel than for steel processed by the conventional method. In addition, in order to ensure the Charpy impact value of approximately 30 J/cm^2 attained for conventionally processed steel tempered at 500°C, for MAF processed steel it is possible to reduce the tempering temperature to approximately 400°C without adversely affecting elongation, and this increases the proof stress 20% to 1,600 MPa as compared to steel processed by the conventional method.

(4) Metallugical structure

Figure 2.87 shows the prior austenite grain structure of steel processed by the conventional method and the MAF method. The crystal grain structure for the MAF steel is more refined than that for the conventionally processed steel. Moreover, in the MAF process, as the degree of plastic deformation increases, the result is stratified crystal grains that are extremely refined and elongated. This grain structure is characteristic of ausforming steel in which dynamic recrystallation occurs [8]. Figure 2.88 shows the result of the use of transmission electron microscope to observe conventionally processed steel and the martensitic structure formed by quenching steel processed using the MAF method. Compared to the conventionally processed steel, the width of the martensite rath was smaller in the MAF steel.

(5) Fatigue strength

To increase the fatigue strength of leaf springs, in general, shot-peening or stress-peening (shot peening with a preload applied in the direction of use)

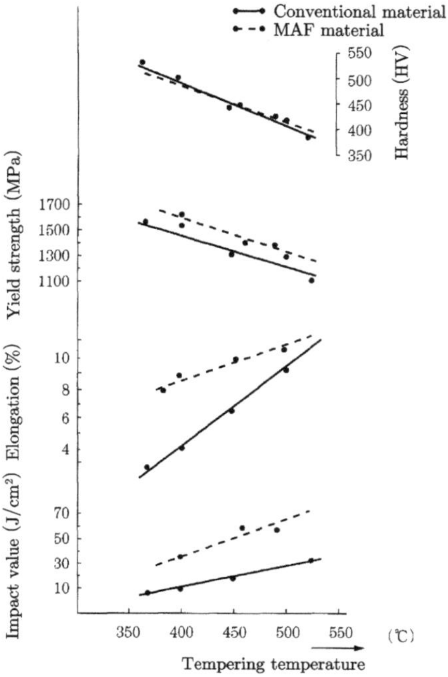

Fig. 2.86. Relation of tempering temperature, hardness and mechanical property

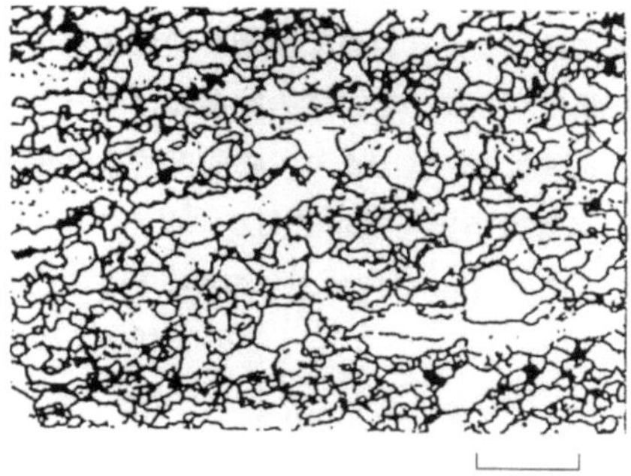

(a) MAF process 25 μm

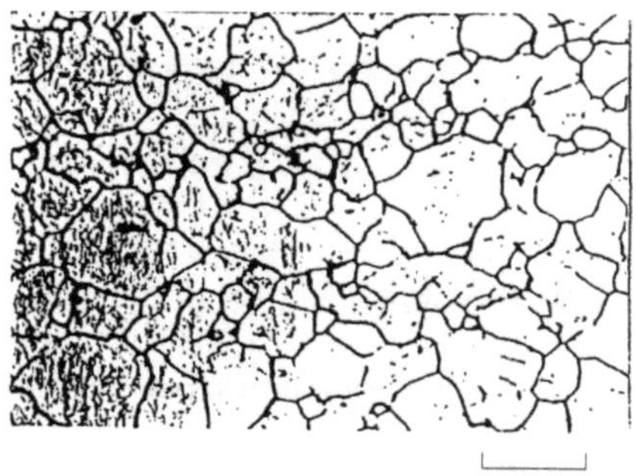

(b) Conventional process 25 μm

Fig. 2.87. Crystal grain structure

(a) MAF process 1 μm

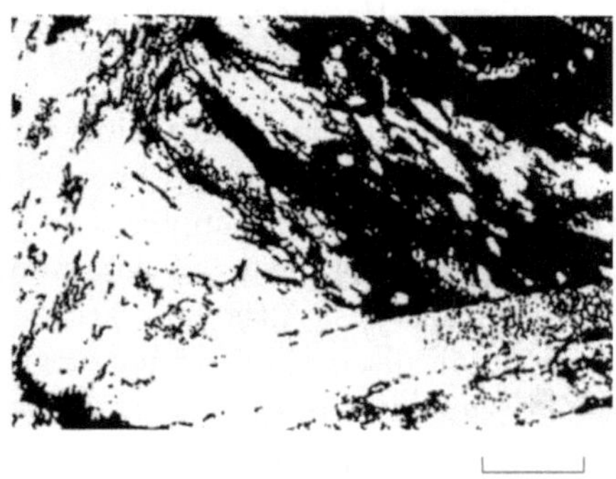

(b) Conventional process 1 μm

Fig. 2.88. Image of quenching structure in MAF process and conventional process

is conducted after tempering. Figure 2.89 shows the S-N carves of steels processed by the stress-peening method. As this result shows, for the same stress, MAF processed steel has a life five times that of conventionally processed steel, and the fatigue limit is also increased.

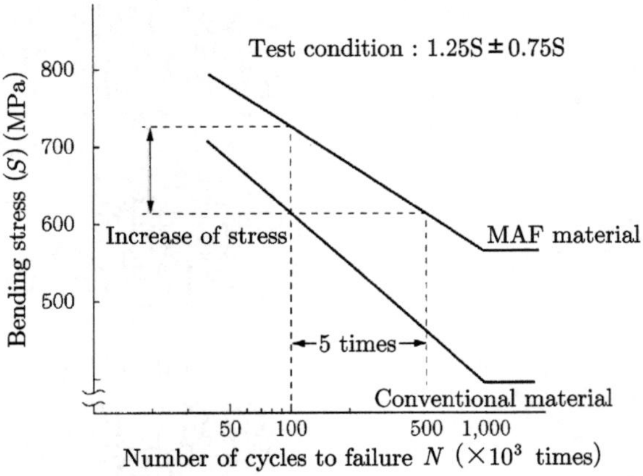

Fig. 2.89. S-N ratio of MAF processed material

References

1. M. Omiya: Zairyou (Japanese) 20 (1971) p. 1024
2. M. Omiya: Zairyou (Japanese) 20 (1971) p. 1101
3. R. F. Harvey: Iron age, 168 (1951) p. 70
4. R. A. Bock and W. M. Justusson: Lightweight Leaf Spring SAE680412 (1968)
5. M. Omiya: Transactions of JSSR 15 (1970) p. 20
6. M. Komatsu: Transactions of JSSR 22 (1977) p. 20
7. I. Tamura: Hagane no netsushori (Japanese) NihonTekkouKyoukai (1969) p. 75
8. M. Maki and I. Tamura: Iron and Steel 70 15 (1984) p. 284

(11) Material evaluation technologies

The required material properties are specified in standards and specifications according to the applications for that material. Various kinds of tests are conducted as means of determining whether the material fulfills the required properties. Although various properties are required for spring steel, this section discusses the technologies used to evaluate mechanical strength, one of the major properties of spring steel.

(i) Hardenability

(1) What is hardenability?

In many cases, hardness is used to evaluate the mechanical strength of spring steel. It is known that hardness affects the fatigue strength and permanent set of the spring. In order to ensure adequate hardness, the structure must be made into martensitic during quenching. Hardenability indicates the depth to which quenching has been conducted – in other words, the depth of the martensitic structure. Previously in the quenching of spring steel, a general guide was to obtain a 50% martensitic structure. Recently, however, from the standpoint of improving permanent set resistance and high stress performance, the ratio of martensitic structure has been increasing. On a provisional basis, the ISO [1] has established a value of 54–56 HRC for quenching hardness.

(2) Method used to evaluate quenching

The end quench method (Jominy end quench test) is a popular method of testing hardenability, and this method has been standardized in JIS [2]. In this method, one end of an austenitic test specimen is water-cooled (in other words, quenched), and then the relationship between hardness and the distance from the water-cooled end of the test specimen is determined. The resulting curve is called a hardenability curve. Figure 2.90 shows a schematic diagram for a hardenability test unit [2].

The region between the upper and lower limits of the hardenability curves obtained for the same grade of steel is known as the hardenability band. Using the hardenability curve, it is possible to determine the quenching hardness for steel of a certain diameter. For example, let us suppose that a ϕ30 mm round bar of spring steel SAE 5155H (equivalent to JIS SUP9) is quenched. Figure 2.91 [3], shows the hardenability band for SAE 5155H. Figure 2.92 [4], shows the method used to determine the cooling speed in the round bar section from the relationship between the diameter of the round bar and the distance from the water-cooled (quenching) end of the Jominy end quench test specimen. In the case of oil quenching, the section hardness distribution can be obtained from Fig. 2.91 and Fig. 2.92 (in other words, as determined from (a) and (b) in Fig. 2.93 and as shown in (c)). Conversely, if the size and

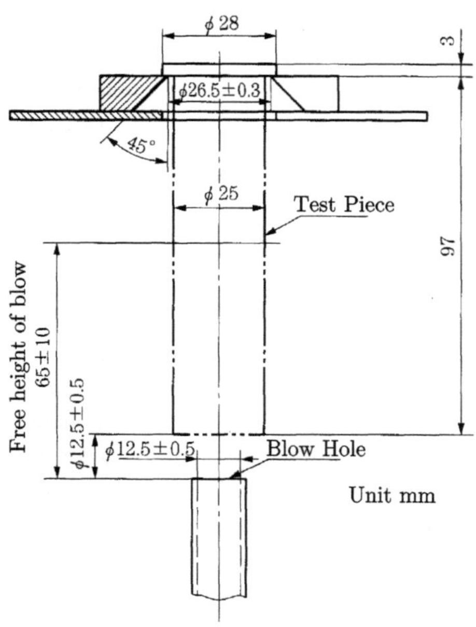

Fig. 2.90. Schematic diagram of apparatus to test hardenability

C	Mn	Si		Cr		
.50/.60	.60/1.00	0.15/0.35		.60/1.00		

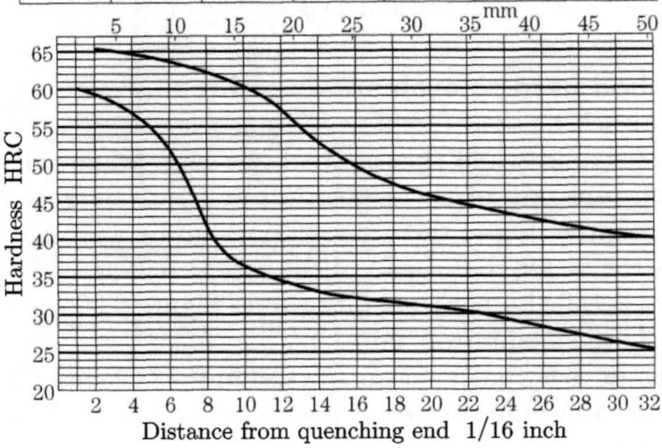

Fig. 2.91. Hardenability band for SAE 5155H (JIS SUP9) steel

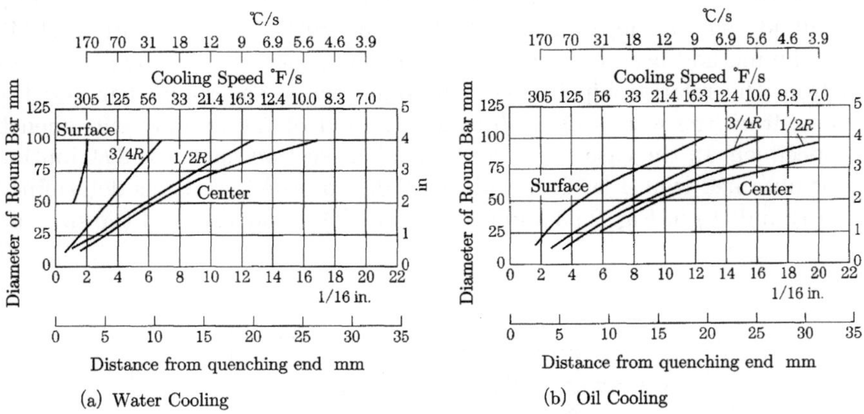

Fig. 2.92. Relation between distance from quenching end and diameter of round bar

hardness are determined in the spring design, the hardenability curve needed for the spring steel can be derived from Fig. 2.93. The hardenability bands for various grades of spring steel have already been determined, making it possible to select the desired spring steel. Recent progress in steel melting and refining technologies has made it possible to reduce the variations in components. As shown in Fig. 2.94 [6], there is a close relation between chemical compositions and hardenability, so it has become possible to narrow the hardenability band as well, enabling the variations in hardness after quenching and tempering to be reduced as well.

(ii) Surface flaws

The flaws on the surface of spring have a major impact on fatigue life. With the high stress performance of springs in recent years, even tiny flaws

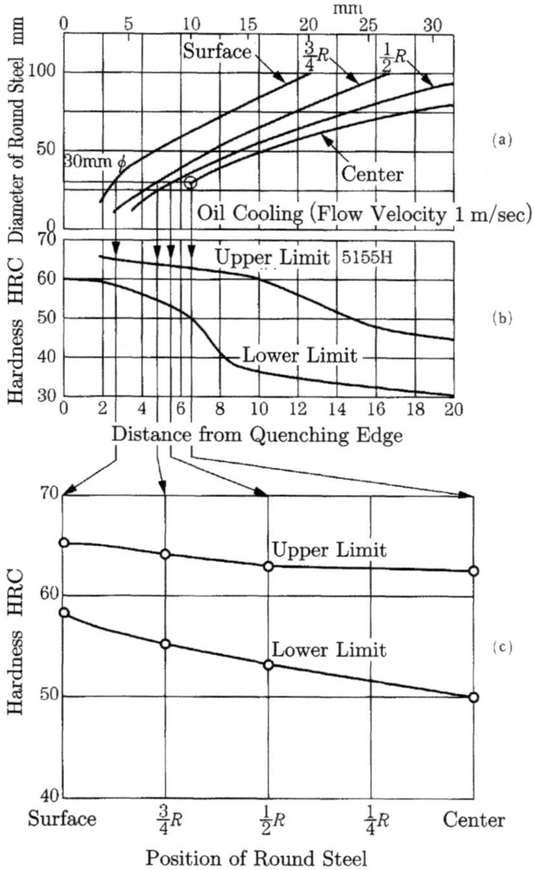

Fig. 2.93. Method for estiblishing the hardenability curve by quenching

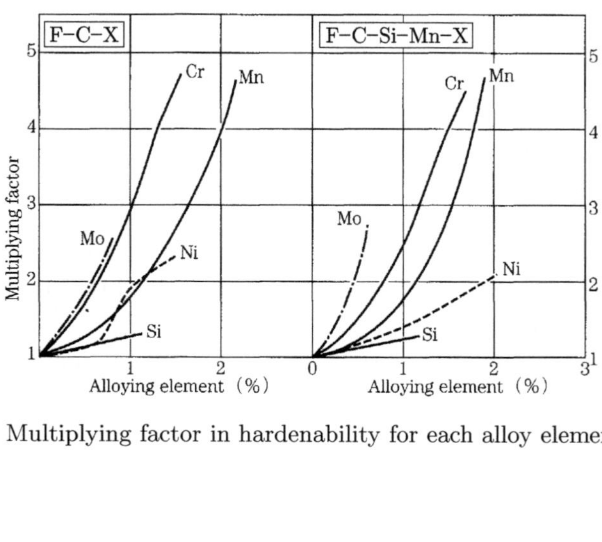

Fig. 2.94. Multiplying factor in hardenability for each alloy element (0.4%C)

	Radiogaraphic Testing	Ultrasonic Testing	Magnetic Particle Testing	Leakage Magnetic Flux Testing	Eddy Current Testing	Liquid Penetrant Testing	Optical Testing
Rough Sketch of crack detection	Radiation Source, Penetrameter, Tone meter, Film	Vertical Method, Vertical Probe, Bebeled Method, Bebeled Probe	Magnetic Powder, Leakage Magnetic Flow	Leakage Magnetic Flow, Sensor, N, S, Magnetic Container	Magnetic Saturated Coil, Defect Coil	Directive Pattern of Defect, Develop-ed Film, Surface Defect	Rotary Mirror, Diffused Light Sensor, Specular Light Sensor, Laser, Scanning Line
Phenomena for test	Electromagnetic Wave	Supersonic Wave	Electro-Magnetism	Electro-Magnetism	Electro-Magnetic flaw	Infiltration Capillarity	Beam of light
Detective Material	Metal, Non-metal	Metal, Non-metal	Ferromagnetic	Ferromagnetic	Conductive	Metal, Non-metal	Metal, Non-metal
Detectable Defect	Inside	Inside	Surface (Subsurface)	Surface (Subsurface)	Surface (Subsurface)	Surface	Surface

Fig. 2.95. Comarison of non-destructive testing methods

may become the starting point for fracture. Moreover, serial productions of spring are almost completely automated, so if a flaw exists on the surface, that flaw may become the origin for fracture during processing, which may result in trouble on the production line. It is important to inspect the surface for flaws; non-destructive testing methods have been standardized by JIS [7–11]. Figure 2.95 shows a comparison of several different non-destructive testing methods [12].

(1) Ultrasonic testing (UT)

An ultrasonic beam is transmitted into the material. If there is a flaw or foreign matter in the material, the ultrasonic beam is reflected or scattered. The size of the defects can be estimated from the reflected wave.

(2) Magnetic particle testing (MT)

When a magnetic material such as steel is magnetized, and there is a flaw on or directly beneath the surface, the flow of magnetic flux is disturbed at that location and leakage magnetic flux is produced. By using magnetic particles (fluorescent or non-fluorescent) to detect this leakage magnetic flux, it becomes possible to tell whether or not there is a flaw at that location. These flaws can be detected either through a visual inspection or automatically by means of image processing.

(3) Leakage magnetic flux testing (LMFT)

In this method, coiled sensors and the like are used (in place of the magnetic particles used to detect flaws in magnetic particle testing) to automatically detect the presence of leakage magnetic flux and the size of flaws.

(4) Eddy current testing (ET)

In this method, an eddy current is produced in the material. If a flaw is present, changes are produced in the eddy current in that location, and this fact is used to detect surface flaws and, to some extent, inner flaws.

(5) Liquid penetrant testing (PT)

This method is commonly known as the color check method. The surface of material is coated with a penetrant (a red or fluorescent liquid) that possesses penetrating force, and this liquid is allowed to thoroughly seep into open flaws in the surface. Then the penetrant on the surface is removed and a developer (a fine white powder) is used to draw the penetrant that has seeped into the flaws, to the surface, in order to detect these flaws.

(iii) Decarburization

When steel materials including spring steel are heated to high temperatures, the carbon in the steel reacts with the oxygen, carbon dioxide, moisture, hydrogen and other gases in the atmosphere and escapes into the air as a result of chemical reactions like those shown in Equations (2.3), (2.4) and (2.5), reducing the carbon content in the surface of the steel. This phenomenon is known as decarburization.

$$2C + O_2 \rightarrow 2CO \tag{2.3}$$

$$C + CO_2 \rightarrow 2CO \tag{2.4}$$

$$C + H_2O \rightarrow CO + H_2 \tag{2.5}$$

In decarburization, the reactive components in the atmosphere react with the carbon in or around the surface of the steel, reducing the carbon concentration. As a result, the carbon in the interior of the steel is dispersed to the surface, advancing the decarburization process. The decarburization reaction is noteworthy in the gamma phase steel, but not very noteworthy in the alpha phase steel. Decarburization of spring steel reduces the hardness of the steel near the surface of the spring, which also reduces its fatigue strength. Accordingly, caution is needed with regard to decarburization when heating spring steel and when heating during the spring forming process. According to JIS [13], the depth of the decarburized layer is measured by observing the steel with an optical microscope and measuring the hardness and carbon concentration. In many cases, the evaluation is made by observation with an optical microscope. Methods of depicting the decarburized layer include total decarburized layer depth and ferrite decarburized layer depth. Total decarburized layer depth refers to the depth from the top surface of the decarburized layer to the base metal, to the point at which it is impossible to distinguish any difference in chemical or physical properties. Ferrite decarburized layer depth refers to the depth from the surface of the level at which the surface of the steel is decarburized and becomes ferrite only. The Japan Spring Manufacturers Association (JSMA) standard is used as the method of measuring the decarburized depth of steel wire for spring manufacturers [14]. There is also a method known as the chord method [15], that is used to measure the hardness at the decarburized layer depth. Other methods include the use of electron probe X-ray microanalysis (EPMA) to perform a carbon analysis and measure the depth of the decarburized layer.

(iv) Non-metallic inclusions

Non-metallic inclusions comprise oxides and sulfides and other compounds produced in the steelmaking process, some of which remain in the steel (endogenous), and those produced by external factors such as by damage from refractory products (exogenous). These types of non-metallic inclusions affect the fatigue life of spring, so their size, properties and quantity should be evaluated. Care is needed particularly when the non-metallic inclusions are near the surface of the steel, as depending upon their size or properties they may tend to become origins for steel fracture.

Non-metallic inclusions are normally measured using an optical microscope. JIS uses optical microscopes to measure the type and number of non-metallic inclusions and judge the level of cleanliness of each steel [16]. Table 2.9 shows the types and shapes of non-metallic inclusions. With regard to the quantity, an optical microscope is used to evaluate the ratio of non-metallic inclusions as the percentage of total area, determined with the point counting method. With increasing the design stress and required quality of springs in recent years, as well as the increased life of valve springs and the like, the effect of non-metallic inclusions is extremely important, and rather than quantity, the size and type of the non-metallic inclusions in particular are seen as more serious problems. This is because the high stress performance of springs increases the susceptibility to notching, and non-metallic inclusions that are harder than the base metal or those with angular shapes in particular tend to become the origins for fatigue fracture, resulting in breakage. Accordingly, to evaluate the size of non-metallic inclusions near the surface, a test method that incorporates the penalty point method has been standardized by the JSMA [17]. In this method, large non-metallic inclusions near the surface in particular are assigned penalty points; the larger the size, the higher

Table 2.9. Types of non-metallic inclusion

Sorts	Characteristic
Type A inclusion	Inclusion (sulfide. silicate, etc.) with viscous deformation in processing. When necessary, subdivide into A_1 inclusions (sulfide) and A_2 inclusions (silicate).
Type B inclusion	Granular inclusions (almina, etc.) forming groups discontinuously in the direction of processing. In steel containing Nb, Ti and Zr (independently or in combinations of two or more), when necessary subdivided into B_1 inclusions (almina and other oxide) and B_2 inclusions (Nb, Ti or Zr carbonitride).
Type C inclusion	Irregularly distributed inclusions (granular oxides, etc.) with no viscous deformation. In steel containing Ni, Ti or Zr (independently or in combinations of two or more), when necessary subdivided into C_1 inclusions (oxide) and C_2 inclusions (Nb, Ti or Zr carbonitride).

the point score. The size is determined by measuring the largest non-metallic inclusions in the direction perpendicular to the material processing direction on the polished surface of the test specimen, prepared parallel to the material processing direction.

In conventional fatigue design, in general stress of approximately 10^7 repeated cycles was used as the fatigue limit. In recent years, however, the service life has been extended, and for some components a value of approximately 10^{10} cycles is used. In the case of such high cycles, it is said that non-metallic inclusions tend to affect the failure [18], so evaluation of non-metallic inclusions will continue to be an important issue. Moreover, research is being pursued into the use of extreme value statistical processing techniques to estimate the size of non-metallic inclusions and conduct a quantitative evaluation of the fatigue limit [19]. Two other methods for identifying non-metallic inclusions are electron beam melting [20] and acid dissolution [21], and these methods are used to evaluate the effect of the type and size of non-metallic inclusions on fatigue strength.

(v) Surface roughness

JIS B 0601 [22], (Geometrical Product Specifications (GPS) – Surface texture: Profile method – Terms, definitions and surface texture parameters) defines three parameters for use in depicting surface roughness: average arithmetic roughness (Ra), maximum height (Ry) and ten-point average roughness (Rz).

These parameters are used to quantify surface irregularities. Among these parameters, maximum height (Ry) expresses the rectilinear interval between two parallel lines enclosing the section curve. As shown in Fig. 2.96 (22), ten-point average roughness expresses the sum of the average of the absolute values for the height from a standard height average line to the height of the highest through fifth highest peak, and the average of the absolute values for the depth of the deepest through the fifth deepest trough.

(vi) Corrosion fatigue test

Automotive suspension springs are painted before use. However, the paint may peel off after being struck by small stones and the like while driving. Therefore, the evaluation of corrosion fatigue of not only springs but spring steel itself as well can be an important issue. Two methods are used to perform this evaluation. In one method, at the spring steel stage, test specimens are placed in a corrosive environment and fatigue tests are conducted after

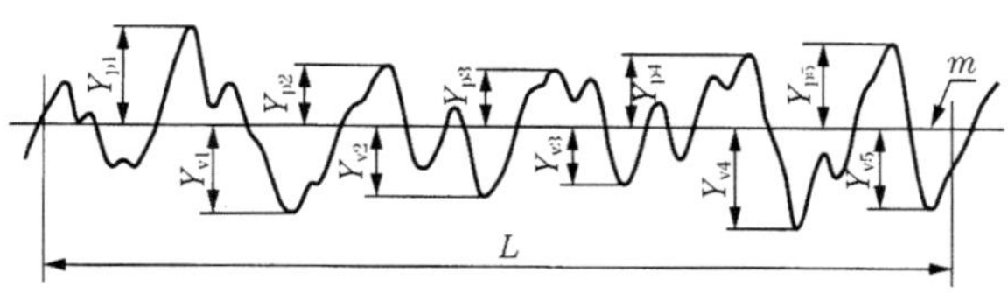

Fig. 2.96. Method for measuring ten-point average roughness

corrosion has progressed to some degree. In the other method, the test is performed for the test specimens during the corrosion process. For actual springs, standard accelerated test methods have not yet been established to measure corrosion fatigue. In recent years, studies [23], [24] have used extreme value statistical processing to evaluate the size of the corrosion pits produced in corrosion tests and the like, and the relationship between shape and fatigue strength. In another study, a laser microscope was used to provide three-dimensional measurements of the size of the corrosion pits [25].

(vii) Delayed fracture test

Delayed fracture is a phenomenon in which a material subjected to static stress suddenly experiences fracture with apparently brittle appearance after a certain period of time has passed. For example, steel materials with comparatively high strength such as high strength bolts are thought to be subjected to high tightening stress, and in some environments the intrusion of hydrogen may produce material degradation at that location. Depending upon the concentration of hydrogen in steel, delayed fracture is a concern in the case of springs as well if hydrogen intrudes in corrosive environments, as the spring has high strength from the outset. At present, there is no standard method for evaluating delayed fracture in spring steel. Figure 2.97, [26] shows an example of the shape of a test specimen used for a delayed fracture test used for bolts. Figure 2.98 [27] shows examples of the test equipment used for these tests.

(viii) Microstructure

The mechanical properties of spring steel are closely related to microstructure. Especially, the microstructure must be observed to determine whether the spring has received proper heat treatment. Microstructures are observed primarily through the use of optical microscopes. To observe finer microstructures, scanning electron microscopes and transmission electron microscopes

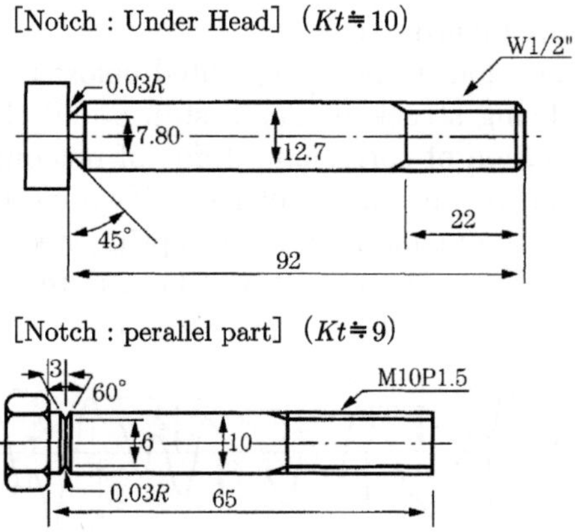

Fig. 2.97. Example of delayed fracture test

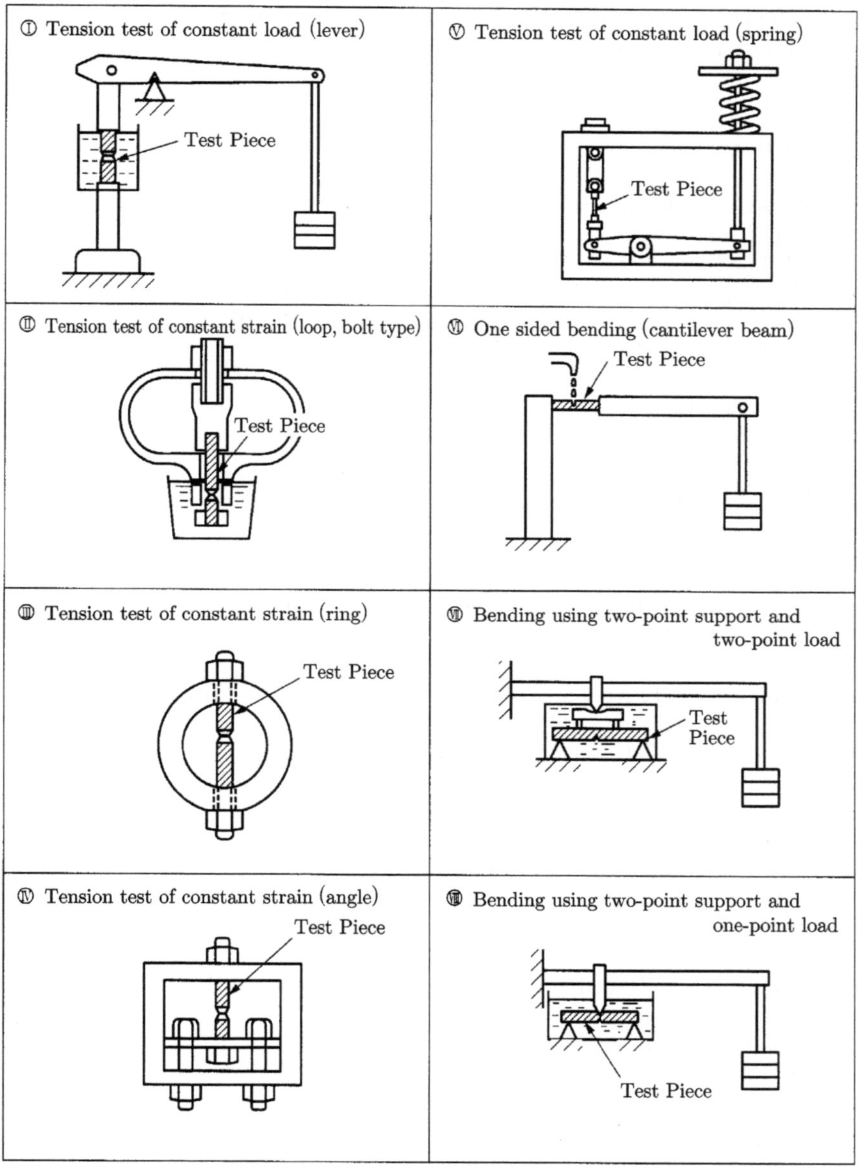

Fig. 2.98. Test equipments for delayed fracture

and the like are used. Transmission electron microscopes capable of performing analyses are used to identify fine precipitates.

(ix) Austenite grain size

The austenite grain size will affect the hardenability, toughness and other properties of spring steel. The methods used to evaluate the austenite grain size of steel are specified in JIS G 0551 (Steels – Micrographic determination of the apparent grain size) [28].

References

1. ISO683-14 Hot-rolled steels for quenched and tempered springs (1992)
2. JIS G 0561
3. SAE handbook Vol. 1 Materials (1983) p. 6
4. ASM Handbook Vol. 4 Heattreatment (1991) p. 86
5. Tokushukou Club: Tokushukou guide (Japanese) 4 (1977) p. 86
6. SAE handbook Vol. 1 Materials (1983) p. 6
7. JIS G 0565-1993
8. JIS G 0568-1983
9. JIS Z 2319-1991
10. JIS Z 2343-1992
11. JIS Z 2344-1993
12. T.Okamura: Tokushukou 46 (1997) 6, p. 4
13. JIS G 0558-1998
14. JSMA Standard No. 10 (1974)
15. R. H. Gassner: Metal progress (1978) March p. 59
16. JIS G 0555-1977
17. JSMA Standard No. 13 (1988)
18. K. Kanazawa, T. Abe, S. Nishijima: Kin-Zai-ken Data Sheet (Japanese) No. 9 (1995)
19. Y. Murakami, K. Kawakami, M. Saito: Trans. of JSSR 35 (1990) p. 1
20. T. Hagihara et al.: Wire Journal Inter. 34 (1991) 4, p. 29
21. Y. Yamada et al.: Trans. of JSSR 31 (1986) p. 1
22. JIS B 0601-1994
23. Y. Murakami et al.: Trans. of JSSR 39 (1994) p. 7
24. Y. Ito, A. Yoneguchi: Trans. of JSSR 43 (1998) p. 37
25. J. Inada, M. Shimotsusa, N. Ibaraki, T. Nakagawa, S. Miyauchi: Trans. of JSSR 42 (1997) p. 15
26. T. Hasegawa, T. Nakahara, Y. Yamada, M. Nakamura: Wire Journal Inter. 35 (1992) 8, p. 49
27. Nihon Gakujutsu Sinkoukai No. 129 (1975) p. 19
28. JIS G 0551-1998

2.1.3 Cold Formed Springs

Springs are machine elements where the elastic strain energy is utilized. In principle, any material satisfying this can be applied to springs. However, when choosing material on a commercial basis, the material designated by Japanese Industrial Standard, JIS and Japan Spring Manufacturers Association, JSMA can be normally used. Because, those materials have many experiences and results technically, and can show a competitive cost.

The spring material under the manufacturing processes can be classified into two groups: hot forming and cold forming. The hot forming is suitable for large size springs. The cold forming can be more effective when processing small size springs. According to the statistical data from JSMA, the ratio of the production weight of hot formed springs to that of cold formed springs is six to four. However, in terms of production number, the cold formed spring is substantially large, and in terms of the sales amount the cold formed spring exceeds the 70% of the total amount.

Cold formed springs can be classified further to wire springs including coil springs, sheet springs for automobiles, flat springs, power springs and spring hanger clips (fastener springs). Among them, the most numerous springs are wire springs and flat springs.

With respect to spring material, the wire spring mostly use ferrous material, while many flat springs do non-ferrous material. The cold formed spring can show an excellent dimensional accuracy by using pre-hardened material, except precipitation hardening springs and some others. Therefore, the cold formed spring material should have excellent formability to meet with the mass productions.

(1) Piano wire and hard drawn steel wire

(i) Grades and applications of piano wire

The piano wire can be applied for engine valve springs, mechanical springs, and music instruments. The piano wire for springs can be merely called "piano wire". As shown in Table 2.10, JIS G 3522 standardizes the piano wire as three grades A, B, and V. In each grade, the specifications for tensile strength, winding property, torsion property, bending property, wire diameter and its tolerance, appearance, surface flaws, decarburization, and surface conditions are designated. Apart from JIS, the preformed straight piano wire which is used for forming and for a helical torsion spring with a long arm, can be also included in the category of the piano wire.

As for the application of piano wire, the V-grade piano wire can be used for springs with high rate of repetition, especially for valve springs of automobiles, ships, and agricultural implements. The A-grade and B-grade piano wire can be used for important parts such as clutch springs and brake springs of automobiles, springs for electric components, machine tools, and construction equipment, aggregate batcher springs, and high quality springs for sporting goods and others. The A-grade piano wire has high value for allowable maximum stress and excellent properties for fatigue against repetitive loads. The B-grade piano wire is suitable for springs that are required, in design, to have high tensile strength and sufficient permanent set resistance, yet they are in-

Table 2.10. Grade, character and applicable diameter range of piano wire [1]

Grade	Symbol	Applicable diameter range
Piano wire A type	SWP-A	over 0.08 mm and under 10.0 mm inclusive
Piano wire B type	SWP-B	over 0.08 mm and under 13.00 mm inclusive
Piano wire V type	SWP-C	

ferior to A-grade piano wires on toughness which becomes an issue in severe forming.

(ii) Manufacturing process for piano wire

The piano wire requires uniformity of chemical composition, and absence of harmful segregation, impurities and non-metallic inclusions. Therefore, piano wire rod, the raw material for the piano wire, is resourced from high-grade steel scrap, which contains little of harmful impurities such as P, S and Cu, all of which are specified in JIS, or high-grade pig iron. The manufacture of rod starts from melting at an electric arc furnace or converter, followed by secondly-refining, where impurities can be reduced from the melts and nonmetallic inclusions in the melt can be controlled on their compositions or decreased. Then, billet is made through continuous caster, where the melt is cast, and blooming mill. Surface flaws of the billet can be removed elaborately by grinding or scarfing. The billet can be heated and rolled under sufficient controls, and finally finished to excellent piano wire rod free from defects such as harmful flaws and decarburization.

Depending upon the chemical compositions, especially, carbon and manganese contents, the piano wire rod (JIS G 3502) is classified into 18 classes ranging from SWRS62A to SWRS92B. Among them, the most popular rod is SWRS 82A.

Table 2.11 shows a list of the chemical compositions of piano wire rods. Although the precise manufacture processes of piano wire differ depending on the grades, in basic terms they look alike. In each process, rigorous quality control is required for all the piano wire. Figure 2.99 shows representative manufacturing process charts of piano wire, starting with piano wire rod. The wire rod, given heat treatment called patenting, are pickled, provided with surface treatment and finally finished through wire drawing at room

Table 2.11. Chemical composition of JIS piano wire rod [5]

Grade	Chemical composition mass%					
	C	Si	Mn	P	S	Cu
SWRS 62 A	0.60–0.65	0.12–0.32	0.30–0.60	≦ 0.025	≦ 0.025	≦ 0.20
SWRS 62 B	0.60–0.65	0.12–0.32	0.60–0.90	≦ 0.025	≦ 0.025	≦ 0.20
SWRS 67 A	0.65–0.70	0.12–0.32	0.30–0.60	≦ 0.025	≦ 0.025	≦ 0.20
SWRS 67 B	0.65–0.70	0.12–0.32	0.60–0.90	≦ 0.025	≦ 0.025	≦ 0.20
SWRS 72 A	0.70–0.75	0.12–0.32	0.30–0.60	≦ 0.025	≦ 0.025	≦ 0.20
SWRS 72 B	0.70–0.75	0.12–0.32	0.60–0.90	≦ 0.025	≦ 0.025	≦ 0.20
SWRS 75 A	0.73–0.78	0.12–0.32	0.30–0.60	≦ 0.025	≦ 0.025	≦ 0.20
SWRS 75 B	0.73–0.78	0.12–0.32	0.60–0.90	≦ 0.025	≦ 0.025	≦ 0.20
SWRS 77 A	0.75–0.80	0.12–0.32	0.30–0.60	≦ 0.025	≦ 0.025	≦ 0.20
SWRS 77 B	0.75–0.80	0.12–0.32	0.60–0.90	≦ 0.025	≦ 0.025	≦ 0.20
SWRS 80 A	0.78–0.83	0.12–0.32	0.30–0.60	≦ 0.025	≦ 0.025	≦ 0.20
SWRS 80 B	0.78–0.83	0.12–0.32	0.60–0.90	≦ 0.025	≦ 0.025	≦ 0.20
SWRS 82 A	0.80–0.85	0.12–0.32	0.30–0.60	≦ 0.025	≦ 0.025	≦ 0.20
SWRS 82 B	0.80–0.85	0.12–0.32	0.60–0.90	≦ 0.025	≦ 0.025	≦ 0.20
SWRS 87 A	0.85–0.90	0.12–0.32	0.30–0.60	≦ 0.025	≦ 0.025	≦ 0.20
SWRS 87 B	0.85–0.90	0.12–0.32	0.60–0.90	≦ 0.025	≦ 0.025	≦ 0.20
SWRS 92 A	0.90–0.95	0.12–0.32	0.30–0.60	≦ 0.025	≦ 0.025	≦ 0.20
SWRS 92 B	0.90–0.95	0.12–0.32	0.60–0.90	≦ 0.025	≦ 0.025	≦ 0.20

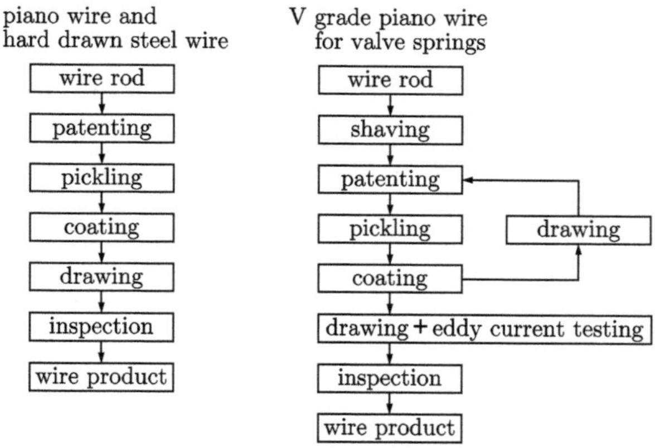

Fig. 2.99. Manufacuring flow chart of piano wire and hard drawn steel wire

temperature. Depending upon the wire size, these processes can be repeated. In the case of V-grade piano wire, shaving (continuous peeling for removal of surface flaws and decarburization) or continuous centerless grinding are given in the middle of the process.

Furthermore, in order to detect flaws formed in the wire-making process nondestructively, eddy current tests are performed on the overall length of the wire. In this case, if the flaws are harmful, they are marked and removed later, to ensure that the material with the surface flaws cannot be used for coiled springs. There are two types of eddy current testing: a through type in which the wire is passed through a fixed eddy current coil and a rotation type in which an eddy current probe coil rotates around wire at a high speed (see Fig. 2.128 and Fig. 2.129).

① Patenting

Patenting is heat treatment process, where high-carbon steel wire is continuously transformed to microstructure of fine pearlite, either by isothermal cooling or continuous cooling. In practice, wire rod, traveling continuously, is first held above the A_3 point, and subsequently cooled down below the A_1 point to be transformed to pearlite. Figure 2.100 shows a schematic of TTT diagrams of this heat treatment. Several kinds of cooling media, used in the patenting for the transformation of wire, include lead, air, and sand that is floated and fluidized by a mixture of air and combustion gas. They are called lead patenting, air patenting, and fluidized bed patenting, respectively.

Recently, a variety of direct patenting methods, which can control the cooling speed of rod, has been developed so that they can replace the first patenting in wire making. The direct patenting is mainly of three kinds: Stelmor type with air which is blasted on a high temperature as-hot-rolled rod thus cooling down to gain fine peartlite microstructure, EDC type with boiled water in which the hot rod is continuously soaked, and DLP type with molten salt held around 550°C in which the rod is continuously soaked as well.

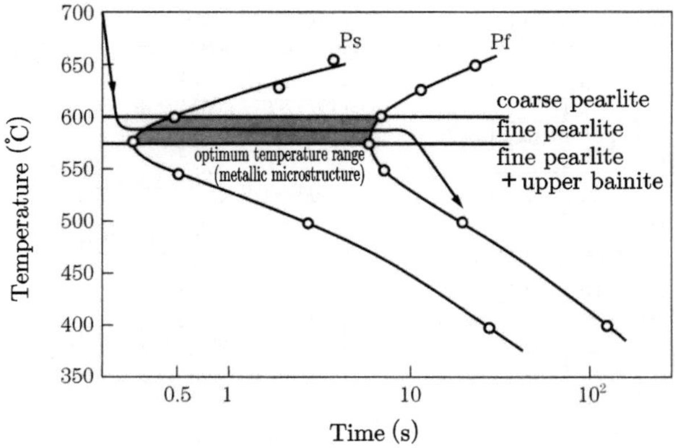

Fig. 2.100. Schematic of patenting in TTT diagram [2]

The last patenting in wire making process, having direct influence on the properties of the piano wire, is of significant importance. To ensure uniform and fine microstructure of pearlite, rigorous quality control is demanded on heating temperature and period, isothermal transformation temperature and period, and so on.

A patenting furnace is so carefully designed that the inner atmosphere can be rigorously controlled for prevention of the decarburization of wire. The tensile strength of patented wire depends upon its size, composition; especially C and Mn contents, and transformation temperature. The gained tensile strength due to drawing also depends upon the drawing reduction and the composition of rod. Therefore, it is necessary to properly control the chemical composition of rod, the size of wire which is patented last, and the transformation temperature, so that the tensile strength may be adjusted within prescribed ranges.

② Pickling

Pickling, either by hydrochloric acid or sulfuric acid solution, is generally provided on patented wire and hot-rolled rod to remove their surface scale (ferrous oxide film). The pickling must be carried out to the desired degree. Both overpickling and insufficient pickling may result in degraded quality of the wire and cause some troubles in wire drawing.

Operations in the pickling include the control of acid concentration, immersion time, and iron ion concentration. After being pickled, depending upon the use of the product, the wire is coated by lime, borax, and/or phosphate, which help as lubrication film in the following wire drawing. Those films act, in wire-drawing, as an agent, not only preventing contact between the die and the wire but also carrying lubricants. In general, the lime or borax film results in white as-drawn surface, whereas the phosphate film results in black as-drawn surface. Table 2.12 shows the comparison between the white and black surfaces.

Table 2.12. the comparison between the white and black surfaces

	White surface	Black surface
Surface color	metallic	grey to black
Tempered color	show temper color	little temper color
Spring formability	normal	better than the white
Corrosion resistance	need anti-rust oil	better than the white
Compatibility for plating	better than the black	need attensions

③ Drawing

Wire-drawing is performed through a die in a single-draft drawing machine or through dies in a continuous one. Lately, the continuous drawing machine has been mostly used. The drawing lubricants used are metallic soap for the dry drawing and aqueous or oil lubrication liquid for the wet drawing. The die generally used in the wire drawing is a carbide die, but for ultra-fine wire and one that requires strict size allowance, a natural/artificial diamond die is used. Basically, the mechanical properties of as-drawn wires depend upon the total reduction of the drawing after the last patenting. Even with the same total reduction, the difference in the drawing condition affects the wire's mechanical properties largely. When the drawing condition becomes inappropriate – for instance, cooling of the wire in the drawing process is insufficient – the elongation and toughness of the drawn wire deteriorate, thereby bringing about an abnormal torsional fracture. For this reason, in the wire drawing, it is necessary to control the temperature rise of the wire as small as possible and also cool down the wire as fast as possible. Especially in the case of continuous drawing machine where heat accumulates in the wire, it is required to cool the wire sufficiently by increasing the number of turns around each block. A direct water-cooling method, by which water is poured on the wire just drawn, is also applied [3]. In the case of the drawing of large size wire, a straight type drawing machine can be primarily used, which enables to draw the wire without any twists. The straight type drawing machine has also been spread for small size wire. A slip type wet drawing machine is also applied to small size wire.

Piano spring wire requires high degree of wire casting at the finish. It is important to control the angle of the die and adjust the condition of the straightening rolls, thereby keeping the free coil size of a ring taken from the edge of the as-drawn wire and the lift of the edge of the ring, which is set on flat surface, within the normal specifications.

(iii) Grades and use of hard drawn steel wire

Hard drawn steel wire is named variously depending upon its carbon content, size, mechanical property and applications. As shown in Table 2.13, JIS G 3521 divides them into three grades, A, B, and C, and designates them as SW-A, SW-B, and SW-C, respectively. JIS specifies its tensile strength, coiling performance, torsional property, bend property, diameter and its tolerance,

Table 2.13. Grade, character and applicable diameter range of hard drawn steel wires [4]

Grade	Symbol	Applicable diameter range
Hard drawn steel wire A type	SW-A	over 0.08 mm and under 10.0 mm
Hard drawn steel wire B type	SW-B	over 0.08 mm and under 13.00 mm
Hard drawn steel wire C type	SW-C	

and appearance. The applications of hard drawn steel wire range widely in various industries, that is, springs for safety pins, switches, weights, chairs and beds, and toys, springs used for the lifting of a shutter and saddle of a bicycle. Shaped wire, which is manufactured from hard drawn steel wire through flat-rolling, is used for washer springs, springs for a bicycle transmission, springs used for the lifting of a seat belt, and so on. The other applications of hard drawn steel wire, which are not formed to springs but applied as they are, include reinforced steel wire for an overhead transmission wire, and control cable. Besides, SW-A is applied to a wire gauze, fence, frame for a seat, and so forth.

(iv) Manufacturing method of hard drawn steel wire

The raw material from which hard drawn steel wire is made is specified in JIS G 3506 (high carbon steel wire rods). This specification classifies them, according to the carbon contents, into 12 categories, among which 9 categories are further grouped as A and B depending upon the manganese contents. The total amounts to 21 categories. Table 2.14 shows a list of the chemical compositions. In general, the hard drawn steel wire is categorized to 0.30%C, 0.40%C, 0.50%C, 0.60%C, 0.70%C, and 0.80%C, depending upon the carbon content. The designation is such that the group of 0.30%C is called 30 carbon.

In comparison with JIS G 3502 (Piano wire rods), the levels of quality assurance differ between high carbon steel wire rod and piano wire rod; that of piano wire rod is severer than the other. The manufacture of hard drawn steel wire primarily starts from wire rod with high drawability. The rod is manufactured through steel making at an electric arc furnace or converter, casting to billets or blooms at a continuous caster, hot-rolling, and finally direct patenting such as Stelmor, EDC, and DLP types. The manufacturing process of hard drawn steel wire is shown in Fig. 2.99, mentioned above.

Hard drawn steel wire is manufactured from high carbon steel wire rods through patenting, descaling such as pickling or shot blasting, surface treatment such as coating with lime, borax, or phosphate, and finally drawing with dies.

(v) Differences between piano wire and hard drawn steel wire

① Difference in rod quality

The difference between the chemical composition of piano wire and that of hard drawn steel wire is exemplified in Table 2.15, contrasting the piano wire

Table 2.14. Chemical composition of JIS high carbon steel wire rods [5]

Grade	Chemical composition mass%				
	C	Si	Mn	P	S
SWRH 27	0.24–0.31	0.15–0.35	0.30–0.60	≦ 0.040	≦ 0.040
SWRH 32	0.29–0.36	0.15–0.35	0.30–0.60	≦ 0.040	≦ 0.040
SWRH 37	0.34–0.41	0.15–0.35	0.30–0.60	≦ 0.040	≦ 0.040
SWRH 42 A	0.39–0.46	0.15–0.35	0.30–0.60	≦ 0.040	≦ 0.040
SWRH 42 B	0.39–0.46	0.15–0.35	0.60–0.90	≦ 0.040	≦ 0.040
SWRH 47 A	0.44–0.51	0.15–0.35	0.30–0.60	≦ 0.040	≦ 0.040
SWRH 47 B	0.44–0.51	0.15–0.35	0.60–0.90	≦ 0.040	≦ 0.040
SWRH 52 A	0.49–0.56	0.15–0.35	0.30–0.60	≦ 0.040	≦ 0.040
SWRH 52 B	0.49–0.56	0.15–0.35	0.60–0.90	≦ 0.040	≦ 0.040
SWRH 57 A	0.54–0.61	0.15–0.35	0.30–0.60	≦ 0.040	≦ 0.040
SWRH 57 B	0.54–0.61	0.15–0.35	0.60–0.90	≦ 0.040	≦ 0.040
SWRH 62 A	0.59–0.66	0.15–0.35	0.30–0.60	≦ 0.040	≦ 0.040
SWRH 62 B	0.59–0.66	0.15–0.35	0.60–0.90	≦ 0.040	≦ 0.040
SWRH 67 A	0.64–0.71	0.15–0.35	0.30–0.60	≦ 0.030	≦ 0.030
SWRH 67 B	0.64–0.71	0.15–0.35	0.60–0.90	≦ 0.030	≦ 0.030
SWRH 72 A	0.69–0.76	0.15–0.35	0.30–0.60	≦ 0.030	≦ 0.030
SWRH 72 B	0.69–0.76	0.15–0.35	0.60–0.90	≦ 0.030	≦ 0.030
SWRH 77 A	0.74–0.81	0.15–0.35	0.30–0.60	≦ 0.030	≦ 0.030
SWRH 77 B	0.74–0.81	0.15–0.35	0.60–0.90	≦ 0.030	≦ 0.030
SWRH 82 A	0.79–0.86	0.15–0.35	0.30–0.60	≦ 0.030	≦ 0.030
SWRH 82 B	0.79–0.86	0.15–0.35	0.60–0.90	≦ 0.030	≦ 0.030

Table 2.15. Difference in chemical composition between piano wire and hard drawn steel wire [5]

Grade	Chemical composition mass%					
	C	Si	Mn	P	S	Cu
SWRS 82 A	0.80–0.85	0.12–0.32	0.30–0.60	under 0.025	under 0.025	under 0.20
SWRH 82 A	0.79–0.86	0.15–0.35	0.30–0.60	under 0.030	under 0.030	–

rod SWRS82A (JIS G 3502) with the high carbon steel wire rod SWRH82A (JIS G 3506).

Impurities like P, S, and Cu are controlled more strictly for the piano wire rod than for the high carbon steel wire rod. In addition, it is required that the piano wire rod should be free from any flaws of more than 0.1 mm in corrosion tests and the depth of total decarburized layer (ferrite layer) should be less than 0.07 mm, whereas the high carbon steel wire rod does not have such specifications. Accordingly, the level of quality of piano wire and that of the hard drawn steel wire differ from the beginning; that is, the wire rod as raw material.

② Difference in specification of products

Table 2.16 shows the differences of JIS specifications between piano wire and hard drawn steel wire. The Tensile strength, torsion value, flaws, decar-

Table 2.16. Differences in JIS specifications between piano wire and hard drawn steel wire [1], [4] and [5]

Type			Piano wire			Hard drawn steel wire		
Designation			SWP-A	SWP-B	SWP-V	SW-A	SW-B	SW-C
applicable diameter range mm			0.08–10.0	0.08–7.00	1.00–6.00	0.08–10.0	0.08–13.0	
Diameter tolerance and ovality	1.00 mm		±0.010 mm, ≦ 0.010 *mm*			±0.020 mm, ≦ 0.020 *mm*		
Tensile strength MPa	1.00 mm		2060–2260	2260–2450	2010–2210	1470–1720	1720–1960	1960–2210
Diameter tolerance and ovality	2.00 mm		±0.015 mm, ≦ 0.015 *mm*			±0.030 mm, ≦ 0.030 *mm*		
Tensile strength MPa	2.00 mm		1810–2010	2010–2210	1770–1910	1270–1470	1470–1720	1720–1960
Diameter tolerance and ovality	4.00 mm		±0.030 mm, ≦ 0.030 *mm*			±0.050 mm, ≦ 0.050 *mm*		
Tensile strength MPa	4.00 mm		1670–1810	1810–1960	1670–1810	1180–1370	1370–1520	1520–1720
Apperance			must be free from detrimental defects					
Specification of rod: Diameter tolerance and ovality			±0.40 mm, ≦ 0.40 mm			±0.50 mm, ≦ 0.50 mm		
Specification of rod: flaw depth	method		Flaw detection test shall be carried out by appropriate means such as magnetic particle inspection and pickling test. Flaw depth shall be measured by an instrument with sufficient accuracy.			none		
Specification of rod: flaw depth	specifications		depth of flaw must not exceed 0.10 mm					
Specification of rod: decurbarization depth	method		see JIS G 0558, 4			none		
Specification of rod: decurbarization depth	specifications		total decarburization layer must be under 0.07 mm					
winding property	diameter		≦ 0.70 *mm* (excl.)		none	≦ 0.70 *mm* (excl.)		
winding property	method		test piece wound around core wire with the same diameter at least 4 times must be inspected whether or not fractured and its state of flaw occurance			test piece wound around core wire with the same diameter more than 4 times must be inspected whether or not fractured and its state of flaw occurance		
winding property	specifications		wire must not occur detrimental flaw on its surface or fracture			wire must not occur detrimental flaw on its surface or fracture		
bending property	diameter		over 6.00 mm		none	over 6.00 mm		
bending property	method		two points of test piece are checked for presence of flaw or fractured or not after bended to different 90° direction under the radius of wire diameter			two points of test piece are checked for presence of flaw or fractured or not after bended to different 90° direction under the radius of wire diameter		
bending property	specifications		must be free from detrimental flaw and fracture			must be free from detrimental flaw and fracture		

Specification of wire	twist property	diameter	0.70–6.00 mm	1.00–6.00 mm	0.70–6.00 mm
		method	seize both ends of test piece firmly with interval of 100 times of wire diameter and rotate one end until fracture. should count number of twists and inspect state of cross section of fracture		
		specifications	number of twists 0.70–2.00 inclusive over 25 times inclusive 2.00–3.50 inclusive over 20 times inclusive 3.50–6.00 inclusive over 15 times inclusive cross section of fracture must be vertical to wire axis and free from flaw or crack, twist state must be umiform in full length, free from flaw and local twist and shall not be extraordinaly spiral shape.	number of twists over 25 for all size. Cross section of fracture must be vertical to wire axis and free from flaw or crack, twist state must be umiform in full length, free from flaw and local twist and shall not be extraordinary spiral shape	number of twists over 0.70 mm and under 2.00 mm over 25 times over 2.00 mm and under 3.50 mm over 20 times over 3.50 mm and under 6.00 mm over 15 times cross section of fracture must be vertical to wire axis and free from flaw or crack, twist state must be umiform in full length, free from flaw and local twist and shall not be extraordinaly spiral shape
	Apperance	specification	surface must be smooth in surface and free from detrimental defects like flaw		
	flaw depth	diameter	over 1.00 mm		none
		method	test piece with appropriate length relieved residual strain in advance must be soaked in adequate boling solution which is the mixture of hydrochloric acid water, by 200 mm. After wire loses it diameter 1%, test piece is checked for the presence of flaw or not. Depth of flaw is generally measured with micrometer after grinding the flaw to be disappeared.		
		specifications	flaw depth 1.00–2.00 mm inclusive under 0.02 mm inclusive 2.00–3.00 mm inclusive under 0.03 mm inclusive 3.00–4.00 mm inclusive under 0.04 mm inclusive 4.00–5.00 mm inclusive under 0.05 mm inclusive 5.00–6.00 mm inclusive under 0.06 mm inclusive 6.00–8.00 mm inclusive under 0.07 mm inclusive 8.00–10.0 mm inclusive under 0.08 mm inclusive	flaw depth under 0.01 mm inclusive under 0.02 mm inclusive under 0.02 mm inclusive under 0.03 mm inclusive under 0.03 mm inclusive	
	decarburization depth	diameter	over 0.70 mm inclusive	over 1.00 mm inclusive	none
		method	obey in JIS G 0558, transverse cross section of tested sample polished and etched is inspected for its decarburization state by means of microscope equiped with magnification of 100 or 500 times		
		specifications	must be free from detrimental decarburization layer	must be free from ferritic decarburization. Total decarburization layer must not exceed 1.5% of diameter and 0.05 mm.	

burization, and wire diameter are more strictly specified for piano wire than hard drawn steel wire.

(vi) General properties of piano wire and hard drawn steel wire

① Drawing reduction and mechanical properties

As mentioned above, piano wire is given strength and toughness through wire drawing preceded by patenting. An example of the relation between the mechanical properties of piano wire and the drawing reduction is illustrated in Fig. 2.101. While the tensile strength and the hardness rapidly increase due to work hardening with increasing in drawing reduction after patenting, the reduction in area slightly rises and over the reduction of 70 to 80% drops conversely. The elongation drops abruptly at the first stage of drawing, and then stays almost the same until 70 to 80% reduction. It decreases again over a reduction of 80%. Similarly, the number of bends and the number of torsion both decline in the range of drawing reduction over 85%.

As specified in JIS G 3522, piano wire excels as material for springs, because of their significantly high tensile strength and appropriate toughness. Yet, as-drawn piano wire has large residual stress due to wire drawing, and low elongation. The yield point and the elastic limit of piano wire are low for their high tensile strength. This tendency prevails for the torsional property.

A torsion test is performed to evaluate the tosional property of steel wire. As the strength of wire increases and its ductility is decreased, longitudinal

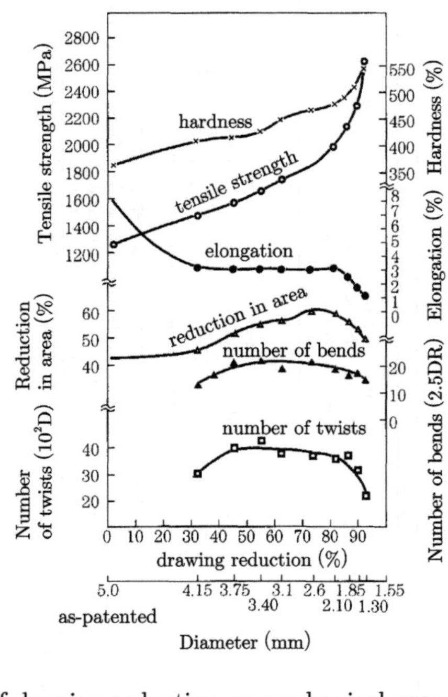

Fig. 2.101. Effect of drawing reduction on mechanical properties of piano wire [6]

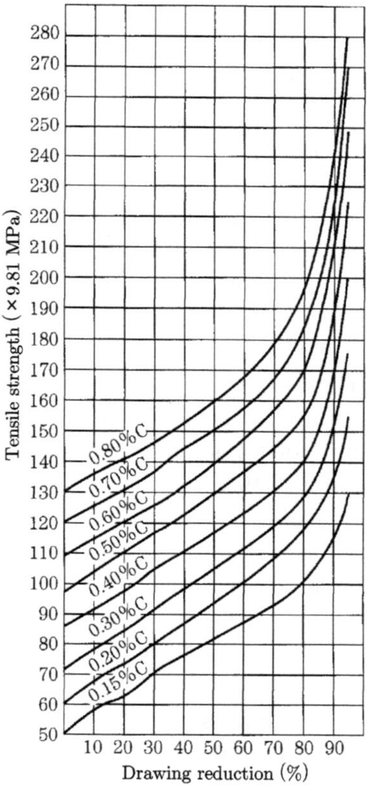

Fig. 2.102. Relation between drawing reduction and tensile sterength of various hard drawn steel wires [8]

cracks along the drawing direction – called delamination – appears in the early stage of torsional deformation. Steel wire causing a delamination is not of practical use. Accordingly, preventing delamination is the most important factor in strengthening steel wires. Each steel wire maker is devising various kinds of original ideas to make the strengthening of steel wires compatible with the prevention of delamination. Besides, the specified tensile strengths of both piano wire and hard drawn steel wire have a dependence on wire diameter: the larger the diameter is, the lower the specified strength becomes. This is because with the increasing diameter of patented wire the drawn steel wire becomes less ductile thus bringing about delamination in a torsion test at low reduction of wire-drawing.

Figure 2.102 shows the effect of the carbon content of as-drawn steel wires on their tensile strength. This relationship holds for piano wire. Higher carbon contents lead to higher tensile strength after patenting, thereby resulting in higher tensile strength after drawing at the same drawing reduction. Therefore, while the material of carbon content being 0.75 to 0.85 is used for piano wire and C-grade hard drawn steel wire that require high strength, the material of carbon contents being 0.60 to 0.75 is used for A/B-grade hard drawn steel wire that requires low strength.

② Properties at elevated temperature

The properties and variations of piano wire at elevated temperatures are indicated in Fig. 2.103 which is the case of 4.0 mm diameter piano wire annealed at 350°C for 15 minutes. As the Fig. 2.103 explains, being held at an elevated temperature the piano wire loses its strength and regains the reduction in area and elongation. In particular, the yield point and torsional yield stress deteriorate significantly in comparison with the corresponding tensile strength and torsional strength.

③ Fatigue strength

A large volume of research has been conducted on the fatigue strength of piano wire. An example of torsional Schenck type fatigue tests at room temperature is shown in Fig. 2.104.

④ Modulus of rigidity

The properties of torsion are of great significance in the knowledge of the spring performances of piano wire and hard drawn steel wire. Figure 2.105 shows results obtained by measuring modulus of rigidity (Shear modulus) G for high-carbon steel wire. Figure 2.106 shows the temperature dependence of the modulus of transverse elasticity of various steel spring wires.

⑤ Relation between hardness and tensile strength

The relation between hardness and tensile strength for piano wire and hard drawn steel wire is shown in Fig. 2.107. In general, it is considered that the fraction of a value in Vicker's hardness against that in tensile strength (HV / Tensile Strength in MPa) is approximately 0.3 for quenched and tempered material, and in the case of oil tempered wire, the notion tends to be held. In the case of material strengthened through cold working, like piano wire, hard drawn steel wire, and stainless spring wire, however, the fraction is likely to be lowered.

(vii) Low-temperature annealing properties of piano wire and hard drawn steel wire

① Objective

Springs made of piano wire and hard drawn steel wire are generally given low-temperature annealing from 200 to 350°C (also called annealing, blueing, or stress relieving) after spring forming. The aims of this heat treatment can be shown below.

1) To upgrade mechanical properties such as elastic limit and improve permanent set in fatigue.
2) To remove residual stress due to spring forming and improve fatigue strength.

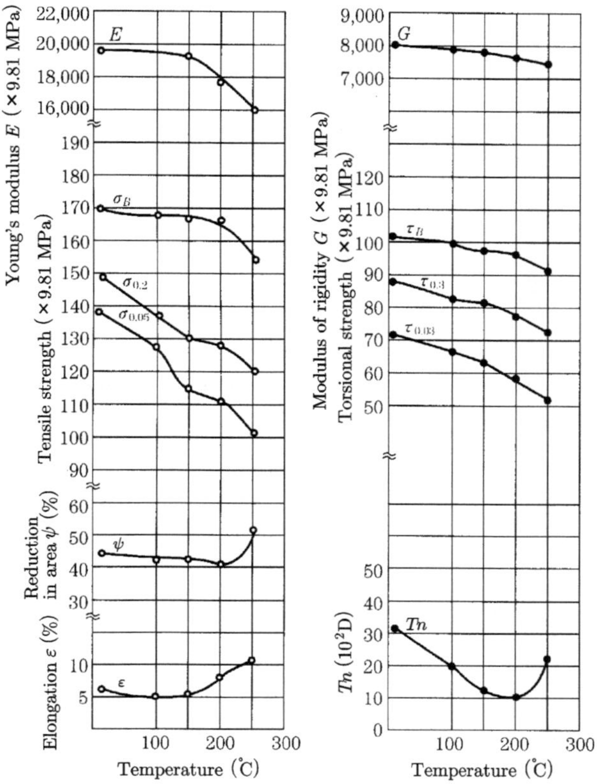

Fig. 2.103. Variations of mechanical properties on heating (RT–250°C) for 4.0 mm diameter piano wire [9]

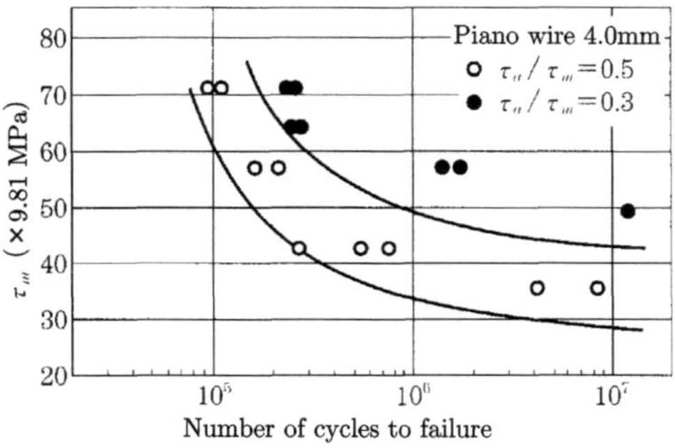

Fig. 2.104. An example of schenck type torsional fatigue test results at room temperature [10]

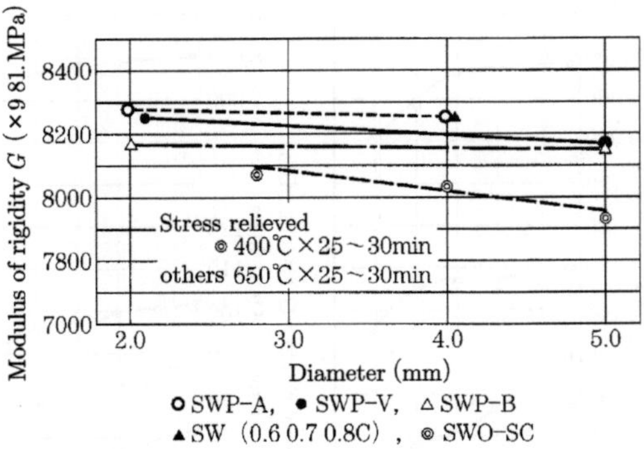

Fig. 2.105. Relation between diameter and modulus of transverse elasticity of various steel wires [11]

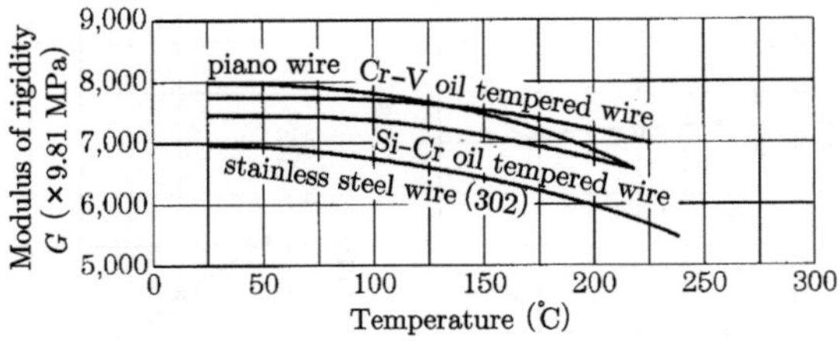

Fig. 2.106. Temperature dependence of the modulus of transverse elasticity of steel wires for springs [11]

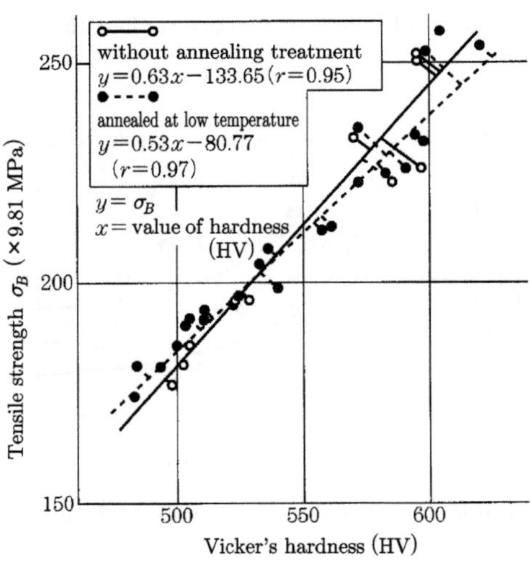

Fig. 2.107. Relation between tensile stress and hardness of hard drawn steel wire [12]

② Change of mechanical properties

Springs are mechanical parts that take advantage of elastic deformation; therefore, it is to be required that the elastic limit of the materials used for springs should be as high as possible. Piano wire and hard drawn steel wire possess high elastic limit even just after drawing. If the high elastic limit alone is desired, heat treatment between 200 and 300°C for several minutes enables the elastic limit to be even higher. This would be due to that the precipitation of carbides in steel wire and the formation of Cottrell atmosphere make dislocations hard to move. Figure 2.108 shows the relationship between the annealing temperature and the mechanical properties.

③ Change of permanent set resistance

An example of the variation of permanent set resistance with low temperature annealing is shown in Fig. 2.109. The behavior of permanent set resistance tends to be inversely proportional to that of tensile strength, and its minimum is located around 200 to 300°C. The fact that the permanent set remains low after annealing between 350 and 400°C where the tensile strength returns to almost the same value before annealing can be considered to be correlated with the high ratio of the yield point to the tensile strength.

④ Variations of residual stress and fatigue property

When piano wire or hard drawn steel wire are formed into springs in the cold forming, tensile residual stress develops on the inner surface of the coils.

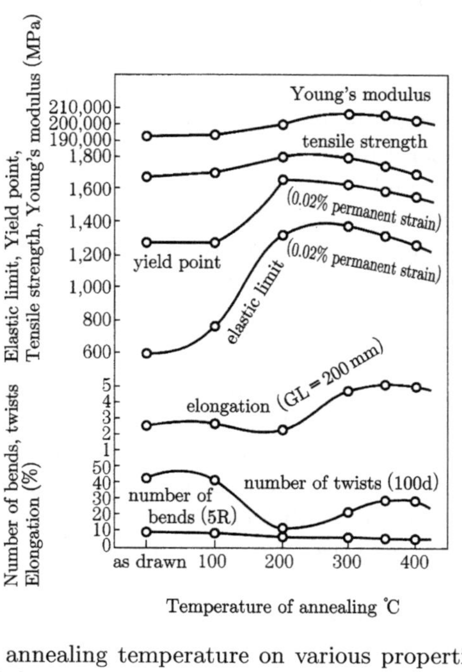

Fig. 2.108. Effect of annealing temperature on various properties of hard drawn steel wire (SW-C 4.0ϕ) [13]

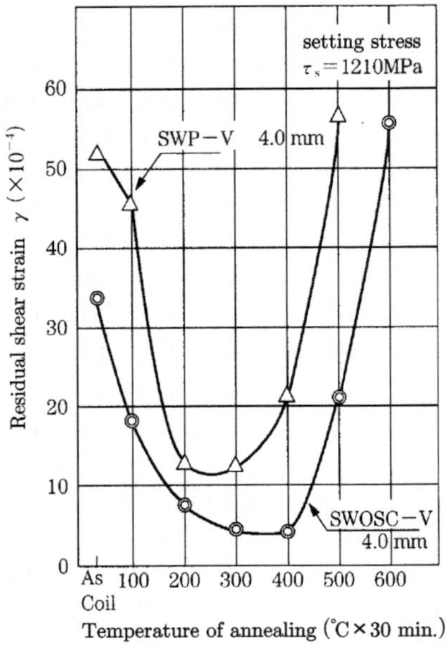

Fig. 2.109. Variation of permanent set resistance with low temperature annealing [14]

This residual stress, which can give bad effects on the permanent set resistance and fatigue property of the springs, should be removed. Figure 2.110 shows an example as to how the residual stress due to spring forming can be removed by low temperature annealing. The temperature of annealing, at which the residual stress is removed, should be as high as possible, but an excessively high temperature would weaken the material, leading to decreases in the fatigue strength and permanent set resistance. As for annealing time, too short annealing time should be avoided. The duration within one hour seems to be sufficient. Figure 2.111 shows the relationship between the annealing temperatures and fatigue strength of wires.

⑤ Changes with low temperature annealing after shot peening

Shot peening process, which is aimed at raising the fatigue property of material, generates beneficial compressive residual stress on the wire surface. On the other hand, however, it is also a sort of plastic working which deforms the material's surface layer and induces working strain, thereby generating a sort of lattice defects called mobile dislocations. If a spring is used as shot peened condition, the mobile dislocations move to form plastic strain, which results in the permanent set of the spring. In order to subdue the tendency of this permanent set, the low temperature annealing is carried out after the shot peening to fix mobile dislocations by stain aging, provided that the annealing temperature should be under the limit so as not to lower the compressive residual stress. Figure 2.112 shows the relation between the residual stress and

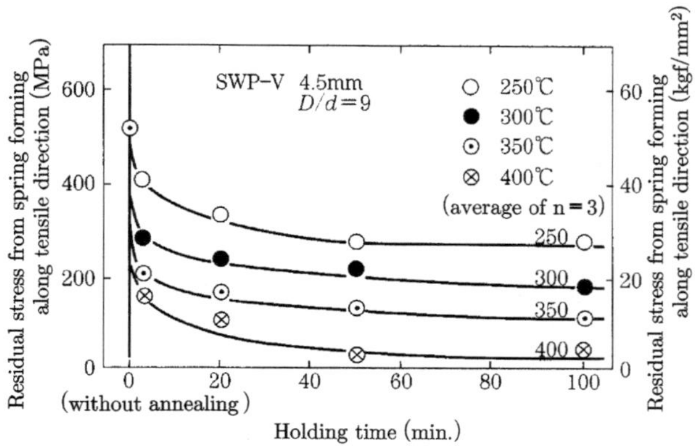

Fig. 2.110. Change in residual stress of SWP-V wire with low temperature annealing [15]

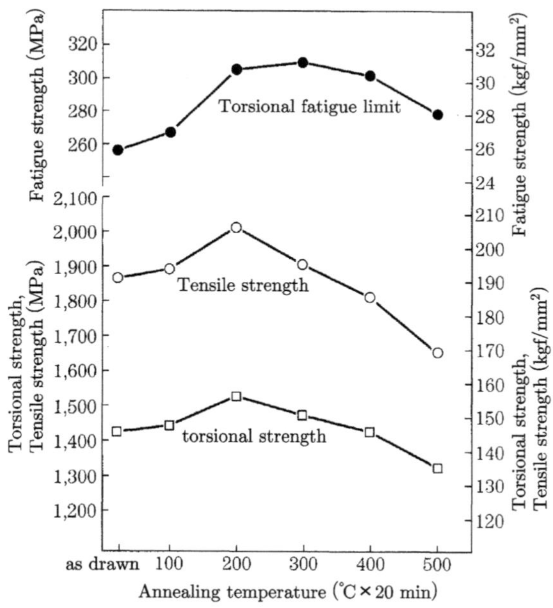

Fig. 2.111. Relation between annealing temperature and fatigue strength of piano wire [16]

fatigue limit of shot-peened springs, and low temperature annealing. It can be seen from the Fig. 2.112 that the preferable temperature of low temperature annealing should be 200 to 250°C, although this temperature can slightly decrease the compressive residual stress.

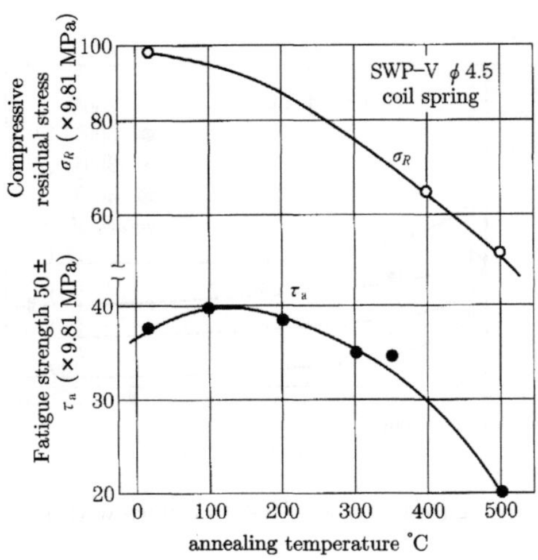

Fig. 2.112. Relation between heating temperature after shot peening, and residual stress and fatigue strength [7]

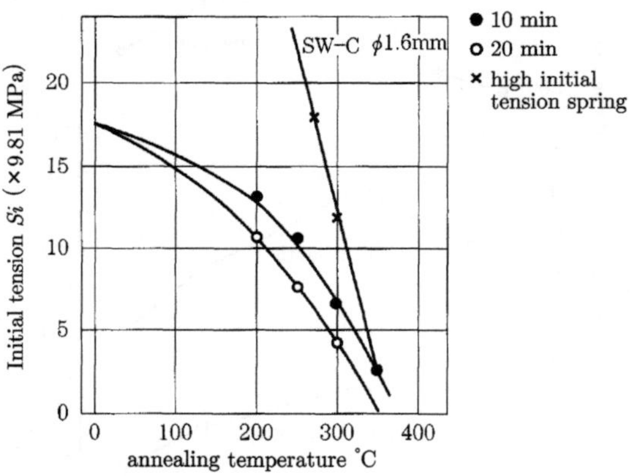

Fig. 2.113. Relation between annealing temperature and initial tension [18]

⑥ Change of Initial Tension

In the case of tension springs, initial tension is an important spring property. However, it can be lowered with increasing temperature through low temperature annealing. Figure 2.113 shows the results supporting this phenomenon. Accordingly, as for the tension springs, the most preferable temperature for the annealing is 200 to 250°C, which is the highest within such a range as not to decrease the initial tension significantly. What should be noted is that in the case of tension springs that are mostly given hook forming, low temperature annealing can deteriorate bending workability and torsional workability of spring. Especially, since the annealing around 200°C can substantially deteriorate the torsional properties, the hook forming after the low temperature annealing would be likely to cause the breakage of spring and bring about an unexpected trouble. Therefore, it would be preferable that the hook forming should be done before the low temperature annealing. When the hook forming is required to be done after the low temperature annealing, the care such as loosening the configuration of rise of the hook or slowing down the speed of hook forming should be taken.

⑦ The optimum conditions for low temperature annealing

The optimum conditions for the low temperature annealing corresponding to each material vary with the conditions of spring applications, properties required for springs, and type of annealing furnaces. As far as piano wire and hard drawn steel wire are concerned, the optimum conditions can be 300 to 350°C temperature and 20 to 30 minutes for springs being the importance to fatigue strength, and 200 and 250°C temperature and 20 to 30 minutes for tensile springs being the importance to permanent set resistance or high initial tension. Table 2.17 shows the recommendable conditions of low temperature annealing.

(viii) Anisotropy of piano wire / hard drawn steel wire and changes with low temperature annealing

While the microstructure and mechanical properties of high carbon steel wire as patented are nearly isotropic, cold drawing develops fiber structure thereby prominently making the mechanical properties anisotropic. Figure 2.114 [20] shows the experimentally measured results of 0.1% tensile and

Table 2.17. Recommended temper conditions for cold formed wire spring for respective materials (°C × min.) [19]

(a) anti-fatigue application: springs especially required higt fatigue strength

material		heat treatment temperature, °C		time, mins	comment
SWP	piano wire	A	300–350	20–30	dynamic high stress springs
SW	hard drawn steel wire	B	200–250	20–30	tension springs requiring initial tension, static high stress springs

(b) general use: springs used with low amplitude or statically

material		heat treatment temperature, °C	time, mins*1	comment
SWP	piano wire	200–350	15–20	general coil springs, static high stress springs
SW	hard drawn steel wire			

*1 time for around 1.0 mmϕ

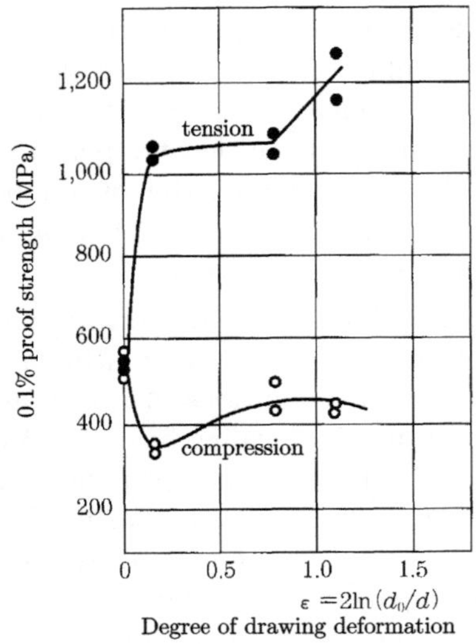

Fig. 2.114. Variation of proof stress of 0.78% carbon steel wire with drawing reduction in both tension and compression directions along with the wire axis [20]

compressive proof stress for 0.78% carbon steel wire. The compressive proof stress of as-drawn wire in the axial direction is lower than that of as-patented wire. Based on the fact that in similar experiments both tensile and compressive proof stresses of full annealed pure iron wire increase with drawing, the decrease in the compressive proof stress of the 0.78% carbon wire can be presumably related with the presence of lamellar cementite (Fe_3C). The proof stress in the radial direction increases with increasing drawing reduction. Figure 2.115 shows measurements on the tensile strength – both in the axial and radial directions – of 0.76% carbon steel wire that is patented and cold drawn. The tensile strength in the radial direction rises with wire-drawing less than that in the axial direction, and it can be seen that the difference between those becomes wider as the drawing reduction increases.

The torsional proof stress, $\tau_{0.3}$ of drawn high carbon steel wire is less than or as same as that of the as-patented wire, when the drawing reduction is small. However, it tends to increase with further wire drawing [21], [22]. For example, experiments done by L. Godecki, where 0.9% carbon steel wire was tested, say following conclusions.

(i) In the drawing with the reduction of area for every die being around 12%, $\tau_{0.3}$ decreases with as-patented wire being drawn through the first die, and then slightly increases up to the level of the as-patented wire until the total reduction of area reaches 50–60%. Above that point of reduction, it increases further with wire drawing.

(ii) In the drawing with the first reduction being 21.5% and average reduction on and after the second being 23%, $\tau_{0.3}$ shows smaller after the first die,

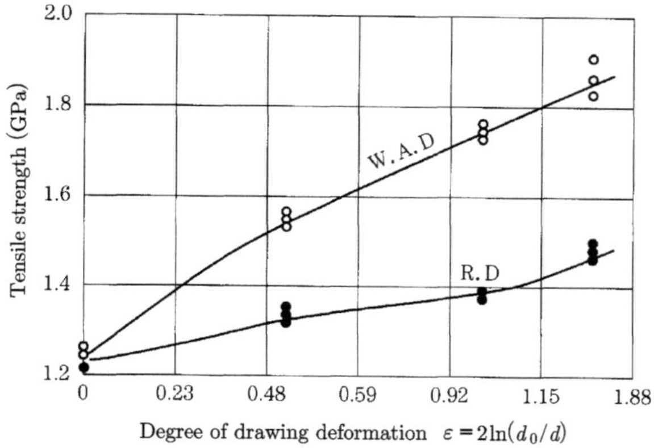

Fig. 2.115. Tensile strength in the axial (W.A.D) and radial (R.D) direction of 0.76% carbon cold drawn steel wire [20]

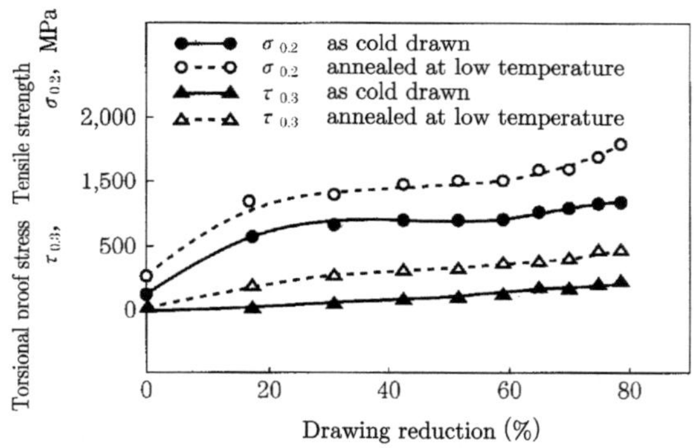

Fig. 2.116. Relation between 0.2% proof stress and torsional yield point, and drawing reduction [22]

but shows larger after the second die and subsequently increases with further wire drawing.

According to experiments done by Kawabata et al. [22], in the early stage of wire drawing, $\tau_{0.3}$ of as-drawn wire is equal to that of as-patented wire (Figure 2.116). With regard to piano wire and hard drawn steel wire, the smaller the diameter and the larger the reduction per each die even with the total reduction being the same, the tensile strength tends to become higher. This is presumably due to that smaller wire can make the total reduction larger, that a larger reduction per each die makes the temperature of the drawn wire higher thus inducing age-hardening, and that the residual stress gets smaller [23]. Accordingly, the anisotropy of the mechanical properties of wire can depend upon the size, tensile strength, and drawing conditions. Low temperature annealing reduces the anisotropy of such drawn wires (Fig. 2.116, Fig. 2.117). However, the normal low temperature annealing conducted in the

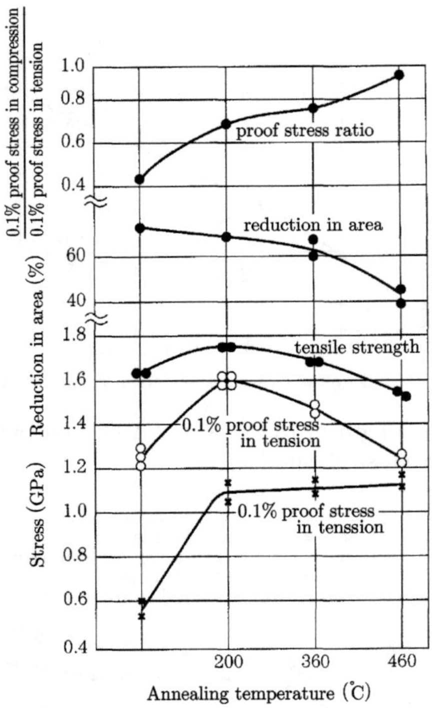

Fig. 2.117. Decrease of anisotropy by low temperature annealing [20]

Table 2.18. Results from rotating bending fatigue tests

	over 1,900 MPa class	over 2,000 MPa class	over 2,100 MPa class	over 2,200 MPa class
Tensile sterength after low temperature annealing	1912	2020	2089	2177
Average value of 10^7 times strength	705	721	739	764
Standard deviation of 10^7 times strength	19.1	6.8	4.6	9.7
Average − (3 × standard deviation)	648	700	726	737

temperature range between 200 and 350°C would not eliminate this anisotropy completely, thus presumably leaving the effect of the direction of loading on the fatigue strength and permanent set. Table 2.18 and Figure 2.118 show the results of rotating bending fatigue tests and spring fatigue tests, respectively, for high carbon steel wires of four kinds of strength level, which were drawn from different rod sizes with the drawing reduction ranging from 75 to 85%. In the case of the rotating bending fatigue tests with loading being repeated tension/compression in the axial direction, the fatigue limit increases with increasing tensile strength. On the other hand, in the case of the spring fatigue

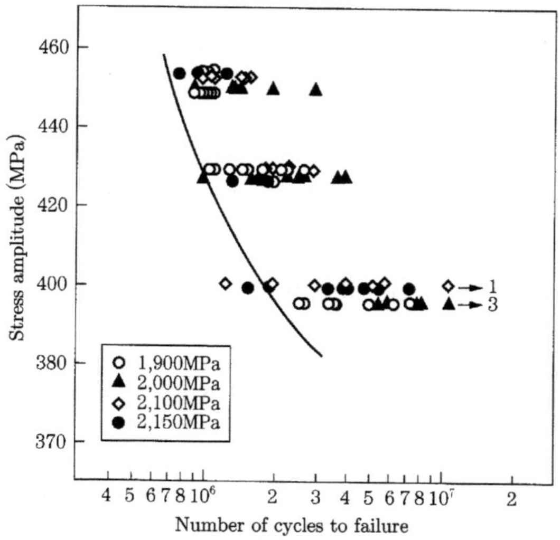

Fig. 2.118. Results of spring fatigue tests for 1.0% Si high carbon steel wire

tests under the partially pulsating torsional stress, the fact that the increase of the tensile strength did not show any improvements in the fatigue strength, can be considered to be due to such anisotropy.

(ix) Remarks in practical use

① Hydrogen embrittlement and delayed fracture

Delayed fracture is the phenomenon where material breaks suddenly after a certain period with brittle fractures without apparent plastic deformation, under the stress which would not cause any fatigue fractures. The cause for this phenomenon is considered to be related with the presence of hydrogen. The occurrence of hydrogen embrittlement due to hydrogen should be taken place under some degree of tensile strength and such amount of hydrogen that corresponds to the hardness and tensile strength of material. The delayed fracture in steel wire for springs can be classified into the following two cases,

(i) Hydrogen is generated at cathodes in pickling or metal plating (electroplating) – as pretreatment before chemical treatment –, penetrates into the steel, and cause the delayed fracture of spring in use.

(ii) Another embrittlement is the spring cracks, which are generated when low temperature annealing is not performed immediately after spring cold forming. This kind of cracks is especially easy to be generated when spring processing flaws called "tool marks" are produced along inner surface of cold wound springs. It can be assumed that the section around such processing flaws, which are largely deformed and become more brittle and harder than the normal section, originates delayed fractures at the crack, conjointly with an extremely small amount of hydrogen contained in the original steel. Therefore, it is essential to remove the tensile residual stress due to coiling immediately by low temperature annealing and to prevent such processing flaws.

It is said that piano wires and hard drawn steel wires are less susceptible to hydrogen embrittlement than oil tempered wires. This is due to the difference in microstructure. While the oil tempered wire consists of tempered martensite microstructure, the piano wire and hard drawn steel wire consist of pearlite microstructure deformed through drawing. The deformed pearlite is considered to be less susceptible to hydrogen than tempered martensite. Preventive measures for the hydrogen embrittlement (delayed fracture) are discussed below.

As for piano wire and hard drawn steel wire, plating and chemical treatment can be applicable because of its low susceptibility to hydrogen. However, after such chemical treatments, it is essential to reduce the amount of hydrogen in the steel by providing baking process immediately. The baking condition chosen for zinc plating is 4 to 6 hours at $191 \pm 14°C$ within 4 hours after the plating, and that for chemical surface treatment is approximately one hour at 120°C. In addition, plating and chemical treatments that generate a large volume of hydrogen such as thick plating, should not be applied to piano wire and hard drawn steel wire for minimizing the possibility of delayed fracture. It is also advisable to avoid using the products in corrosive environments. As mentioned above, it is also necessary to immediately remove the tensile residual stress caused at coiling process, by providing low temperature annealing.

② Variation of coil diameter with low temperature annealing

It is well known that the diameter of coil varies with low temperature annealing. Figures 2.119–2.121 show the variation of coil diameter with low temperature annealing for piano wire, stainless steel wire, and oil tempered wire [25]. According to these results, it is found that the low temperature annealing decreases the diameter of coil for piano wire and oil tempered wire, and increases that of stainless steel wire for springs.

There are various considerations concerned with the cause for these phenomena. One is that the phenomena are affected by the residual stresses developed during the drawing process and spring forming process. That is, due to the difference in the mechanism of hardening between piano wire and hard drawn steel wire, and stainless steel wire, their directions of distribution of the residual stress which generate during drawing are reversed. This leads to the reversed directions of coil diameter change caused by low temperature annealing. Therefore, coiling process should be made, taking the diameter change after low temperature annealing into consideration.

③ Variation of wire diameter due to rotary straightening

The degree of straightness and the variation of diameter for various steel wires processed through straightening are shown in Fig. 2.122 to 2.123. It is found that for the straightness of wire prior to the straightening process

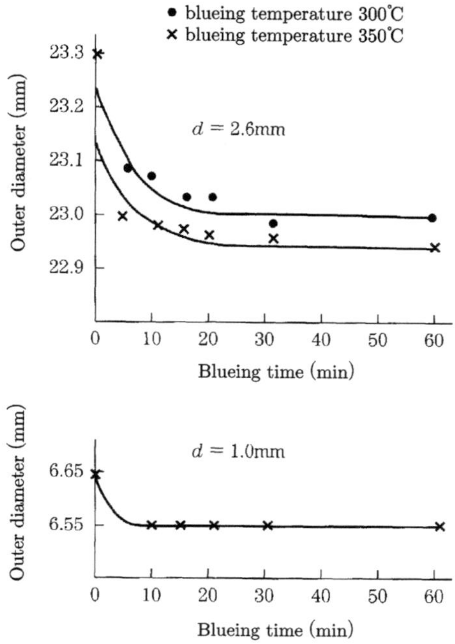

Fig. 2.119. Variation of coil diameter of piano wire (SWP-B) spring with low temperature annealing [25]

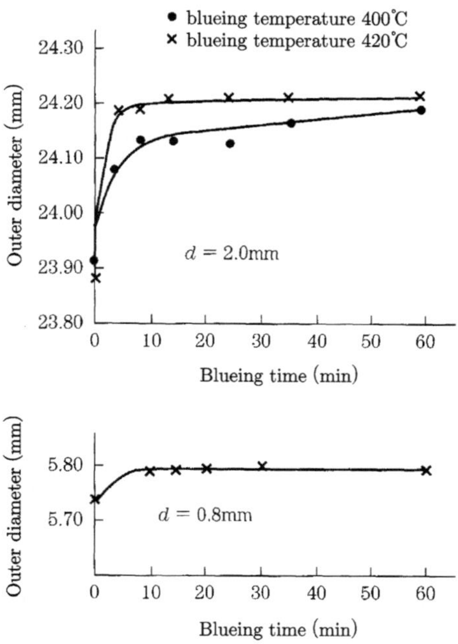

Fig. 2.120. Variation of coil diameter of stainless steel wire (SUS304-WPB) spring with low temperature annealing [25]

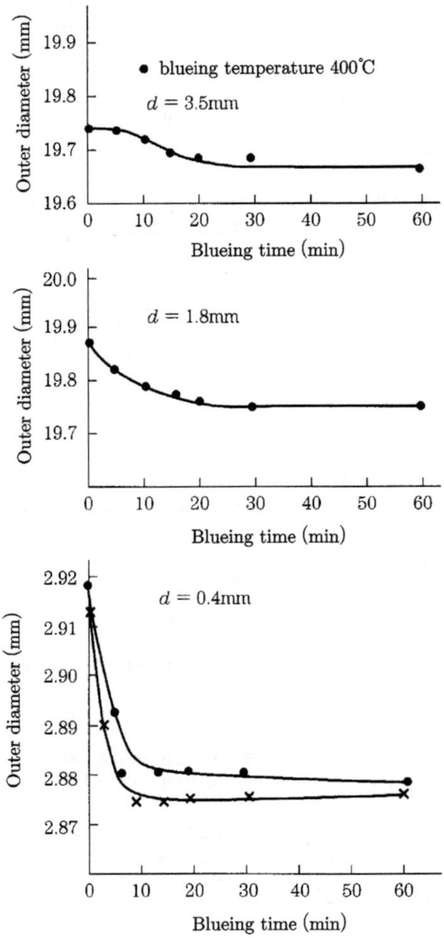

Fig. 2.121. Variation of coil diameter of oil tempered wire (SWOSC-V) spring with low temperature annealing [25]

the cold drawn steel wire such as SWP-B and SUS304-WPB has much more variations than oil tempered wire. As for the change in diameter, while the diameter of oil tempered wire does not change with the straightening process, the diameter of cold drawn steel wire increases with increasing the degree of straightening and settle down a certain value after some degree of straightening. This can be considered due to the change of residual stress caused by Bauschinger effect.

(x) Grades and features of plated piano wire and hard drawn steel wire

For the purpose of preventing the hydrogen embrittlement caused by the plating to spring, and distortion problems such as entanglements in spring

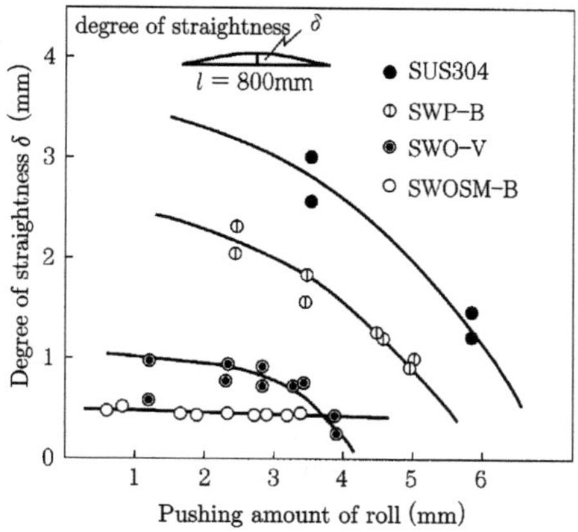

Fig. 2.122. Relationship between pushing amount of straightening roll and maximum flexture at 800 mm length of various steel wires [26]

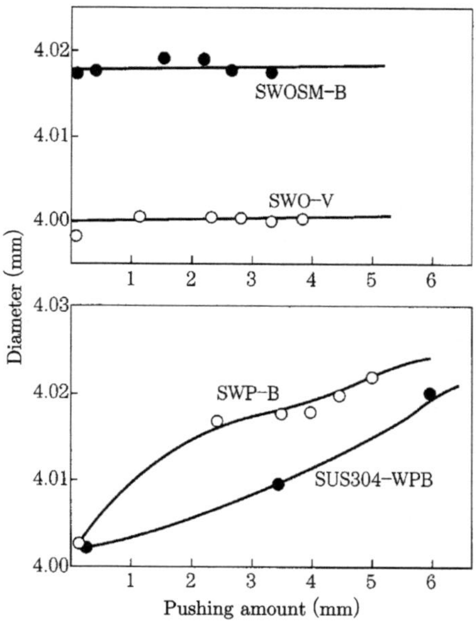

Fig. 2.123. Relation between pushing amount of straightening roll and diameter of wire [26]

plating process, the pre-coated wire plated with nickel, zinc, aluminum, etc., can be sometimes applied. There is also another type of pre-coated wire, which is plated with thick layers of copper for improvement of electrical conductivity. Compared with phosphor bronze, this wire has high modulus of elasticity and high allowable stress to be applied for conductive springs. Table 2.19 shows the grades and features of main plated steel wire for springs.

Table 2.19. Grades and characteristics of major plated steel wires for springs

kind of plating	characteristic	naming	wire maker
Zinc	protect iron from corrosion in nature by corroding Zn itself as a sacrifice (white rust). molten zinc plating (large diameter) has thick plating layer, resulting in superior corrosion resistance. electro zinc plating (small diameter) has thin plating layer, resulting in inferior corrosion resistance.	SWIC-G preco-Z delta wire	Shinko Steel Wire Inc. Suzuki Metal Inc. Metax
Nickel	good corrosion resistance and good deformability. The thicker the plating layer, the better corrosion resistance	SWIC-F preco-N	Shinko Steel Wire Inc. Suzuki Metal Inc.
Nickel-Zinc alloy	alloy plating which contains more than 3% Ni in Zn matrix has characteristics of good appearance or glare of Ni plating and protective feature of Zn plating and reduced pinhole.	mac wire	Marubishi Metal Inc.
Zinc-Aluminium	Zn–Al alloy plated wire, having both sacrifice anti-corrosion effect of Zn and high corrosion resistance of Aluminum, has corrosion resistant film more than twice of normal Zn plating.	delta-AZ	Metax
Zinc-Copper	take on gold color by low temperature annealing	"deru" color gold	Shinko Steel Wire Inc.
Zinc-Copper Chromium	take on black color by low temperature annealing	"deru" color black	Shinko Steel Wire Inc.

References

1. JIS G 3522 Piano wire
2. M. Ochiai: Suzuki Metal Industry internal report
3. Y. Nakamura et al.: Kobe steel technical report 26 (1976) p. 65
4. JIS G 3521 Hard drawn steel wire
5. JIS G 3502 Piano wire rod, JIS G 3506 Hard drawn steel wire rod
6. S. Komura: Spring (Japanese) Ser. 3 Maruzen (1982) p. 54
7. K. Tarui and N.Ibaraki: Journal of the JSTP, 39-447 (1998)
8. K. Takeo: Heat-treatment of steel (Japanese) Ser. 5 Maruzen (1962) p. 519
9. Joint researches Transactions of JSSR 13 (1967) p. 1

10. Joint researches Transactions of JSSR 13 (1967) p. 43
11. Iron and steel Institute of Japan: Iron & Steel Handbook (Japanese) (1967) p. 1039
12. K. Shigeno and K. Nonoyama: JSSR Technical report (1985) p. 13
13. Society of wire products: Wire products (Japanese) Ser. 4 p. 232
14. T. Ozone and H. Kato: JSSR Autum technical report (1986) p. 6
15. H. Toshino and M. Kato: JSSR Autum technical report (1981) p. 1
16. JSMA Spring Text book small spring (1990) p. 36
17. JSMA Heat-treatment of small spring SP-3 (1982) p. 108
18. JSMA Heat-treatment of small spring SP-3 (1982) p. 108
19. JSMA Heat-treatment of small spring SP-3 (1982) p. 146
20. Y. Yamada and T. Yokoyama: JSTP wire drawing working group No. 18 (1983)
21. L. Godecki: Wire Industry, February (1969) p. 151
22. Y. Kawabata, T. Wakamiya, Y. Yamaoka, S. Yamashita: Trans. of JSSR 23 (1978) p. 1
23. Y. Yamada, T. Yokoyama, M. Toyama: JSTP wire drawing Working group No. 20 (1984)
24. H. Hayashi, M. Onoda, H. Yarita: JSSR Autum technical report (1988) p. 5
25. JSMA Easy spring technology (Japanese) (1991) p. 83
26. K. Okata, S. Nakata, S. Takenouchi, K. Hirano: Trans. of JSSR 26 (1980) p. 27

(2) Oil tempered wire

Although oil tempered wire, from a perspective of manufacturing process, should be called "Oil Quenched and Tempered Wire", it is considered that the word is shortened due to the lengthiness. In Japan, the name of "Quenched Wire" was used before World War II. After the war, the introduction of American specifications for oil tempered wires into Japan by the investigating commission for American spring technology, organized by Japan Spring Association, which is presently JSMA, had the name "oil tempered wire" settled down in Japan. The "Quenched" is the expression in the U.S.A., while it is called "Oil Hardened and Tempered wire" in the United Kingdom.

(i) Grade, applications and manufacturing processes

① Grade and applications

It is no exaggeration that most oil tempered wires are used for springs. Oil tempered wire is manufactured according to the specifications corresponding to the applications for springs. Oil tempered wire can be classified as follows, depending upon their applications, grade, and the shape of cross-section.

(a) Classification by applications

- Oil tempered wire for springs
- Oil tempered wire for valve springs
- Oil tempered wire for suspension springs
- Oil tempered wire for press die springs

The oil tempered wire for valve springs is, as described below, spring steel wire from which extremely strict quality assurance is required. Thus, the manufacturing processes include a steelmaking process with the control of compositions of non-metallic inclusions, shaving process for removal of surface flaws, eddy current inspection process for removal or marking of flaws, and so on. In the case of oil tempered wire for springs, these processes are not applied, but occasionally a shaving process is applied depending upon the applications.

The oil tempered wire for suspension springs is oil tempered wire for automotive suspension spring. The large diameter ranged from 8.0 to 14.0 mm, can be mostly applied.

The oil tempered press die spring wire is used for the press die springs designated in JIS B 5012. Oil tempered wire with circular section is used for round coiled springs, and those with trapezoid section or rectangle section are used for flat wire coiled springs. The trapezoid section wire is designed such that, after being coiled, the thickness of inside coil can be equivalent to that of outside coil on account of spring index. When the spring index is large, a rectangular section can be applied. Since the press die spring is applied under a large load, the non-circular cross-section is used in order to obtain a large load as much as possible under the small space.

(b) Classification by Steel Grade

• Carbon Steel	SWO-A, B, V
• Low Alloy Steel	
Si–Mn Steel	SWOSM-A, B, C
Cr–V Steel	SWOCV-V
Si–Cr Steel	SWOSC-V
High-Carbon Si–Cr Steel	Non-JIS
Si–Cr–V Steel	Non-JIS
Si–Cr–Ni–V Steel	Non-JIS
Si–Cr–V–Mo Steel	Non-JIS

Table 2.20 [1], [2] shows the grades and chemical compositions of oil tempered wires designated by JIS. Recently, the percentage of the applications of oil tempered wire which is not designated by JIS for automobile valve springs and suspension springs has been increasing. Such new material can be discussed in Section (vii) and (viii).

(c) Classification by cross-sectional profile

- Round Section
- Non-circular Section
  - Egg-shaped profiles and similar profiles
  - Trapezoid (mainly for press die)
  - Rectangle (mainly for press die)

Egg-shaped profiles for non-circular section springs and other similar profiles in oil tempered wire have been applied for the valve spring and clutch

Table 2.20. Grades and chemical compositions of JIS standardized oil tempered wire [1], [2]

	oil tempered wire			wire rod		chemical compositions (mass %)							
	grade	standard	Character	specification	Mark	C	Si	Mn	P	S	Cr	Cu	V
general use	Carbon steel oil tempered wire for mechanical springs	G 3560	SWO-A	high carbon steel wire rod JIS G 3506	each within SWRH 57(A, B)–SWRH 82(A, B)	0.53–0.88	0.10–0.35	0.30–1.20	under 0.04	under 0.04	–	–	–
			SWO-B			0.53–0.88	0.10–0.35	0.30–1.20	under 0.030	under 0.030	–	–	–
	Silicon-Chromium steel oil tempered wire for mechanical springs		SWOSC-B	same as high carbon steel wire rod JIS G 3506 except for chemical composition	chemical composition coresponds to SUP 12 of spring steel	0.51–0.59	1.20–1.60	0.50–0.90	under 0.035	under 0.035	0.055–0.90	–	–
	Silicon-Manganese steel oil tempered wire for mechanical springs		SWOSM-A, B, C		chemical composition coresponds to SUP 6 of spring steel	0.56–0.64	1.50–1.80	0.70–1.00	under 0.035	under 0.035	–	under 0.030	–
valve spring use	Carbon steel oil tempered wire for valve springs	G 3561	SWO-V	same as piano wire rod JIS G 3502 except for chemical composition	SWRH 62B, 72B	0.60–0.75	0.12–0.32	0.60–0.90	under 0.025	under 0.025	–	under 0.020	–
	Chromium-Vanadium steel oil tempered wire for valve springs		SWOCV-V		chemical composition coresponds to SUP 10 of spring steel	0.45–0.55	0.15–0.35	0.65–0.95	under 0.025	under 0.025	0.80–1.10	under 0.020	0.15–0.25
	Silicon-Chromium steel oil tempered wire for valve springs		SWOSC-V		chemical composition coresponds to SUP 12 of spring steel	0.51–0.59	1.20–1.60	0.50–0.80	under 0.025	under 0.025	0.50–0.80	under 0.020	–

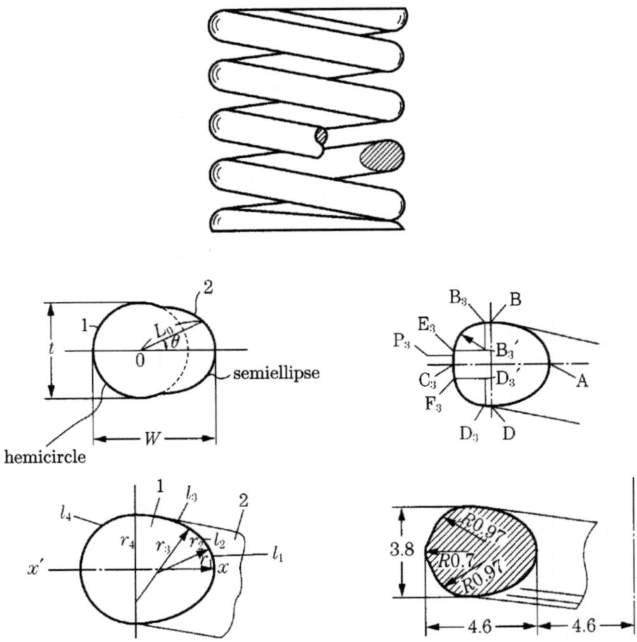

Fig. 2.124. Example of cross sectional profiles of egg-shaped spring wire [3]

spring required the reduction of solid height and weight saving on the occasion of commercial applications of non-circular piano wire around 1982. These shapes of spring wire have been employed by worldwide automotive manufacturers. Examples of cross-sectional profiles of non-circular wire are shown in Fig. 2.124 [3]. Figure 2.125 [4] shows examples of commercial applications of non-circular sections with the aspect ratio, the ratio of long diameter to small diameter.

② Manufacturing methods for oil tempered wire

(a) Manufacturing methods of wire rod for oil tempered wire

The microstructure of oil tempered wire is tempered martensite obtained by quenching and tempering. Accordingly, the segregation of C and Mn in the vicinity of the center, segregation of P in the grain boundary, and other impurities segregation in oil tempered wire give larger effects on degradation of ductility and toughness than in piano wire and hard drawn steel wire which have fibrous structure formed by wire drawing. Compared with piano wire, oil tempered wire is more susceptible to non-metallic inclusion due to its high tensile strength in the large diameter side as well as the difference of microstructure. Thus, steel for oil tempered wire is manufactured through an electric arc furnace or converter from carefully selected high grade scrap with little amount of impurities or high grade pig iron. Especially for oil tempered valve spring wire with strict requirements concerning the cleanliness of non-metallic inclusion, a special steelmaking called composition control of

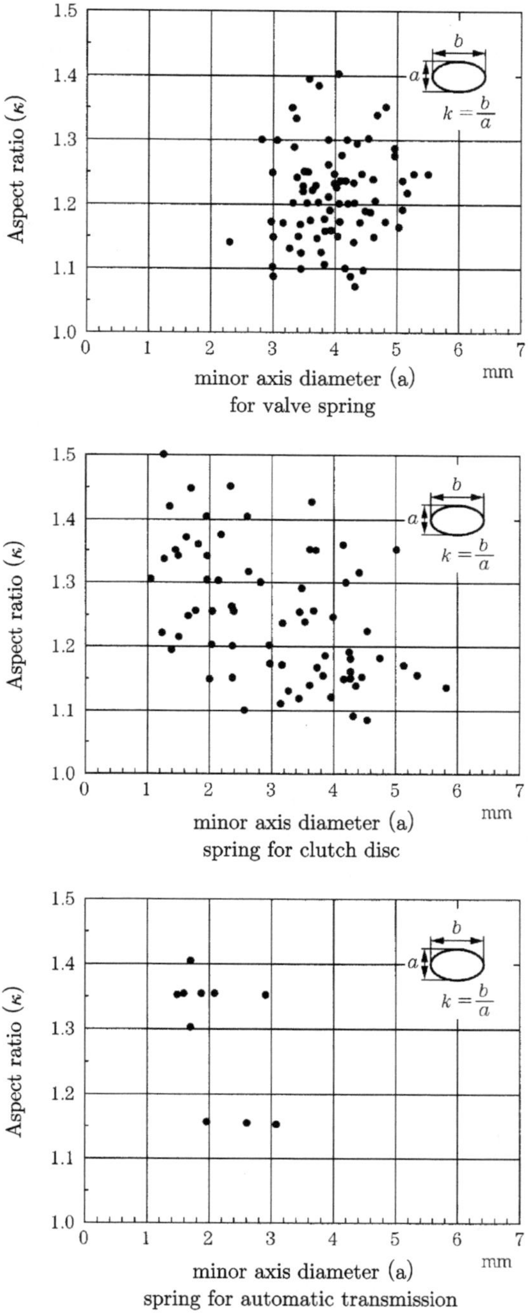

Fig. 2.125. Aspect ratios of non-circular cross sectional wires for various use [4]

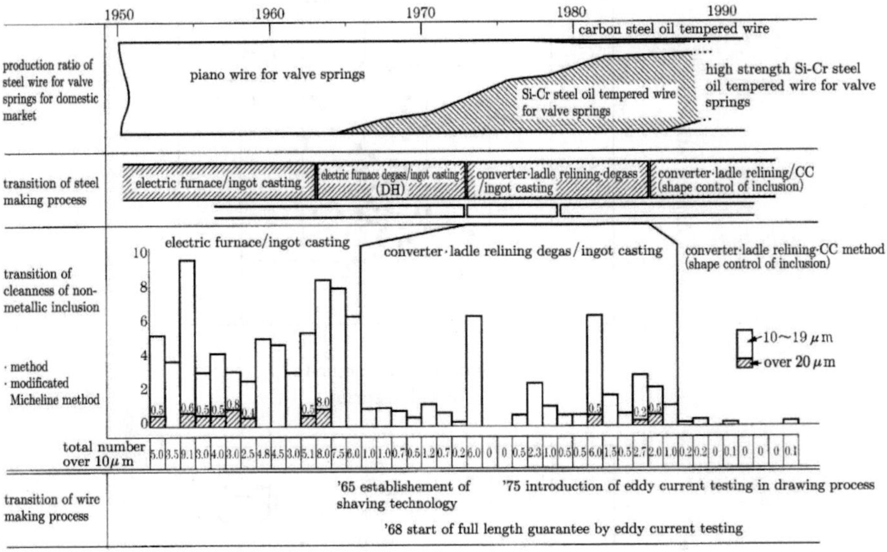

Fig. 2.126. Transitions of steel wire for valve springs and technical improvements [5]

non-metallic inclusion can be carried out by secondary refining (ladle refining). Means such as reducing impurities and non-metallic inclusions with a combination of secondary refining and degassing process are used for other steel for oil tempered wire.

The steelmaking technology combining the composition control of non-metallic inclusion and continuous casting described later is substantially effective for oil tempered valve spring wire rod, thus having been spreading over the world. As indicated in Fig. 2.126, the steelmaking process for spring wire rod has been advancing as summarized below:

1) 1960's
   Electric Furnace → Degassing → Ingot casting
2) 1970's
   Converter → Ladle refining (Al deoxidation) → Degassing → Ingot casting
3) Since the middle of 1980's
   Converter → Ladle refining (Si deoxidation / Composition control of non-metallic inclusion) → Continuous casting

The composition control of non-metallic inclusions consisting of oxides is based on the concept that non-metallic inclusions, if present, are made soft and harmless. In this technology, artificial slag is formed by adding special flux on the occasion of ladle refining and the non-metallic inclusions consisting of oxides are controlled such that the melting point is maintained low between 1200–1500°C. The non-metallic inclusion, whose melting point is lowered, is deformed thin and long during hot rolling after casting and divided into small pieces through wire-drawing.

The continuous casting method is characteristic primarily of being capable of producing large amount of higher quality steel more steadily than ingot casting and is even effective for the reduction of non-metallic inclusion. This is due to the unlikeliness of molten steel being oxidized by the isolation from air during casting and the acceleration of floatation of non-metallic inclusions in a tundish. This effect has been confirmed in fatigue tests with oil tempered wire [6]. After continuous casting, the bloom or billet is hot-scarfed. After the billet inspection process where the surface flaws are strictly inspected and removed by grinding, the billet is hot-rolled to wire rod. Although the flaws which occasionally generate in the hot-rolling process could be inspected by an eddy current flaw detector during the hot rolling operation, it is difficult to mark and remove the flawed area in operation due to the high speed of rolling and high temperature. Therefore, most steel mills study the frequency of flaws to apply to the reference of shipping decisions. Recent technology makes it possible to inspect flaws and remove them by operating both eddy current flaw detector and flaw removing machine simultaneously via microcomputers during the off-line rewinding and to remove the flawed parts alone [7].

(b) Manufacturing processes for oil tempered wire.

Although manufacturing processes for oil tempered wire depend upon the grades, the basic process is the heat treatment of rod, followed by wire-drawing in which the rod is deformed to a predetermined diameter, and finally oil tempering treatment. In the case of manufacturing oil tempered valve spring wire with strict requirements in quality, a peeling process called "shaving" and flaw detecting process against surface flaws can be added to the basic processes. Figure 2.127 shows representative manufacturing processes for oil tempered wire.

The heat treatment before wire-drawing, is a process normally called patenting, where the obtained microstructure is fine pearlite possessing high workability. In the case of oil tempered wire of small diameter, the process between patenting and wire-drawing is repeated to make the wire diameter small. The purpose of the patenting, in this case, is to make the wire-drawing easier; therefore, this patenting can be excluded and may be replaced by annealing when the drawing reduction is small. Normally, pickling treatment is used for the removal of scale after patenting.

The rod for oil tempered valve spring wire is, as mentioned above, given the countermeasures against the surface defects including various flaws after the blooming mill process. However, it is extremely difficult to eliminate all surface flaws over the total length of rod. Since more or less amount of flaws can remain, a method of continuous removal of all surfaces in the subsequent wire-making process is employed. In 1960's, a rotating type grinding machine was used. However, due to the circumferential grinding marks left after the grinding, since about 1965, it has become popular to apply a shaving type surface removal machine, which takes the skin of 0.1 to 0.15 mm in thickness off the entire wire surface with a peeling die. Since work-hardened layer, which is hard to be drawn, is formed in the surface layer of shaved rod, patenting or an-

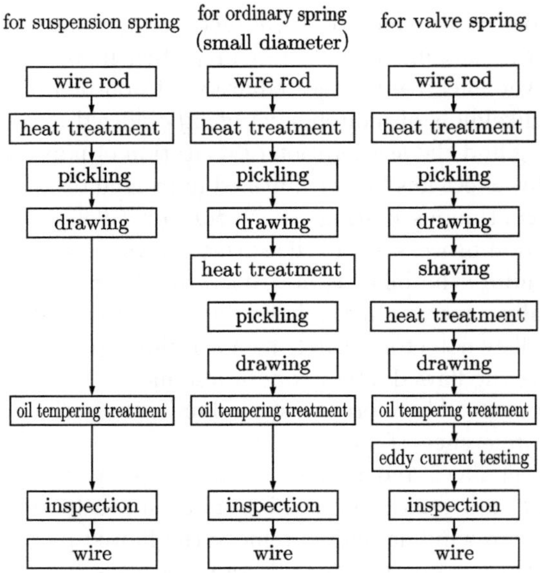

Fig. 2.127. Example of production flow chart of various oil tempered wire

nealing at a temperature below A_1 transformation is provided, thus making the layer transformed to favorable microstructures for wire drawing. Occasionally flaws can occur in the subsequent wire drawing process, therefore, an eddy current flaw detector can be applied either during the wire drawing process and/or after the quenching and tempering process. The eddy current test has two types, that is, a through coil type with differential method and rotation probe type with a probe coil rotating in a spiral manner. Each has its own characteristics. In general flaw detection, reliability is improved by applying these two methods independently or in combination. The features of the through coil type and the rotation probe type are shown in Table 2.21, and schematics of both types are shown in Fig. 2.128 and Fig. 2.129 [8].

With regard to wire drawing, just like piano wire, a continuous drawing machine and a single head drawing machine called single block can be used for large reduction and small reduction, respectively.

Oil quenching and tempering, the focal process of manufacturing oil tempered wire, is heat-treatment where drawn material is continuously quenched and tempered. Figure 2.130 shows an example of oil quenching and tempering treatment equipment. From the viewpoint of metallurgy, this quenching and tempering can be classified into the following three processes.

i) heating process for dissolving carbides into steel, whose microstructure becomes austenite, ii) quenching process for cooling the wire rapidly below Ms point (starting temperature for martensite transformation), and iii) tempering process for giving appropriate toughness and strength to hardened martensite.

Table 2.21. Features of through coil type and rotating probe type

	through coil type	rotating probe type
handling	easier with fewer setting points compared with rotating probe type	more difficult with much setting points compared with through coil type
relation between wire speed and detectability	the higher the speed of wire, the higher the flaw detection, and the lower the speed, the lower the flaw detection	tested area makes spiral shape with non-detected area when tested at high speed, since probe rotates around straightly running wire
relation between kind of defects and detectability	hard to detect continuous flaw like seam	higher detectability to continuous flaw over spot flaw
required space	smaller space, compared with rotating probe type	require larger space, compared with the penetration coil type
other remarks	require demagnetization instrument after the detection	may cause destruction of probe at passage of wire end or wire breakage to be required automatic shelter system

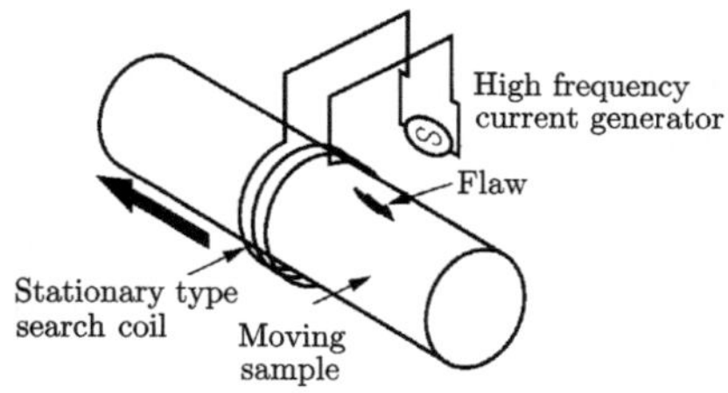

Fig. 2.128. Prinpiple of through coil type eddy current testing machine [8]

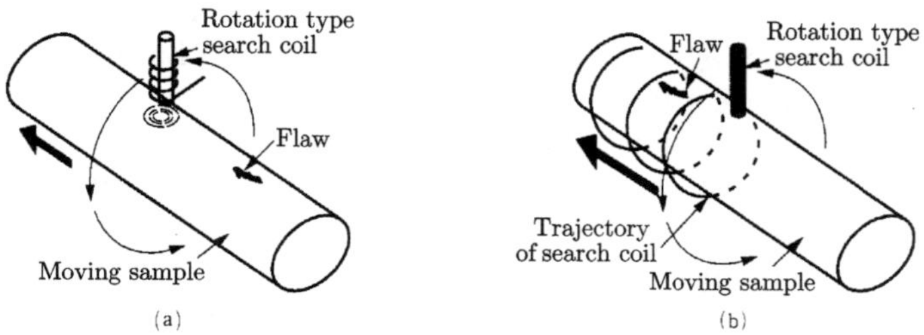

Fig. 2.129. Prinpiple of rotation probe type eddy current testing machine [8]

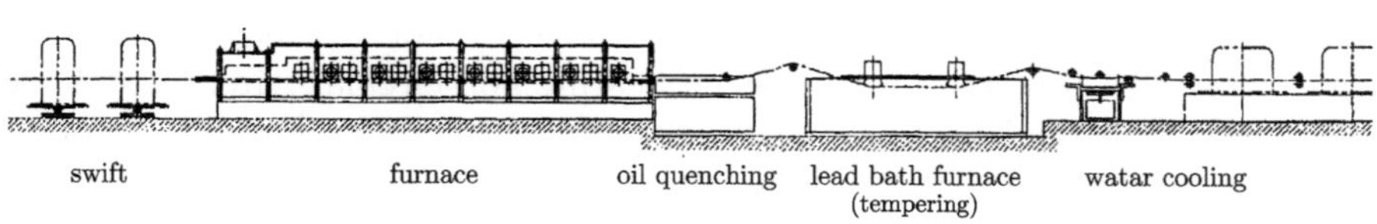

Fig. 2.130. Example of equipment for oil temper process

The aim of the first heating process is to austenitize the steel thereby dissolving additional elements. If the heating temperature and heating time become excessive, it will lead to the coarsening of austenite grain size thereby degrading the mechanical properties after the subsequent quenching and tempering, and also it will increase the possibility of decarburization. There are various methods for the heating, each of which has its own characteristic. At present in Japan, direct fire heating, indirect heating with a radiation furnace, and high frequency induction heating can be applied [9], [10]. Globally, there are even some places applying resistance heating and molten lead heating [11].

The next quenching process is aimed at quenching the austenitized steel below Ms point thereby making the steel transformed to martensite. Mostly oil has been used as the medium for quenching, but lately water-soluble quenching media are also being developed. Thus now, there is no metallurgical reason why oil should be used for the quenching of oil tempered wire. As for high frequency induction heating, water is normally used as the quenching medium. The temperature of the quenching media should be kept below Ms point, and should be such that the cooling capability of the media can be maintained. Concerning with the quenching speed, as for carbon steel, it should be cautious that inadequate speed may cause pearlite transformation. The aim of the tempering process is to change the hard and brittle martensite obtained through the quenching process into the tempered martensite which has mechanical properties suitable for spring wire. Figure 2.131 shows the relation between tempering temperatures and mechanical properties of Si–Cr Steel oil tempered wire. In the case of spring steel, the strength and toughness are mostly evaluated by hardness and impact value, respectively. In the cases of oil tempered wire, although the strength can be determined by a tensile test, it is difficult to sample a specimen for Charpy impact test from oil tempered wire. Toughness is replaced by ductile evaluations such as elongation, reduction in area and bending.

A molten lead furnace which has stable temperature control is frequently used for the tempering treatment, just like patenting. Lately, in the sense of environmental issues, a fluidized bed furnace, dispensing with molten lead, is being used in some divisions. When induction heating is used for austenitizing heating, in which case the feed speed is high, induction heating is also used for the tempering process.

As for the oil tempered wire heat-treatment line, there are multi-line which processes plural wires simultaneously and single-line which processes a single wire at high speed. Although the former is more popular due to its high productivity, the latter can be represented by induction heating process. As for induction heating, as the heating distance is short it becomes possible to prevent the distortion of wire when heating non-circular wire. Due to the smallness of energy loss, since it becomes possible to make intermittent operations it can be said that the induction heating is suitable for small quantity production.

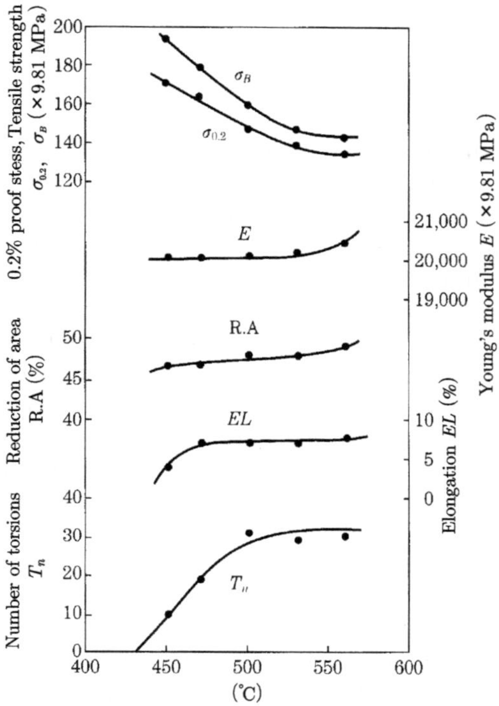

Fig. 2.131. Tempering property of Si–Cr steel

(ii) Mechanical properties of oil tempered wires

① Diameter, Tensile strength, and Inspection item of oil tempered wire

Table 2.22 shows JIS diameter ranges for varieties of spring steel wires. The overall range of oil tempered wire is shifted to the larger size than that of cold drawn steel wire. Si–Cr oil tempered spring wire has a large diameter range owing to their wide commercial applications. As far as valve spring Si–Cr oil tempered wire is concerned, there was no specification below a diameter of 1.6 mm in the early stages of JIS. Since a diameter less than 1.6 mm had been put to practical use in the western countries, and diameters as small as 0.5 mm had been used in large quantity in Japan, the limit on the smaller diameter side was extended to 0.5 mm in 1994, adjusting to ISO. Table 2.23 shows a list of the differences among the specifications – except chemical composition – of varieties of oil tempered wires. Comparing wire rod for valve spring wire and that for normal spring wire, there is a difference of flaws and decarburization to be assured. This difference can be also seen between both wires. A coiling test, bending test, and torsion test are all aimed at the detection of flaws. While the coiling test and bending test focus on rough surface and spot-like flaws, the torsional test focuses on longitudinal cracks that are hard to detect in the coil winding and bending tests as well as seams. For valve spring wire, a decarburization test for determining the depth of a decarburized layer can be additionally carried out.

② Mechanical properties of oil tempered wire and low temperature annealing

Figure 2.132 shows a stress-strain diagram obtained from tensile test, where oil tempered wire is compared with piano wire and stainless steel wire. Oil tempered wire has larger elastic limit and 0.2% proof stress than other steel wire. All spring wires, including oil tempered wire, are always given low temperature annealing after spring forming. The purpose of this heat treatment is to improve the mechanical properties and remove the residual stress originated during spring forming. Especially as for oil tempered wire, which has smaller changes of the mechanical properties due to low temperature annealing than piano wire, the purpose of low temperature annealing is the removal of the residual stress. Accordingly, it is to be desired that low temperature annealing for oil tempered wire be carried out at as high temperature as possible, within the extent to which the strength is not lowered significantly. Figure 2.133 [12] shows the mechanical properties of various oil tempered wires with low temperature annealing, and Table. 2.24 [13] shows the heat treatment conditions for cold formed spring material, including oil tempered wire. Figure 2.134 [14] shows the relation between the heat treatment conditions for Si–Cr oil tempered wire and residual stress, and Fig. 2.135 [15] shows an example of the relation between temperature of low temperature annealing and the fatigue life of springs. The relation between tensile strength and hardness for oil tempered wire and that for springs made of oil tempered wire are shown in Fig. 2.136 and Fig. 2.137, respectively.

③ Characteristics of springs made of oil tempered wire

A spring made of oil tempered wire is superior to other steel spring in both fatigue strength and permanent set resistance. Figure 2.138 [12] shows a fatigue limit diagram with an endurance limit of 10^7 cycles for compression springs made of various valve spring steel wires. In the Fig. 2.138, it can be seen that the Si–Cr steel oil tempered wire, which is primarily used for valve springs of an automobile engine shows the highest fatigue limit and highest permanent set resistance. Besides, the appropriate shot peening enables the fatigue limit of the oil tempered wire to be upgraded by 30%. Since the Si–Cr steel oil tempered wire has high heat resistance, low temperature nitriding can be applicable for the spring.

The permanent set resistance can be mostly evaluated in a heat clamping test as a kind of accelerated test, because insignificant differences can be seen at room temperature. Figure 2.139 [12] shows the result of clamping tests at elevated temperature with a variety of steel springs. Compared with the piano wire, as typical cold drawn steel wire, the oil tempered wire can show higher heat resistance. Among them, the Si–Cr steel oil tempered wire springs surpasses the others.

Table 2.22. Available diameter range of oil tempered wire and other spring steel wire

	Types of steel spring wire	Grade	diameter (mm)																									
			0.08	0.10	0.20	0.50	0.80	1.00	1.20	1.40	1.60	1.80	2.00	3.00	4.00	5.00	3.00	4.00	5.00	6.00	7.00	8.00	9.00	10.0	11.0	12.0	13.0	14.0
oil tempered wire	Carbon steel oil tempered wire for mechanial springs	SWO-A, B											░	░	░	░	░	░	░	░	░	░	░	░	░	░		
	Si–Mn steel oil tempered wire for mechanical springs	SWOSM-A, B																░	░	░	░	░	░	░	░	░	░	░
		SWOSM-C																░	░	░	░	░	░	░	░	░		
	Si–Cr steel oil tempered wire for mechanical springs	SWOSC-B						░	░	░	░	░	░	░	░	░	░	░	░	░	░	░	░	░	░	░	░	░
	Carbon steel oil tempered wire for valve springs	SWO-V											▓	▓	▓	▓	▓	▓	▓	▓								
	Cr–V steel oil tempered wire for valve springs	SWOCV-V											▓	▓	▓	▓	▓	▓	▓	▓	▓	▓	▓	▓				
	Si–Cr steel oil tempered wire for valve springs	SWOSC-V				▓	▓	▓	▓	▓	▓	▓	▓	▓	▓	▓	▓	▓	▓	▓	▓	▓						
cold drawn steel wire	hard drawn steel wire	SW-A	░	░	░	░	░	░	░	░	░	░	░	░	░	░	░	░	░	░	░	░	░	░				
		SW-B, C	░	░	░	░	░	░	░	░	░	░	░	░	░	░	░	░	░	░	░	░	░	░	░	░	░	
	piano wire	SWP-A	░	░	░	░	░	░	░	░	░	░	░	░	░	░	░	░	░	░	░	░	░	░				
		SWP-B	░	░	░	░	░	░	░	░	░	░	░	░	░	░	░	░	░	░								
	piano wire for valve springs	SWP-V											▓	▓	▓	▓	▓	▓	▓	▓								
	stainless steel spring wire	SUS 304-WPA	░	░	░	░	░	░	░	░	░	░	░	░	░	░	░	░	░	░	░	░						
		SUS 304-WPB	░	░	░	░	░	░	░	░	░	░	░	░	░	░	░	░	░	░	░	░	░	░	░	░		
		SUS 631-J1-WBP	░	░	░	░	░	░	░	░	░	░	░	░	░	░	░	░	░	░								

░ for mechanical springs
▓ for valve springs

Table 2.23. Comparison of various specifications of varius oil tempered steel wire [1], [2]

Type			oil tempered wire for valve springs			oil tempered spring wire					
Designation			SWO-V	SWOCV-V	SWOSC-V	SWO-A	SWO-B	SWOSC-B	SWOSM-A	SWOSM-B	SWOSM-C
Diameter range, mm			2.00–6.00	2.00–10.00	0.50–8.00	2.00–12.0		1.00–15.0	4.00–14.0	4.00–12.0	
diameter tolerance and out-of-roundness	1.00 mm		±0.020 mm, $\leqq$ 0.020 mm			±0.025 mm, $\leqq$ 0.025 mm					
Tensile strength MPa	1.00 mm		–	–	2010–2160	–	–	1960–2110	–	–	–
Reduction in Area %	1.00 mm		$\geqq$ 45								
diameter tolerance and out-of-roundness	2.00 mm		±0.020 mm, $\leqq$ 0.020 mm			±0.025 mm, $\leqq$ 0.025 mm					
Tensile strength MPa	2.00 mm		–	–	2010–2160	–	–	1960–2110	–	–	–
Reduction in Area %	2.00 mm		$\geqq$ 45								
diameter tolerance and out-of-roundness	4.00 mm		±0.020 mm, $\leqq$ 0.020 mm			±0.025 mm, $\leqq$ 0.025 mm					
Tensile strength MPa	4.00 mm		–	–	2010–2160	–	–	1960–2110	–	–	–
Reduction in Area %	4.00 mm		$\geqq$ 45								
Apperance			must be free from detrimental defects								
Specification of wire rod	diameter tolerance and out-of-roundness		±0.040 mm, $\leqq$ 0.040 mm			±0.050 mm, $\leqq$ 0.050 mm					
Specification of wire rod	flaw depth	method	measure depth of flaw after 30 min's immersion in boiled solution which is the mixture of 7 hydrochloric acid to 3 water specified in JIS K 1310, basically, use new solution			none					
Specification of wire rod	flaw depth	specifications	depth of flaw must not exceed 0.10 mm			none					
Specification of wire rod	decurbarized depth	method	follow to JIS G 0558, 4			none					
Specification of wire rod	decurbarized depth	specifications	total decarburization layer must be under 0.7 mm			none					

Specification of wire

Type		oil tempered wire for valve springs	oil tempered spring wire
Designation		SWO-V SWOCV-V SWOSC-V	SWO-A SWO-B SWOSC-B SWOSM-A SWOSM-B SWOSM-C
winding property	diameter	under 6.00 mm	
	method	in case of diameter 4 mm and less, winding around a mandrel with the same diameter as wire, and in case of wire more than 4 mm, winding a mandrel with twice the diameter of wire, more than 4 turns with solid conditions, whether or not fractured and presence of flaw should be inspected	
	specifications	wire must not occur detrimental flaw on its surface or fracture	
bending property	diameter	over 6.00 mm	
	method	presence of flaw or fracture should be inspected after bended to 90° direction with radius of wire diameter	
	specifications	must be free from detrimental flaw and fractured	
twist property	diameter	0.70–6.00 mm	under 6.00 mm
	method	seize both end of test piece firmly with interval of 100 times of wire diameter and rotate one end until fracture. One should count number of twists and inspect state of cross section of fracture	
	specifications	cross section of fracture: must be vertical to wire axis and free from flaw or crack. Twist condition: must be free from detrimental flaw after fracture	
apperance	specification	surface must be free from detrimental defects like flaw, scale, and rust	
flaw depth	diameter	0.50 mm and above	over 1.00 mm
	method	test piece with appropriate length relieved residual strain in advance must be soaked in adequate boling solution in 200 mm length. After wire loses 1% of its diameter, test piece is checked for presence of flaw or not. Depth of flaw is generally measured with micrometer after ground to edge of flaw.	
	specifications	flaw depth $0.50 \leqq d \leqq 2.00$ under 0.01 mm $2.00 \leqq d \leqq 6.00$ within 0.5% of wire diameter $6.00 \leqq d \leqq 10.00$ within 0.7% of wire diameter	flaw depth $0.50 \leqq d \leqq 2.00$ under 0.02 mm $2.00 \leqq d \leqq 6.00$ within 1.0% of wire diameter $6.00 \leqq d \leqq 10.00$ within 1.4% of wire diameter
decarburization depth	diameter	over 0.50 mm	over 1.00 mm
	method	Follow to JIS G 0558, transverse cross section of tested sample polished and etched is inspected for its decarburization state by means of microscope equiped with magnification of 100 to 500 times	
	specification	must be free from ferritic decarburization	total decarburization layer must not exceed 1.5% of diameter and 0.05 mm

Table 2.24. Optimum heat treatment conditions for various steel spring wires (diameter around 1.0–4.0 mm) [13]

material grade/nomination		heat treatment temperature	holding time	comments
SWP-A, B, V SW-A, B, C	Piano wire Hard drawn steel wire	300–350°C	20–30 min.	dynamic high stress springs
		200–250°C		tension springs that require initial tension static high stress springs
SUS 304-WPA, B	Austenitic stainless steel spring wire	350–400°C		dynamic high stress springs
SUS 631 J1-WPC	Precipitation hardening stainless steel spring wire	475 ± 10°C	60 min.	precipitasion hardening treatment
SWO-V	Carbon steel oil tempered wire for valve springs	300–400°C	20–30 min.	fine wire under diameter of 1.0 mm treated under 400°C
SWOCV-V	Chromium-Vanadium steel oil tempered wire for valve springs			
SWOSM-A, B, C	Silicon-Manganese steel oil tempered wire			
SWOSC-V	Silicon-Chromium steel oil tempered wire for valve springs	400–450°C		
PBW	Phosphor bronze wire	200–250°C	30–45 min.	bronze wire equivalent to phosphor bronze wire
NSW	Copper-Nickel wire	300–350°C		
Be–Cu	Beryllium-Copper wire	315°C	90–120 min.	age hardening treatment

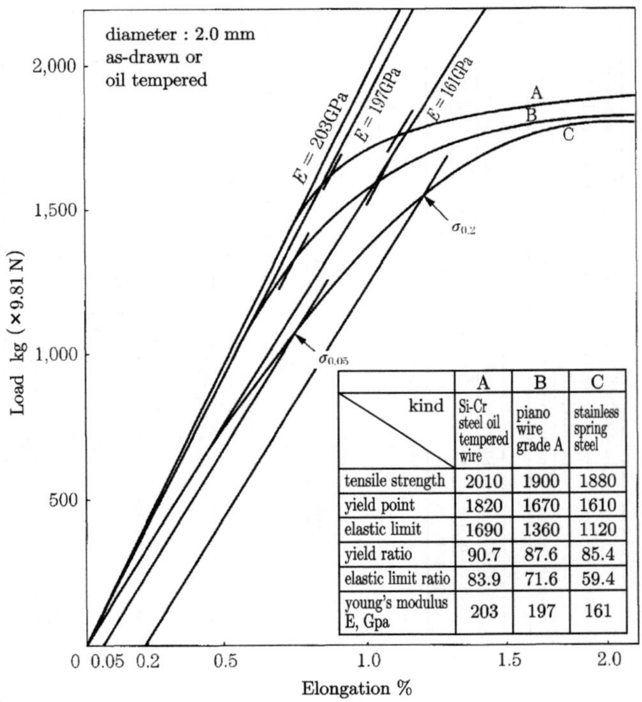

Fig. 2.132. Load-Elongation diagram of various steel spring wires [12]

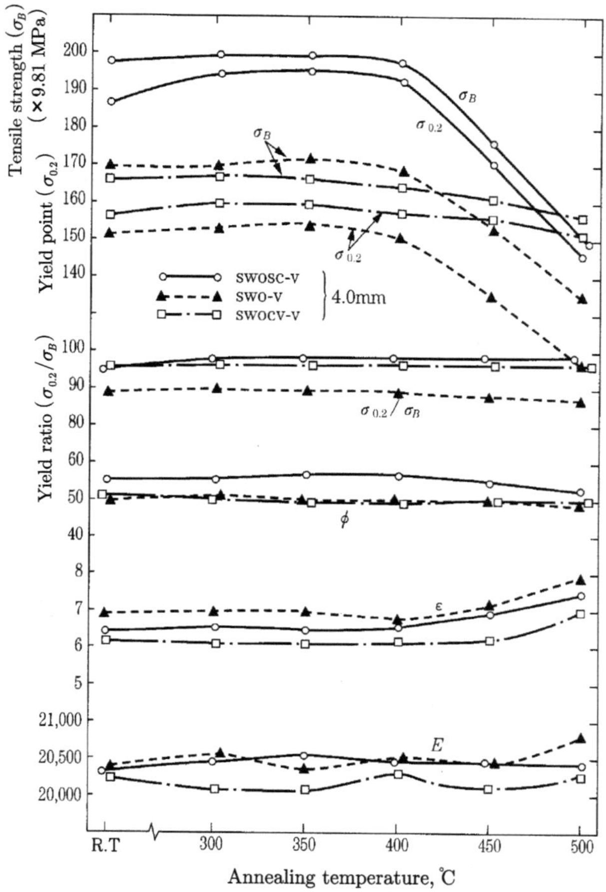

Fig. 2.133. Low temperature annealing property of oil tempered wire for valve springs [12]

(iii) Effects of chemical composition

The grades of oil tempered wire can be classified into two groups: carbon steel oil tempered wire and low-alloy steel oil tempered wire. The carbon steel oil tempered wire has almost the same chemical composition as the piano / hard drawn steel wire. On the other hand, low-alloy steel oil tempered wire includes Si–Mn steel, Cr–V steel, and Si–Cr steel. The effects of Si, Mn, Cr, and V, additional elements in these steel grades, and the effects of C and impurities such as P and S, are described below.

C: Infinitesimal amount of carbon is dissolved into solid iron, but most of carbon exists as carbides, which can alter the microstructure and properties of steel. While steel used for springs contains 0.45 to 1.0% carbon, a majority of oil tempered wires used extensively contain 0.47 to 0.75% carbon. Within this carbon range, higher carbon content leads to higher hardness after quench-

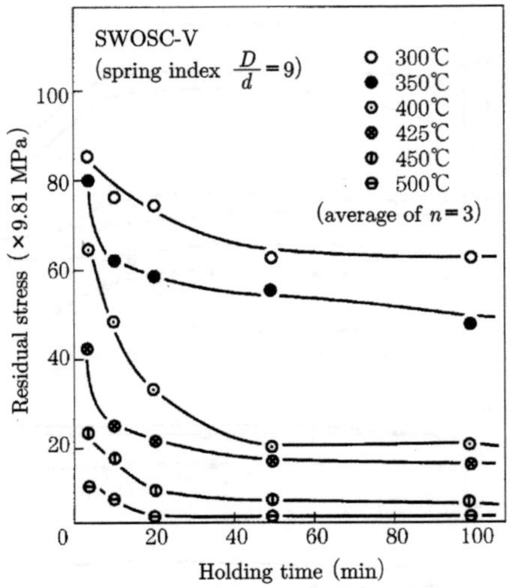

Fig. 2.134. Change of residual stress of SWOSC-V springs with low temperature annealing (without shot peening) [14]

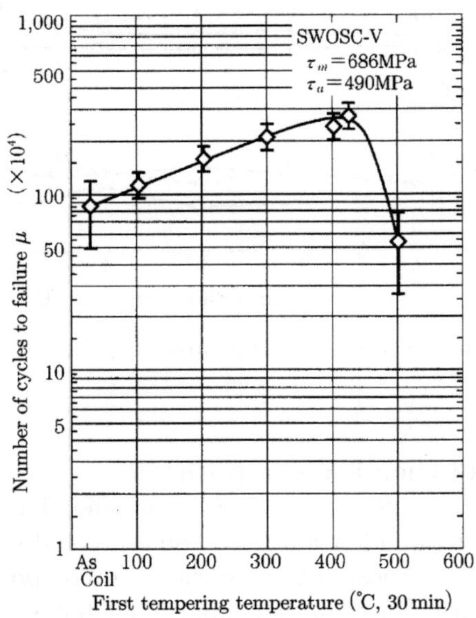

Fig. 2.135. Fatigue property of SWOSC-V springs (with shot peening) [15]

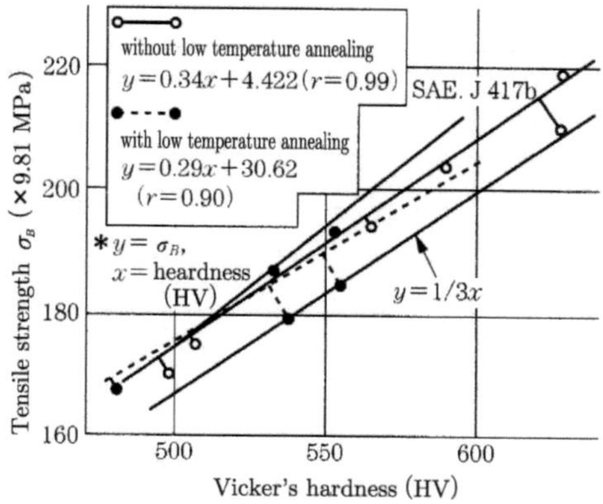

Fig. 2.136. Relation between hardness of oil tempered wire and its tensile strength [16]

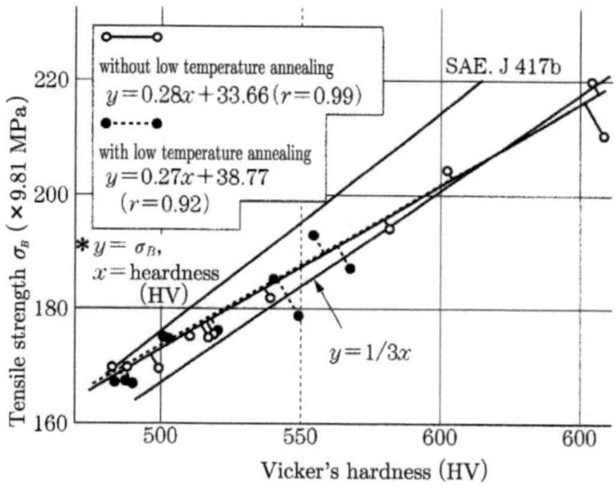

Fig. 2.137. Relation between hardness and tensile strength of cold formed oil tempered springs [16]

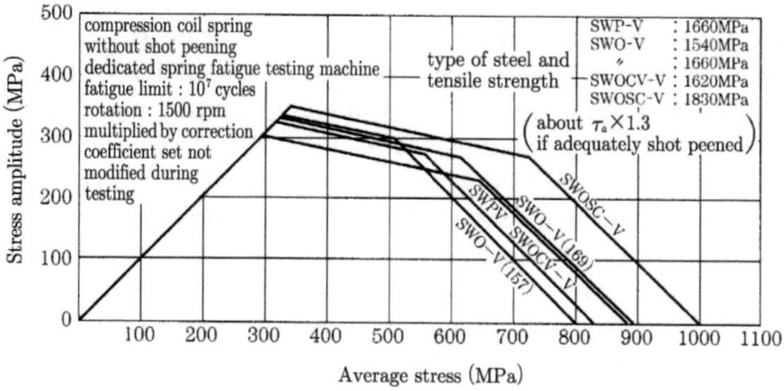

Fig. 2.138. Fatigue limit diagram of various springs made of valve spring quality steel wires [12]

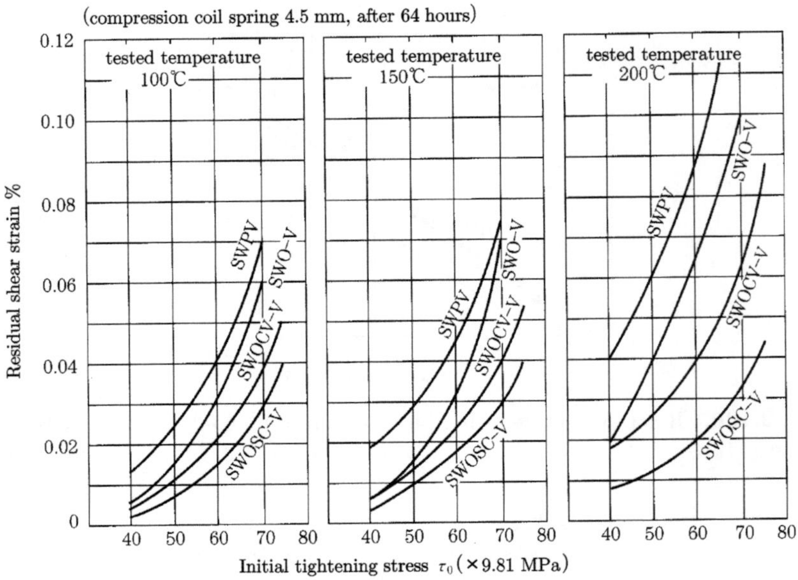

Fig. 2.139. Differences of permanent set properties of springs made of valve spring quality steel wires

ing and allows Cr, V, and Mo to form carbides during tempering, thereby promoting the tempering softening resistance and being likely to lower the toughness.

Si: Being used as deoxidizing agent in steelmaking, Si up to 0.4% is contained in carbon steel oil tempered wire and Cr–V steel oil tempered wire. Si gives ferrite and martensite solid solution hardening and improves the heat resistance and hardenability. Si up to 1.2–2.2% is frequently added to spring steel and oil tempered wire to improve the permanent set resistance and heat resistance. However, it should be noted that high Si steel can be susceptible to decarburization. It has been reported that Si addition of over 2.5% has little effect due to the saturation in the permanent set resistance [17]. Since Si has the effect of delaying decomposition of retained austenite during tempering, in high Si steel some amount of retained austenite tends to remain after quenching and tempering treatment [18].

Mn: Manganese is used as deoxidizing agent in steelmaking. Additionally, Mn fixes sulfur in steel by forming MnS inclusions, thereby preventing initiation of flaws due to hot brittleness related to sulfur. Mn, which has a capability of being dissolved into ferrite and martensite, gives them solid solution hardening, yet has less effect on the permanent set resistance than Si.

Cr: Chromium dissolves into austenite to improve its hardenability and can combine with carbon during tempering to form Cr carbides, which raise the softening resistance during tempering. In high Si steel which is susceptible to decarburization, addition of over 0.5% Cr has the effect of preventing decarburization.

V: Fine non-dissolved vanadium carbides prevent grain boundary migration during austenitizing heating, thereby making grains finer. V, dissolved into steel during austenitizing, precipitates at tempering to bring about secondary hardening. Yet this effect is not usually obtained, since the ordinary tempering temperature for oil tempered wire is lower than that for the secondary hardening. Too much amount of V can increase vanadium carbide in steel, to make the hardness after quenching lower.

P: An insignificant amount of phosphorous segregates on grain boundary, deteriorates the toughness and thus may result in quenching crack. The amount of P in ordinary steels is designated to be controlled under 0.03%, yet, in the case of commercial spring steel, the content of phosphorous is mostly under 0.01%.

S: Although sulfur does not have any harmful influence on the fatigue strength and permanent set resistance, as impurity element, it degrades reduction of area in tensile tests and tends to have bad influence on the corrosion resistance. Thus, in recent years the S content has been decreased to a larger degree than before by carrying out desulfurization during the preliminary treatment of pig iron.

(iv) Effect of microstructure on mechanical properties

The primary factors of microstructure which have effects on the mechanical properties of oil tempered wire can be austenite grain size and retained austenite. The austenite grain size of oil tempered wire depends upon the chemical composition and the heating condition during the quenching and tempering process. Needless to say, fine grain size is desirable for oil tempered wire. Coarse austenite grains would bring about the decrease of the reduction of area and the elongation to make the ductility and toughness lower. In order to keep the austenite grain size fine, the practices of adding a small amount of V, choosing rapid heating such as induction heating, and controlling the heating temperature lower within the extent where sufficient heating be obtained, can be taken. The microstructure of oil tempered wire is basically tempered martensite. However, in the case of alloy steel, or depending upon the heat treatment condition, part of austenite phase generated under solution heating does not transform to martensite by quenching and remains down to room temperature. This is called retained austenite. The amount of retained austenite varies with the chemical compositions of oil tempered wire and the cooling condition from austenitizing. In the case of Si–Cr oil tempered wire, the amount of retained austenite is about several percent. Although it has not been certain in the past, how the retained austenite affects the properties of oil tempered wire, recent researches have been advanced to make it clear [18], [19]. Retained austenite itself does not have any harmful effects on the wire. However, when the amount of retained austenite, the hardness of which is lower than that of martensite, is contained to a large degree, the tempering temperature should be decreased to obtain a predetermined strength, thus making the hardness of the tempered martensite higher than in the case of retained austenite being smaller in quantity. While retained austenite in wire can increase the ductility, simultaneously harder tempered martensite which is rather insufficiently tempered becomes the origin of harmful effects such as deterioration of workability and rise in crack propagation speed. However, it does not always exert harmful effects, but generates strain induced transformation on spring forming, thus raising the hardness of spring inside to improve the fatigue strength. As for permanent set resistance, the retained austenite tends to increase the permanent set slightly. Figure 2.140 and Fig. 2.141 show examples of improvement in the fatigue strength and deteriorations of the permanent set in clamping tests, respectively [19].

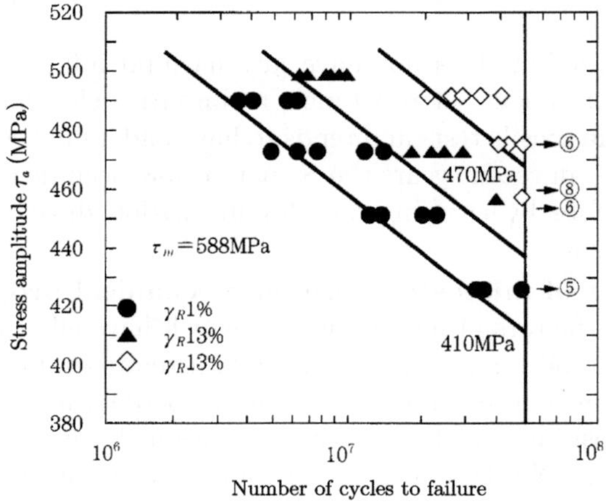

Fig. 2.140. Improvement of spring fatigue strength with strain induced transformation of retained austenite [19]

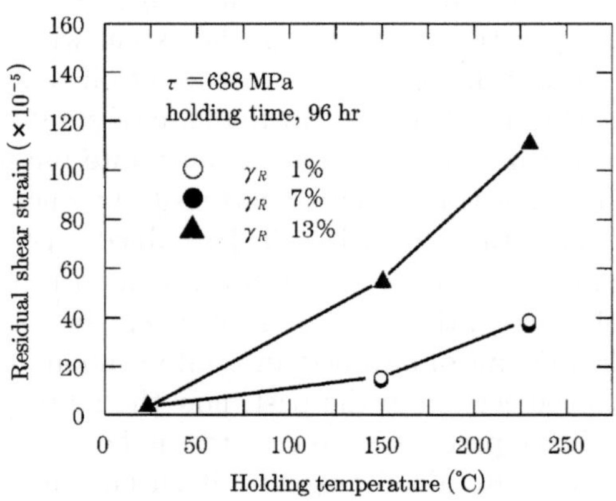

Fig. 2.141. Deterioration of permanent set resistance caused by retained austenite [19]

(v) Variation of mechanical properties with temperature

Figure 2.142 and Fig. 2.143 show the results of tensile tests and torsional tests, respectively, both of which were conducted at elevated temperatures with carbon steel oil tempered wire, Si–Cr steel oil tempered wire and piano wire (4.0 mm in diameter for all). Considering the same process after spring forming, the tested wires were given low temperature annealing, which is

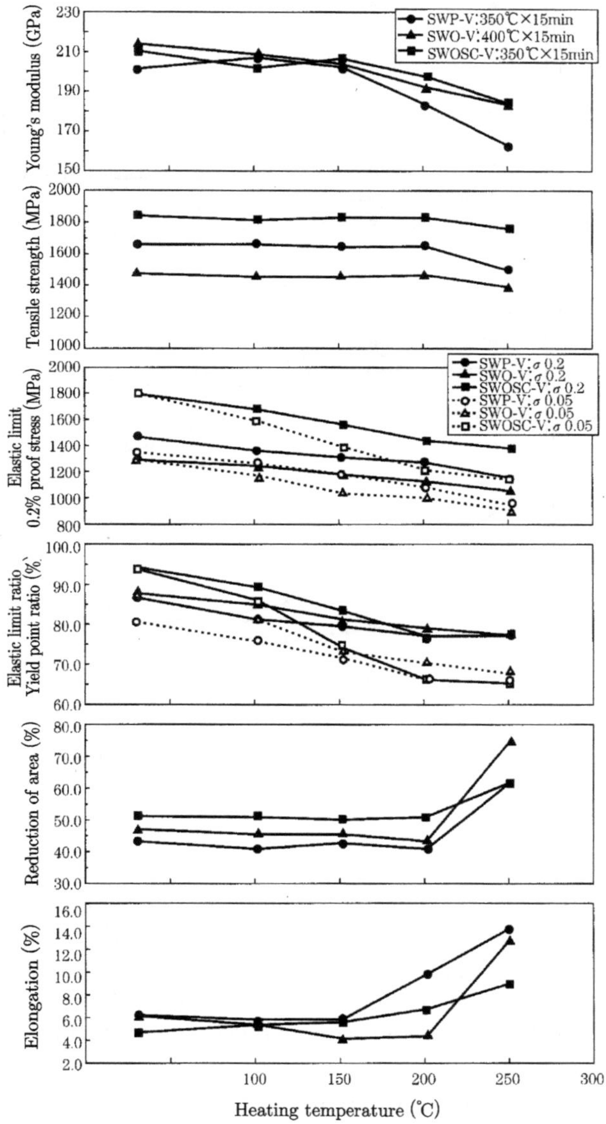

Fig. 2.142. Tensile properties of steel wire for valve springs at high temperature [20]

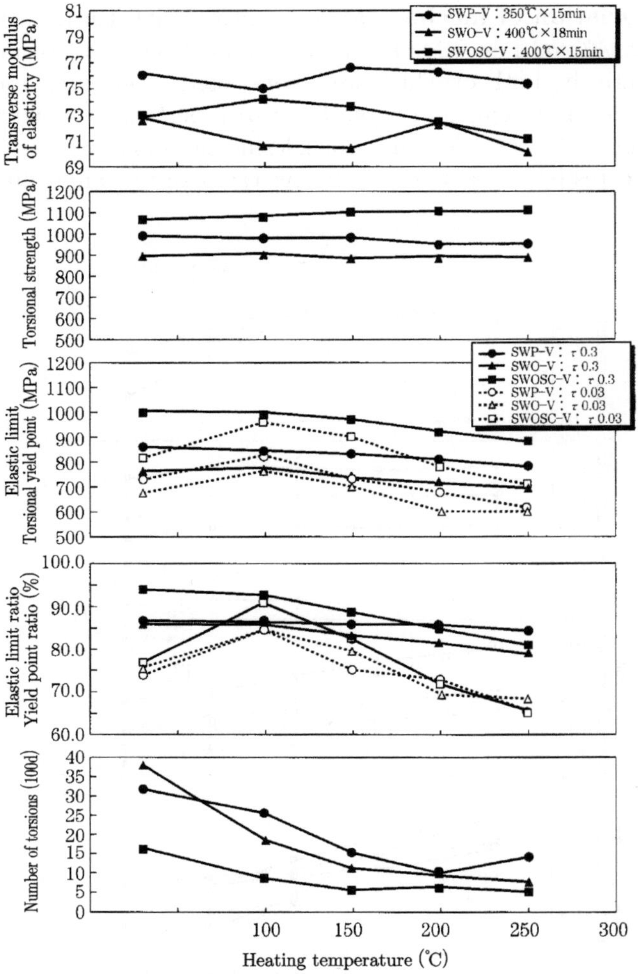

Fig. 2.143. Torsional properties of steel wire for valve springs at high temperature [20]

350°C for 15 minutes for the piano wire, and 400°C for 15 minutes for the oil tempered wire. In both cases, the strength decreases, and the reduction of area and elongation increase with temperature. Comparing with the decrease in tensile strength and maximum torsional strength, the decrease in 0.2% proof stress and 0.05% elastic limit are large. Especially, it should be noted that for Si–Cr steel oil tempered wire, the decrease of yield point and yield ratio is the largest.

(vi) Defects and fatigue strength of oil tempered wire

A valve spring is the indispensably important part for the valve motion system of an automobile engine and its fatigue strength can govern the durability of the engine itself. Accordingly, it is essential that material used for valve springs has high fatigue performance. The stress acting on a spring is normally shear stress which shows the maximum value on the surface. Thus, the defects affecting the fatigue strength of the spring are either surface defects or internal defects just under the surface. Such defects include surface flaws, decarburization, and non-metallic inclusions just under the surface. Since only one defect has the possibility of leading to shutdown of the engine – a fatal accident for automobile –, it is desired that such defects be absent throughout the entire length of oil tempered wire.

① Effect of surface flaws on fatigue strength of oil tempered wire

Since the relation between surface flaws and fatigue strength had been considered as an essential issue, there was joint research organized by JSSR [21], [22]. The research was carried out for both piano wire and oil tempered wire comparatively. The flaws taken as the object of the research are two types such as longitudinal flaws originating from steel billet and in-process flaws originating through spring forming. As shown in Fig. 2.144 with in-process flaws, their harmful effect can be eased by providing shot peening. However, as the depth of in-process flaw becomes larger, the fatigue strength decreases more significantly. As for the longitudinal flaws originating from steel billet, it is reported that a flaw of about 0.05 mm in depth does not affect the fatigue strength when either shot peening or a combination of nitriding and shot peening is provided, see Fig. 2.145 [23]. The effect of surface flaws on the fatigue strength is complicated. In addition, there are other types of flaws which are not taken by these studies. Since it can be thought that the effect of

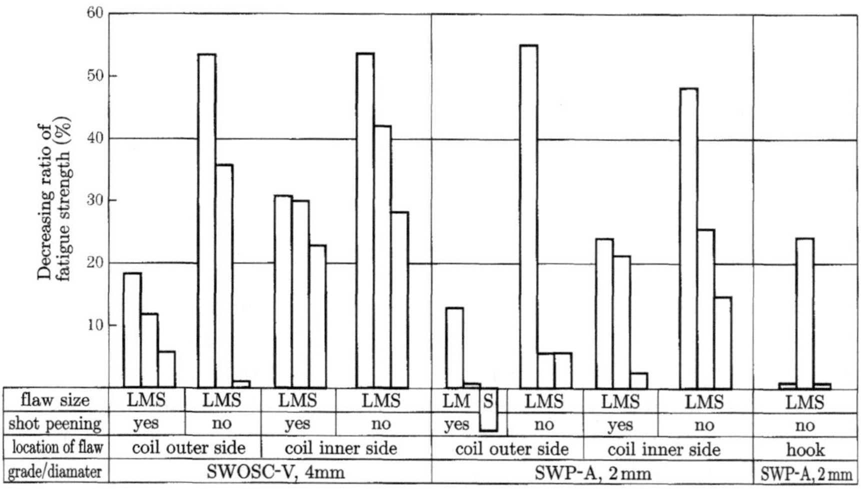

L: large (0.10 mm), M: midium (0.06 mm), S: small (0.02 mm)

Fig. 2.144. Influence of surface flaws occurred in spring processing on fatigue strength [22]

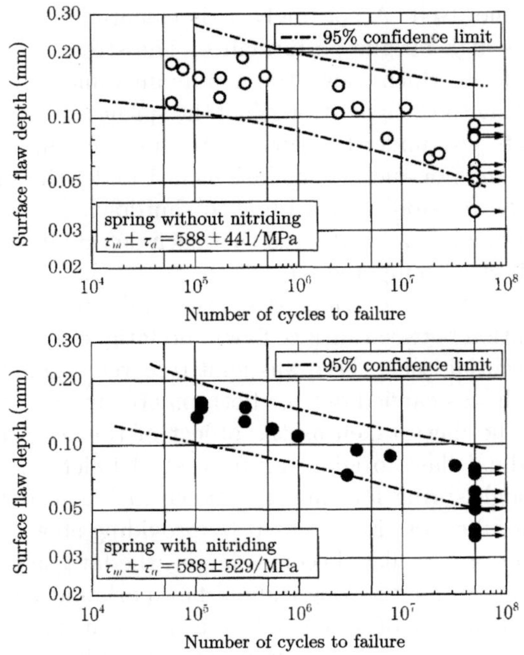

Fig. 2.145. Influence of longitudinal flaw (seam) of wire material on shot peened spring fatigue strength [23]

flaws or defects on fatigue can not be evaluated merely with a unitary index of flaw depth, further study can be required.

② The effect of non-metallic inclusion on fatigue strength of oil tempered wire

It has been known that the fatigue strength of a smooth specimen made of low strength steel or middle strength steel is proportional to the tensile strength or hardness and increases with the strength of material. On the contrary, it has been also known that for high strength material this proportionality becomes unclear and at the same time the fatigue strength varies significantly. It has been considered that one of these reasons is due to non-metallic inclusions [24]. When the oil tempered wire was introduced in Japan after World War II, the strength of carbon steel oil tempered wire and Cr–V steel oil tempered wire was not so high. Therefore, the effect of non-metallic inclusions on the fatigue strength was not apparent. Soon after the Si–Cr steel oil tempered wire appeared in the market, the effect of non-metallic inclusions became obvious due to the higher sensitivity to crack propagation caused by its higher strength. Besides, the spread of scanning electron microscopy and electron prove microanalyzer has made it possible to analyze the surface of fatigue fracture and to investigate the effect of non-metallic inclusions. As a result, as shown in Fig. 2.146, it has been realized that the fatigue fracture of oil tempered wire due to non-metallic inclusions tends to appear in the low-

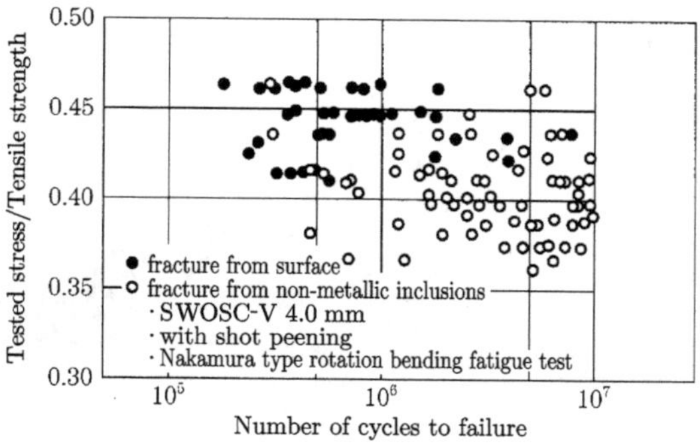

Fig. 2.146. Relation between the ratio of tested stress to tensile strength and fracture mode [28]

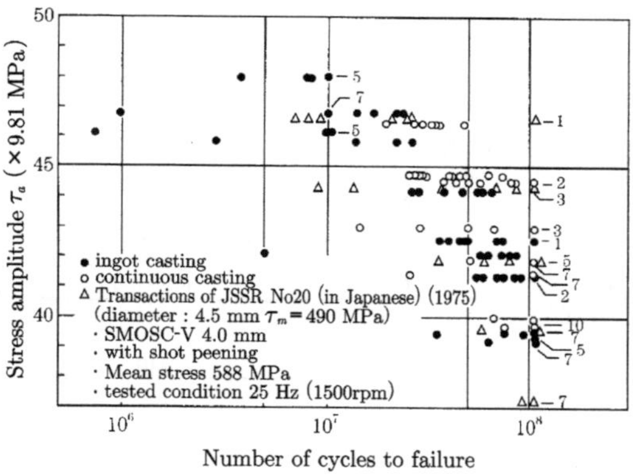

Fig. 2.147. S-N curve of SWOSC-V coil springs on 10^8 cycles [27]

stress long life region [26]. Based on the commercial requirement of fatigue life, the number of cycles to failure has extended from 10^7 to 5×10^7 or even up to 10^8 cycles. Statistical approach has been frequently applied to the testing method. As shown in Fig. 2.147, the fatigue fracture mode changes over 10^7 cycles, making the fatigue strength decline up to 10^8 cycles by 15–16% compared with 10^7 cycles fatigue strength and continue declining further [27–29]. As shown in Fig. 2.148, it is also reported that the S-N curve in a rotating bending fatigue test has a downward slope up to 10^8 cycles without showing any marked fatigue limit [30].

Concerning the relation between the size of non-metallic inclusions and the fatigue strength as mentioned above, there are studies on spring steel and oil tempered wire [31–33]. As the size of non-metallic inclusions is smaller, the fatigue strength or fatigue life can be improved further. In recent years, studies have been conducted from the viewpoint of not only the size, but also what composition of non-metallic inclusions is likely to originate the fatigue fracture. It has been known that the difference in the composition of inclusions,

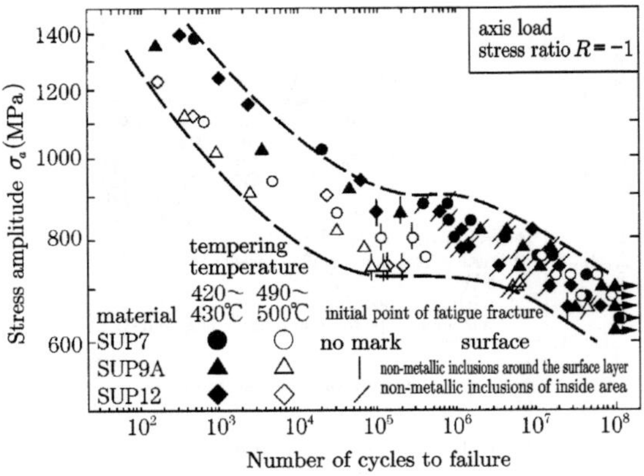

Fig. 2.148. S-N diagram of spring steel coil springs on 10^8 cycles [35]

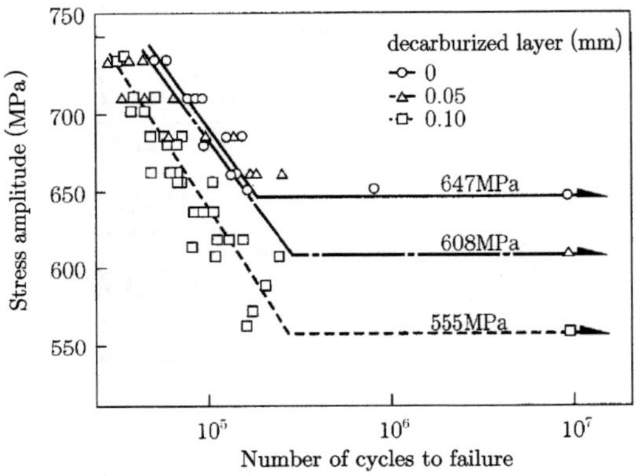

Fig. 2.149. Relation between depth of total decarburized layer and rotating bending fatigue strength [36]

even if the size stays the same, varies the degree of the effect on the fatigue strength [34], [35].

③ Effect of decarburization on fatigue strength of oil tempered wire

Decarburization is not the origin of stress concentration unlike flaws and non-metallic inclusions, but it decreases the surface hardness and deteriorates the fatigue strength on the average. Both ferrite decarburization and total decarburization have effects on the fatigue strength. However, since the ferrite decarburization is critical for the spring fatigue strength, it must be eliminated for spring material and spring manufacturing processes. Figure 2.149 [36] shows the effect of the total decarburized layer thickness (hardness declining depth) of Si–Cr steel oil tempered wire without shot peening on the rotating bending fatigue strength. It can be seen from Fig. 2.149 that as the decarburized layer is deeper the fatigue strength becomes lower. Although the shot peening can improve the fatigue strength for the decarburized wire, its effect on the decarburized wire becomes lower than the non-decarburized wire due to the soft surface layer and rough surface.

(vii) High strength oil tempered wire

In order to promote weight saving, space saving, and higher performance for springs, it is necessary to increase the strength of material. In recent years, this demand has been increasing gradually. To meet with such tendency, primarily in the automotive industry, high-strength oil tempered valve spring wire and high-strength oil tempered suspension spring wire have been developed and made of commercial use.

① High strength oil tempered valve spring wire

Although a valve spring itself is a tiny part in the whole automobile, promoting the mass and space savings with having high performance can improve the freedom of designing valve motion system, to have down-sizing of engine itself and to improve fuel consumptions. In addition, promoting the performance of valve springs improves the motion efficiency of the valve and makes it possible to promote the output of engine power. Based on these reasons, at present, several new steel grades have been developed and made of commercial use as oil tempered valve spring wire. Table 2.25 shows the chemical compositions and characteristics of high-strength oil tempered wires which have been in commercial use so far.

Steel grades adopted extensively for high-strength oil tempered wire include modified Si–Cr steel such as high-carbon Si–Cr steel, high carbon Si–Cr–V steel and high carbon Si–Cr–Ni–V steel, [37–40]. The high strength modified Si–Cr valve spring steel oil tempered wire mentioned above is capable of being given low temperature nitriding treatment due to its high heat resistance, to be applied to high stress valve springs. Since such higher heat resistance can make the nitrided spring quality less scattered, steel whose heat resistance is improved by the addition of vanadium and molybdenum or by the addition of more silicon and cobalt has been in commercial use for oil tempered valve spring wire [41], [42].

② High strength oil tempered suspension spring wire

So far, hot formed springs have been predominantly used for suspension springs. To meet the demands of weight saving around the suspension system, however, suspension springs in small cars tend to become lighter by designing with higher stress. Therefore, high strength oil tempered wire has been developed for suspension springs, to improve the permanent set resistance [43–45]. Since suspension springs contact with aqueous solution during in use, the precautions to prevent the initiation of corrosion pits on surface and hydrogen penetration into steel should be required, for example, by applying the appropriate coating. Especially for the high strength material, it is to be desired that attention should be paid to the material design and spring design corresponding to such environmental brittleness.

Table 2.25. Variations of high strength oil tempered wire developed.

	kind of oil tempered wire	chemical composition (%)									tensile strength grade for valve springs 4.0 mm for suspension springs 10.0 mm		characteristics
		C	Si	Mn	Cr	Ni	Mo	V	Co	other			
for valve springs	Silicon-Chromium steel oil temperd wire for valve springs	0.55	1.40	0.70	0.70	–	–	–		–	1910		JIS specification
for valve springs	High carbon Silicon-Chromium steel oil temperd wire for valve springs	0.64	1.42	0.72	0.76	–	–	–		–	2060		increase of tempering hardness by high carbon
for valve springs	Silicon-Chromium steel oil temperd wire for valve springs	0.65	1.41	0.80	0.75	–	–	0.12		–			refinement of grain size by high carbon and V addition
for valve springs	High strength oil tempered wire for valve springs	0.60	1.48	0.70	0.87	0.31		0.06		–			improvement of toughness by Ni addition
for valve springs	High strength oil tempered wire for valve springs	0.61	1.97	0.70	0.73	–	–	0.10		–			improvement of tempering softning resistance by additional Si
for valve springs	High strength oil tempered carburizing and nitriding wire for valve springs	0.75	1.49	0.49	0.54	–	0.15	0.49		–	2010	nitriding required	refinement of grain size by high carbon and V addition and improvement of tempering resistanse by Mo addition
for valve springs	High strength oil tempered wire for valve springs	0.65	2.20	0.60	1.20	–	–	0.15	0.15	–	2160	nitriding required	improvement of heat resistance by additional Si and Co
for valve springs	High strength Chromium-Molybdenum-Vanadium oil tempered wire for valve springs	0.64	1.43	0.67	1.57	–	0.57	0.26		–	2060	nitriding required	improvement of tempering resistance by addition of high carbon and Cr, Mo, or V
for suspension springs	Silicon-Chromium steel oil tempered spring wire	0.54	1.46	0.70	0.71	–	–	–		–	1740		JIS specification
for suspension springs	High strength oil tempered wire for suspension springs	0.58	1.36	0.70	0.67	–	–	0.21		–	1910		improvement of yield point due to refinement of grain size by V addition
for suspension springs	High strength oil tempered wire for suspension springs	0.59	1.48	0.51	0.59	–	–	0.16		–			
for suspension springs	High strength oil tempered wire for suspension springs	0.60	2.00	1.30	0.15	–	–	0.35		[N] 0.015			improvement of permanent set resistance by high Si and high Mn, and V and N addition

(viii) Remarks to be taken in the use of oil tempered wire

Although oil tempered wire is suitable for springs, it becomes brittle unexpectedly under a certain condition. Considering the fact that its microstructure is tempered martensite and the strength is substantially high, the following points should be noted when using oil tempered wire.

1) Keep from using – to the utmost – pickling and metal plating, which may make wire absorb hydrogen and suffer from hydrogen embrittlement.
2) Keep from carrying out severe bending and coiling with a small spring index (D/d), both of which are more difficult with oil tempered wire than with piano wire and hard drawn steel wire. Especially for oil tempered wire with a large diameter, it should be cautious to be some possibility of breakage.
3) Carry out low temperature annealing soon after the coiling of springs. In the case of high strength Si–Cr steel oil tempered wire, tensile residual stress of more than 1,000 MPa generates on the spring inside surface after the coiling on some occasions, where it can cause delayed fracture or stress corrosion cracking.
4) When nitrate based salt is used for the heat treatment such as low temperature annealing after the coiling of springs, sufficient cleaning is necessary due to the corrosiveness of decomposed salt. To the utmost, keep from using salt baths and use an air heating furnace or electric furnace.

References

1. JIS G 3560: Oil tempered carbon steel wire for mechanical spring
2. JIS G 3561: Oil tempered carbon steel wire for valve spring
3. K. Komura: Small spring (wire spring) Tokushukou (Japanese) 38, 7 (1989) p. 32
4. JSSR joint committee non-circular sectioned coil spring Trans. of JSSR 36 (1988)
5. J. Osawa: Iron and Steel (Japanese) 74, 6 (1988) p. 941
6. H. Tominaga: Symposium for spring fatigue strength (Japanese) JSSR (1989) p. 17
7. H. Kawakami, T. Oshiro, Y. Kawaguchi, Y. Kanesada: Kobe steel technical report 41, 4 (1991) p. 71
8. K. Komura: Spring Edition 3 Maruzen (Japanese) p. 58
9. Y. Seto, E. Soga, K. Uchino, K. Kawane: Trans. of JSSR (Japanese) 27 (1982) p. 28
10. H. Tsubono, K. Nishimura, T. Fujiwara, H. Ioka: Trans. of JSSR (Japanese) 307 (1985) p. 1
11. M. Urbnacic: Double Lead Indemnity, Springs, Fall (1996) p. 95
12. Suzuki metal industry co. Specifications note book
13. Manual Heat-treatment for small springs JSSR (1982) p. 146
14. H. Toshino, M. Kato, K. Shigeno, K. Ariyoshi: JSSR Autumn Technical report (1981) p. 1

15. T. Ozone, H. Kato: JSSR Autumn Technical report (1986) p. 9
16. K. Shigeno, K. Nonoyama: JSSR Autumn Technical report (1985) p. 13
17. J. Sugimoto et al.: Trans. of JSSR 38 (1993) p. 7
18. T. Sakakibara: Trans. of JSSR 38 (1993) p. 1
19. H. Yarita: Trans. of JSSR 43 (1998) p.1
20. JSSR joint research Trans. of JSSR 13 (1967) p. 1
21. JSSR joint research Trans. of JSSR 22 (1977) p. 55
22. JSSR joint research Trans. of JSSR 30 (1985) p. 117
23. M. Terashita et al.: Trans. of JSSR 40(1995) p. 21
24. K. Kanazawa: JSSR Autumn Technical report (1988) p. 53
25. H. Hayashi: JSSR Autumn Technical report (1983) p. 9
26. K. Kanazawa, T. Abe: Trans. of JSSR 37 (1992) p. 15
27. H. Toshino, M. Kato, K. Shigeno: Trans. of JSSR 20 (1975) p. 35
28. H. Hayashi: JSSR Autumn Technical report (1983) p. 9
29. Y. Ishikawa, Y. Izawa, M. Nakanishi: JSSR Autumn Technical report (1997) p. 29
30. T. Abe, T. Omura, S. Matsuoka: JSSR Spring Technical report (1998) p. 17
31. Y. Ito, H. Hayashi, T. Saito, N. Takagil: JSSR Autumn Technical report (1985) p. 17
32. Y. Kobayashi, K. Namiki, M. Takagi, T. Shibata: JSSR Autumn Technical report (1989) p. 13
33. Y. Murakami et al.: Trans. of JSSR 39 (1994) p. 7
34. Y. Yamada et al.: JSSR Spring Technical report (1985) p. 1
35. T. Abe, T. Omura, S. Matsuoka: JSSR Spring Technical report (1998) p. 17
36. N. Ibaraki et al.: ESF International Conference April (1990) p. 1
37. JSSR Valve spring working Group Trans. of JSSR 40 (1995) p. 57
38. D. Matsumoto, T. Murai, T. Yoshioka: Sumitomo electric industries Technical report 147 (1995) p. 66
39. M. Terashita, et al.: Trans. of JSSR 40 (1995) p. 21
40. D. Matsumoto, T. Murai, T. Yoshioka: JSSR Autumn Technical report (1995) p. 27
41. Y. Hagiwara et al.: Trans. of JSSR 35 (1990) p. 13
42. S. Yasuda et al.: Trans. of JSSR 42 (1997) p. 1
43. N. Yokote et al.: Trans. of JSSR 31 (1986) p. 9
44. J. Sugimoto et al.: Trans. of JSSR 38 (1993) p. 7
45. JSSR Working group: Trans. of JSSR 39 (1994) p. 101

(3) Stainless steel wire for spring

(i) Types of stainless steel wire

Material for stainless steel wire is classified from its metallurgical microstructure into mainly three categories that is austenitic, ferritic and martensitic series. Precipitate hardening stainless steel and duplex stainless steel composed of austenite and ferrite phase are additionally improved stainless steel series. JIS G 4308 designates the standard of stainless steel wire material which contains 36 kinds of stainless steel including 22 kinds of austenitic stainless steel, 3 ferritic, 10 martensitic and 1 precipitate hardening. There are many kinds of steel that can be used for stainless steel wire, including heat resistant steel wire designated in JIS G 4311 heat resistant steel bars which contain more than approx. 11% Chromium.

However, there are not so many kinds of steel in JIS G 4314 that designates stainless steel spring wire, containing 4 austenitic and 1 precipitate hardening steel.

Table 2.26 shows chemical compositions of stainless steel wire material. Figure 2.150 [1] also shows the relation between stainless steel wire material and stainless steel wire in JIS standards and Table 2.27 [2] shows types of stainless steel spring wire and its metallurgical classification.

The JIS standard of stainless steel spring wire was revised in 1994, considering the accordance with ISO standard. In this revision, requirement on diameter was adjusted to specifications of ISO/DIS 6931-1 (Stainless steels for Springs Part 1: wire) and straight stainless steel wire was newly defined. However, revision of the specification on mechanical properties consistent to ISO standard was postponed because of ISO/DIS revision of tensile strength to higher level making it difficult for SUS304-WPB to adjust ISO/DIS strength level. The use of SUS301 which has higher work hardening, instead of SUS304, can be required, leaving to further discussions.

Apart from consistency with ISO/DIS, SUS304 straight wire (SUS304-WPDS, SUS304-WPBS) was specified additionally. Its applicable diameter range is 0.29 to 1.60 mm. It should be noted that the tensile strength of WPDS is lower than that of WPBS, as shown in Table 2.28 [3].

Table 2.26. (a) Chemical compositions of austenitic stainless spring steel

in mass%

Type	C	Si	Mn	P	S	Ni	Cr	Mo	N
SUS 302	≦0.15	≦1.00	≦2.00	≦0.045	≦0.030	8.00–10.00	17.00–19.00	–	–
SUS 304	≦0.08	≦1.00	≦2.00	≦0.045	≦0.030	8.00–10.50	18.00–20.00	–	–
SUS 304 N1	≦0.08	≦1.00	≦2.00	≦0.045	≦0.030	7.00–10.50	18.00–20.00	–	0.10–0.25
SUS 316	≦0.08	≦1.00	≦2.00	≦0.045	≦0.030	10.00–14.00	16.00–18.00	2.00–3.00	–

Table 2.26. (b) Chemical compositions of precipitation hardening stainless spring steel

in mass%

Type	C	Si	Mn	P	S	Ni	Cr	Al
SUS 631 J1	≦ 0.09	≦ 1.00	≦1.00	≦0.040	≦ 0.030	7.00–8.50	16.00–18.00	0.75–1.50

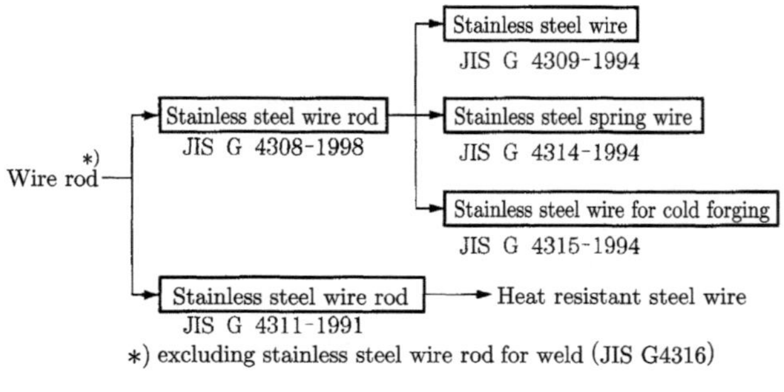

Fig. 2.150. Relation between JIS stainless wire rod and JIS wire

Table 2.27. Types of stainless spring steel and its classification

Type	Grade		Classification
	Division	Code	
SUS 302	A	–WPA	Austenitic series
	B	–WPB	
SUS 304	A	–WPA	
	B	–WPB, –WPBS(1)	
	D	–WPDS(1)	
SUS 304 N1	A	–WPA	
	B	–WPB	
SUS 316	A	–WPA	
SUS 631 J1	C	–WPC	Precipitation hardening

(1) S Means wire required the straightness in the specifications

(ii) Production method of stainless steel wire

In production of stainless steel, molten steel is made first by heating and melting ferroalloy that contain nickel and chromium and scrap as raw material in electric furnace. Next AOD (Argon Oxygen Decarburization) process reduces carbon in the molten steel by blowing argon with oxygen gas into it. Following ladle metallurgy process adjusts chemical composition of stainless steel and adjusts the molten steel temperature to the optimal condition for following continuous casting (CC) process. In the CC process, molten steel refined is continuously cast into billet. Geometry of cross section is usually square or rectangular, but round shape has been also in production recently. Wire rod rolling is basically the same as that for spring steel. Acid pickling of wire rod removes scale layer formed during hot rolling process to finish stainless steel wire rod. In some material, the batch annealing or continuous annealing called DST (Direct Solution Treatment) utilizing the heat of hot rolling process can be carried out before acid pickling. Figure 2.151 [4] shows a typical production process flow of stainless steel wire for springs. Solution treated and acid pickled wire rod after hot rolling undergo flaw inspection and is coated, drawn, and solution treated repeatedly and finally drawn with a large reduction to obtain prescribed tensile strength. As for stainless steel wire for the application to nuclear power plants and automobiles, which should require stringent surface quality, wire rod surface-removed by shaving dies can be used. The solution heat treatment in intermediate process can be normally conducted in continuous bright solution heat treatment furnace in reducing atmosphere. In this case, the following pickling process is not required.

Stainless steel wire for springs tends to be seized by the friction in contact with other substance, because of its low heat conductivity, compared with carbon steel and low alloy steel spring wire. Therefore, in production of spring wire, Ni plating or resin coating can be applied to solution treated wire surface before final drawing, in order to improve coiling ability and prevent wear flaw or poor dimensional precision when cold formed at spring manufacturers.

Table 2.28. Tensile strength of stainless spring steels (JIS G 4314) [3]

	Tensile strength N/mm^2			
	Grade A	Grade B	Grade C	Grade D
Diameter/mm	SUS302-WPA SUS304-WPA SUS304N1-WPA SUS316-WPA	SUS302-WPB SUS304-WPB SUS304-WPBS$^{(1)}$ SUS304N1-WPB	SUS631-WPC$^{(2)}$	SUS304-WPDS
0.080	1650–1900	2150–2400	–	–
0.090				
0.10			1950–2200	
0.12				
0.14				
0.16				
0.18				
0.20				
0.23	1600–1850	2050–2300	1930–2180	–
0.26				
0.29				1700–2000
0.32				
0.35				
0.40				
0.45		1950–2200	1850–2100	1650–1950
0.50				
0.55				
0.60				
0.65	1530–1780	1850–2100	1800–2050	1550–1850
0.70				
0.80				1550–1800
0.90				
1.00				1500–1750
1.20	1450–1700	1750–2000	1700–1950	1470–1720
1.40				1420–1670
1.60	1400–1650	1650–1900	1600–1850	1370–1620
1.80				–
2.00				
2.30	1320–1570	1550–1800	1500–1750	
2.60				
2.90	1230–1480	1450–1700	1400–1650	
3.20				
3.50				
4.00				
4.50	1100–1350	1350–1600	1400–1650	
5.00				
5.50				
6.00				
6.50	1000–1250	1270–1520	1300–1550	
7.00				
8.00				
9.00	–	1130–1380	–	
10.0		980–1230		
12.0		880–1130		

(1) SUS 304-WPBS in diameter ranged from 0.2 to 1.6 mm
(2) After the precipitation hardening, 470 ± 10°C, 1 hour and air cooling, the increase of tensile stress should be over 250 MPa inclusive

In final drawing process, it is normal to use a dry non-slip type continuous drawing machine. In fine wire production under 0.8 mm diameter, a slip type drawing machine with multi stepped capstan corns, can be used with diamond dies in oil lubricant. The wire made by this method has superior surface burnish. In the production of wire required the straightness, the straighten-

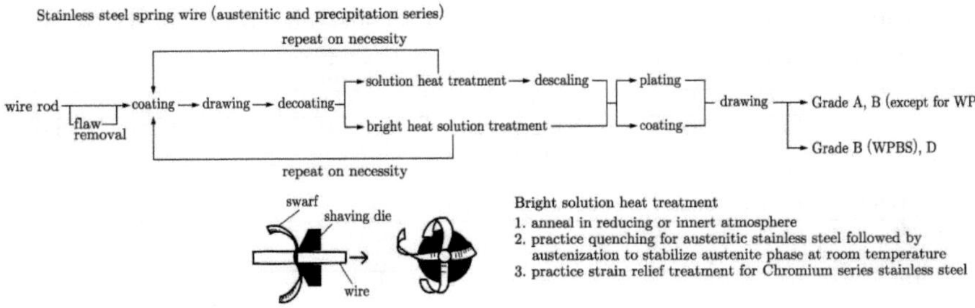

Fig. 2.151. An example of production process flow of stainless steel spring wire

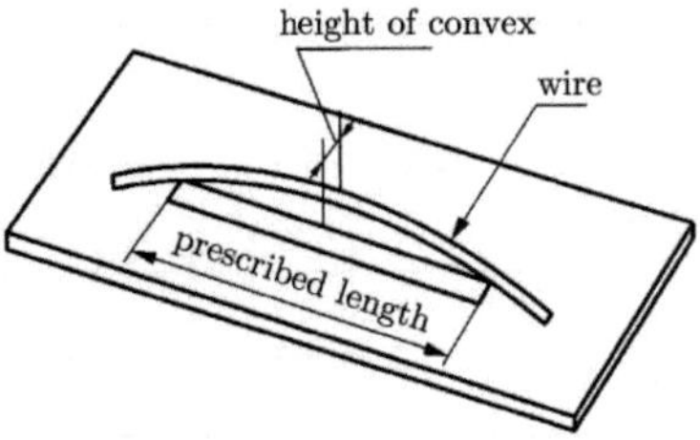

Fig. 2.152. Measuring method of the straightness of wire [2]

Table 2.29. Specification on straightness in JIS straight stainless steel wire [2]

in mm

Diameter	straight	
	height of convex	length of string
0.29 to 0.45	≦ 5	200
over 0.45 to 1.6, incl	≦ 4	200

ing process can be applied after the final drawing. To evaluate this degree of straightness, the straightness testing methods and specifications are designated by JIS, as shown in Fig. 2.152 [2] and Table 2.29 [2].

(iii) Characteristics of Ni plated stainless steel wire for springs

Nickel plated stainless spring steel wire was developed by Sumitomo Electric Industries Inc. (SEI) in 1966 as new stainless spring steel wire, substituting previously used deleterious lead coating. After that, SEI had granted its license to domestic four companies. Owing to improvement of quality and cost by each company, the commercial share of the nickel plated wire has been over 80% of the total stainless steel spring wire market. The major characteristics of Ni plated stainless wire can be summarized below.

1) Its surface burnish is silver white color, and similar appearance like stainless steel. Low friction coefficient and superior spring formability can be shown (Fig. 2.153 [5]).

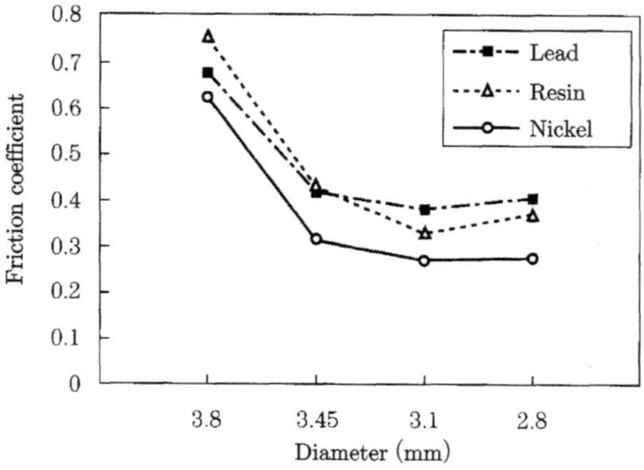

Fig. 2.153. Coefficient of friction between wire and dies in drawing process

2) The oxidization and color change after low temperature annealing is small. However, the removal of plating can be required when used in the corrosion conditions such as food applications.
3) The corrosion resistance in air atmosphere is equal with that of stainless steel wire.
4) No polluted and hazardous material such as toxic lead.

(iv) Characteristics of stainless steel spring wire and its applications

Meta-stable austenitic stainless steel such as SUS302, 304, and 316 shows complete austenitic structure when quenched after solution heat treatment over 1000°C. Under the cold drawing with high reduction during the manufacturing processes of stainless steel wire, part of austenite phase can be transformed into deformation induced martensite, showing high work hardening. This work hardening rate changes with chemical compositions, especially with Ni content as shown in Fig. 2.154 [6]. Therefore, the work hardening rate of autenitic stainless steel spring wire is relatively large for SUS302 and 304, and is lowest for SUS316.

SUS631 J1 is called "semi-austenitic precipitation hardening stainless steel wire" whose metallurgical microstructure after solution heat treatment is meta-stable austenite phase, and can be partly transformed to deformation induced martensite phase in drawing process. The combination of this martensitic transformation and precipitation hardening makes it possible to have high strength. The precipitate hardening heat treatment is normally conducted after spring coiling process.

The SUS631 J1-WPC wire, which is slightly higher Ni in composition than AISI 631 to improve wire drawability, is a unique JIS standard in the world. Typical characteristics and applications of these stainless steel wires are shown in Table 2.30 [7]. The SUS304-WPB wire is widely used for precision machine (watch, scales), optical devices (cameras) and medical devices, besides home electric appliances, and automobiles. The SUS316-WPA wire is applied to

Table 2.30. Major properties of stainless steel wire for springs and its applications [7]

Classifications	Diameter in mm	Properties		Typical applications
SUS 302-WPA SUS 304-WPA	0.080–8.00	minimum tensile strength 1,000–1,650 MPa maximum allowable temperatur 290°C transverse modulus of elasticity 69 GPa normal corrosion resistance, heat resistance, magnetic	in case of 4.0 mm diameter tensile strength 1,230–1,480 MPa	• photo items for darkroom • for reinforcement of blade hoses
SUS 302-WPB SUS 304-WPB	0.080–12.00	minimum tensile strength 880–2,150 MPa maximum allowable temperatur 290°C transverse modulus of elasticity 69 GPa normal corrosion resistance, heat resistance, magnetic	in case of 4.0 mm diameter tensile strength 1,450–1,700 MPa	• springs for various kinds of home electric appliances • automotive applications (for example, radiator cap)
SUS 316-WPA	0.080–8.00	minimum tensile strength 1,000–1,650 MPa maximum allowable temperatur 290°C transverse modulus of elasticity 69 GPa suprerior corrosion resistance, weak magnetic	in case of 4.0 mm diameter tensile strength 1,230–1,480 MPa	• tape recorders • capstan shafts (weak magnetics)
SUS 631 J1-WPC	0.100–6.00	minimum tensile strength 1,300 + 250–1,950 + 250 MPa maximum allowable temperatur 340°C transverse modulus of elasticity 73 GPa superior spring property, strong magnetic	in case of 4.0 mm diameter tensile strength* 1,400 + 250–1,650 + 250 MPa	• springs for engines in automobile • wireless antennas

* SUS 631 J1-WPC shall be increased in tensile strength more than 250 MPa after precipitation hardening

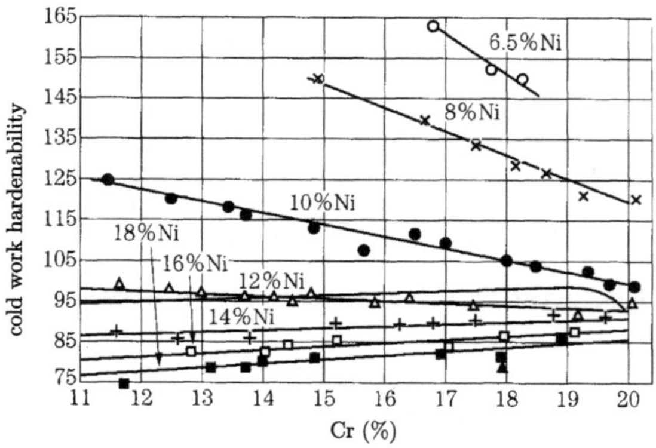

Fig. 2.154. Effect of Cr and Ni on cold work hardenability of austenitic stainless steels [6]

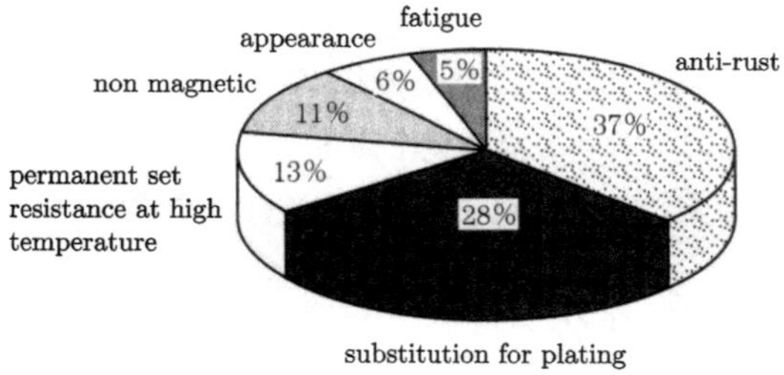

Fig. 2.155. Reasons for using stainless steel wire

such springs required relatively low strength and should be required to have corrosion resistance and to be non magnetic. The SUS631 J1 is used in springs for automotive engines or antenna since it has the highest elastic modulus and superior heat resistance. The reason why stainless steel wire can be used from the view point of user side are summarized in Fig. 2.155 [8]. It can be seen from the Fig. 2.155 that most applications can be for corrosion resistance property and two-third of the applications are the replacement from rust prevention or plating. It is 5% for fatigue strength and even under 20%, including for permanent set resistance at high temperature. While Fig. 2.156 [8] shows the types of spring, compression coil spring, tension coil spring, torsion coil spring are in order. From forming side, SUS304-WPBS and WPDS straight wire, which have been recently designated in JIS, are suitable for long-armed torsion springs and long straight wire forming.

(v) Relation between drawing reduction and mechanical properties of stainless steel wire

Figure 2.157 [9] shows relation between drawing reduction and mechanical properties of stainless steel wire. As described before, SUS304 wire which has high work hardening characteristic has been commonly used for spring wire in Japan, while AISI 302 and other similar material which show more work

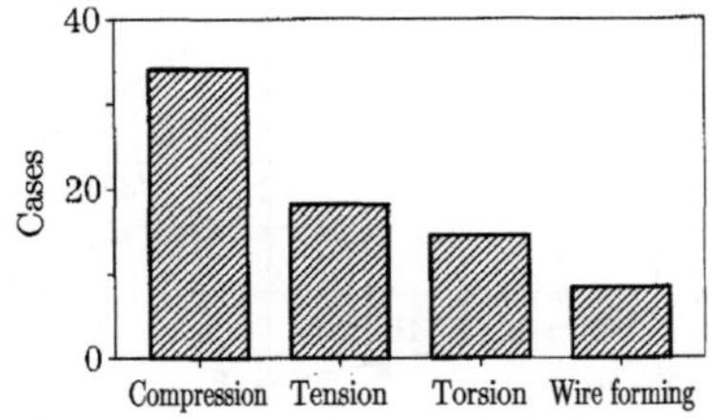

Fig. 2.156. Kind of stainless steel wire spring [8]

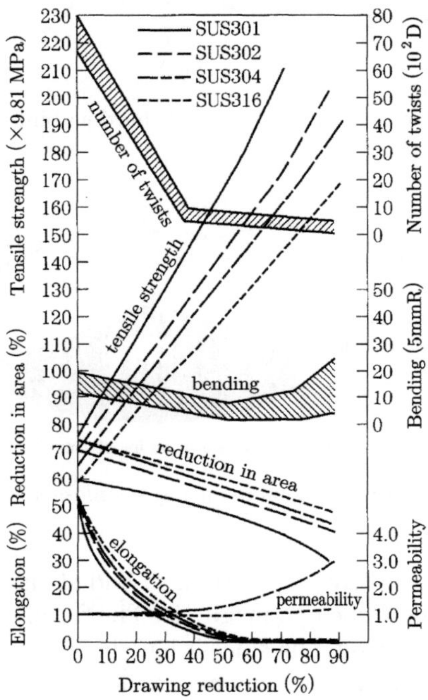

Fig. 2.157. Relation between drawing reduction and mechanical properties of stainless steel wire [9]

hardening, have been used in North America and Europe. These stainless steel spring wires have weak magnetism with strain-induced martensite generated in drawing process. The SUS316 wire with low work hardening rate shows the minimum change in permeability. Figure 2.158 shows magnetic permeability change of stainless steel spring wire with drawing reductions.

The SUS631 J1 is so-called 17-7 PH stainless steel (SUS631), in which Al is added to Cr–Ni steel as a precipitate hardening element. This steel can be hardened with the precipitation of Ni–Al compound by the heat treatment after strain-induced martensite transformation. Figure 2.159 [9] shows the relation between reduction in area and mechanical properties of 17-7 PH series stainless steel wire. The JIS G 4314 designates that the increase of tensile

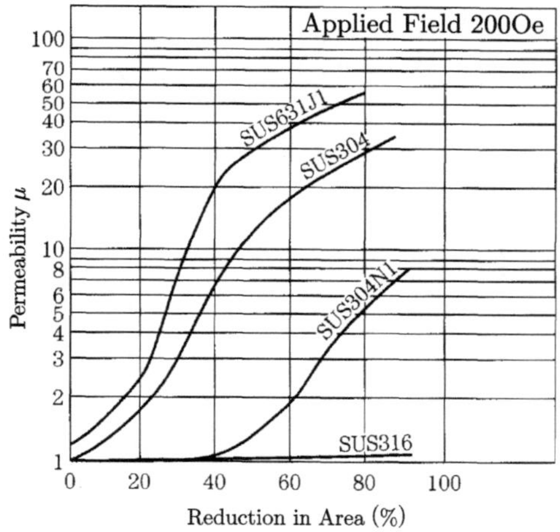

Fig. 2.158. Relation between drawing reduction and magnetic permeability of stainless steel spring wire

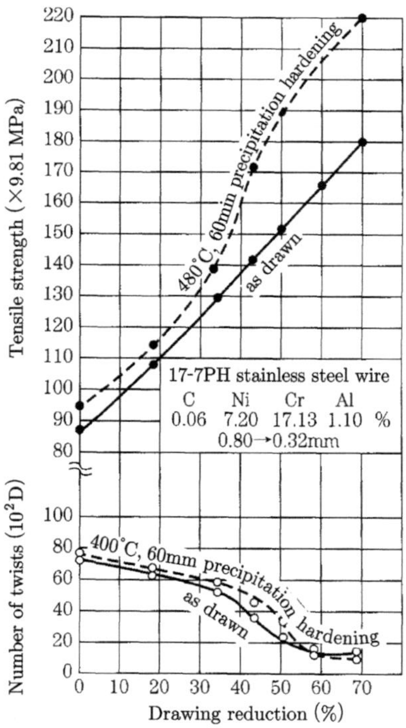

Fig. 2.159. Relation between drawing reduction and mechanical properties of 17-7 PH stainless steel wire [9]

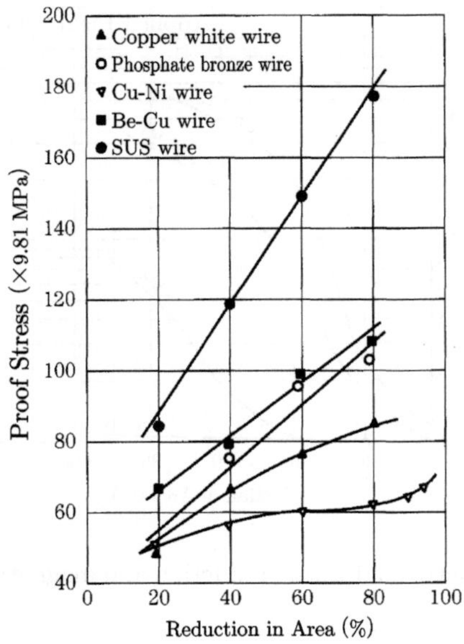

Fig. 2.160. Effect of cold work on proof stress of wire spring material (1.0 mm in diameter) [10]

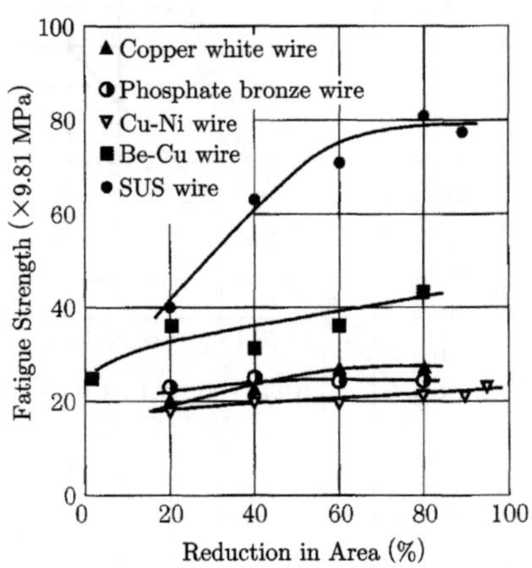

Fig. 2.161. Effect of cold work on fatigue strength of wire spring material (1.0 mm in diameter) [11]

strength of SUS631J1 after precipitation hardening heat treatment at 475 ± 10°C for 60 minutes shall be more than 250 MPa for springs. As for these stainless steel wires, static mechanical properties and fatigue strength can be improved with cold drawing process.

Figure 2.160 [10] and Fig. 2.161 [11] show the comparison of proof stress and fatigue strength between stainless steel wire and non-ferrous spring wire

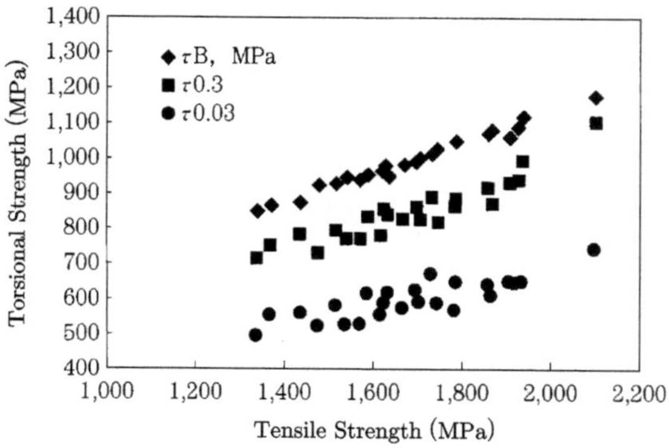

Fig. 2.162. Relation between tensile strength and torsional strength of stainless steel spring wire [12]

(1.0 mm in diameter) such as nickel silver, phosphor bronze, copper beryllium and Cu–Ni alloy. The Fig. 2.161 shows that fatigue strength of stainless steel wire is two times of copper alloys. On the other hand, fatigue strength of stainless steel spring wire is lower than that of steel wire under the same strength level. Since this reason is not clear, further studies should be required. There would be some possibility in future where stainless steel wire springs be replaced by plated high carbon steel wire springs because of its good corrosion resistance and cost performance.

Under these backgrounds, JSSR technical committee on fatigue strength of stainless steel wire [12], have proposed the needs of data accumulation which can be utilized to clarify predominant factors of fatigue and permanent set properties of springs for design reference. The following study committee on fatigue strength of stainless steel wire has carried out systematic studies and analyses of SUS304 wire of 2 mm diameter. The committee reported the relations between 50 to 90% drawing reduction and the characteristic such as martensite phase, residual stress, tensile properties, torsional properties, or fatigue properties.

These studies made it clear that both torsional strength and martensite phase increase with the increase of tensile strength, as shown in Fig. 2.162, and 2.163 [12].

On the other hand, fatigue properties of wire spring are complicating due to many influential factors. Abundant previous research works on spring material have been collected in wire spring calculation system "BANEX" by JSSR database committee. Fatigue data on stainless steel spring wire have been also included in BANEX. Based on this database, the relation between tensile strength, and rotary bending fatigue strength or fatigue strength of springs is shown in Fig. 2.164 and Fig. 2.165 [13]. The ratio of fatigue strength to tensile strength of stainless steel spring wire is known to be lower than that of piano wire and Si–Cr oil tempered wire (solid lines in Fig. 2.165 mean allowable stress amplitude recommended by JIS). It has been said that as for stainless

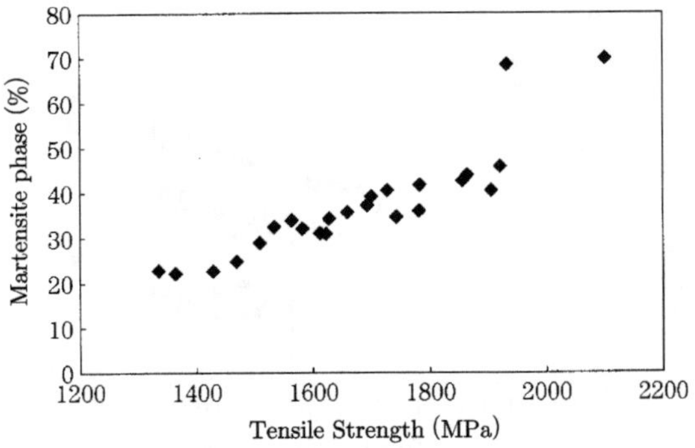

Fig. 2.163. Tensile strength and amount of martensite of stainless steel spring wire

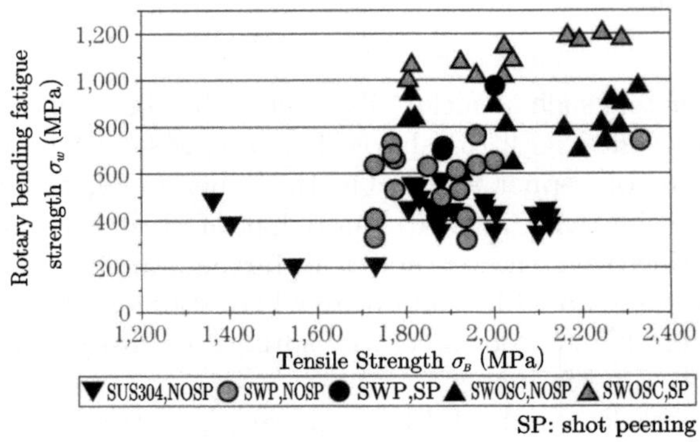

Fig. 2.164. Relation between tensile strength and rotary bending fatigue strength of various steel spring wires (under 3 mm in diameter inclusive) [13]

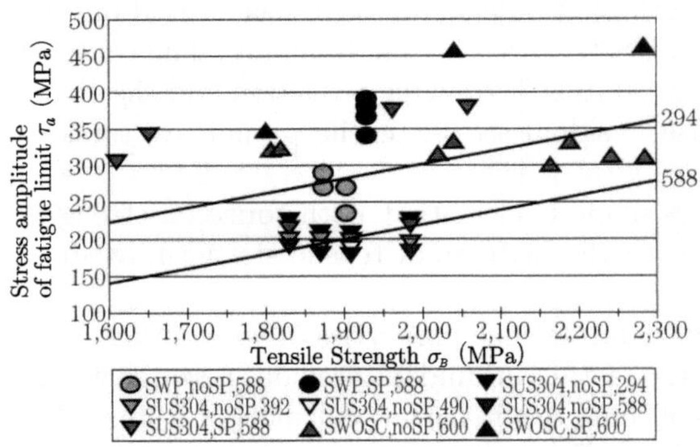

Fig. 2.165. Relation between tensile strength and spring fatigue limit of various steel spring wire (under 3 mm in diameter inclusive, stress value described right indicates mean stress) [13]

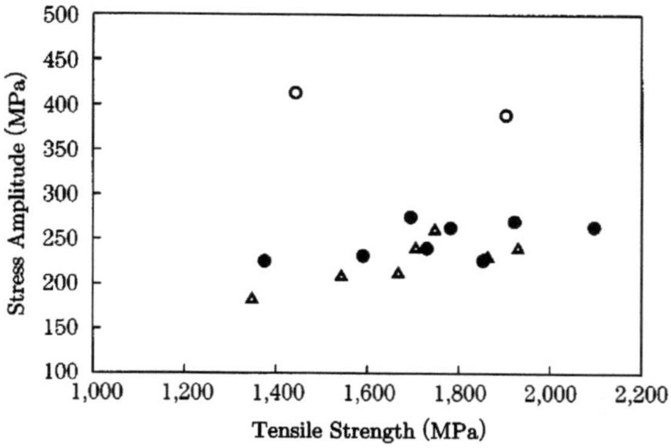

Fig. 2.166. Relation between tensile strength of stainless steel spring wire and fatigue strength of springs made of the same wire [14]

△ as coiling

● low temperature annealing

○ low temperature annealing and shot peening

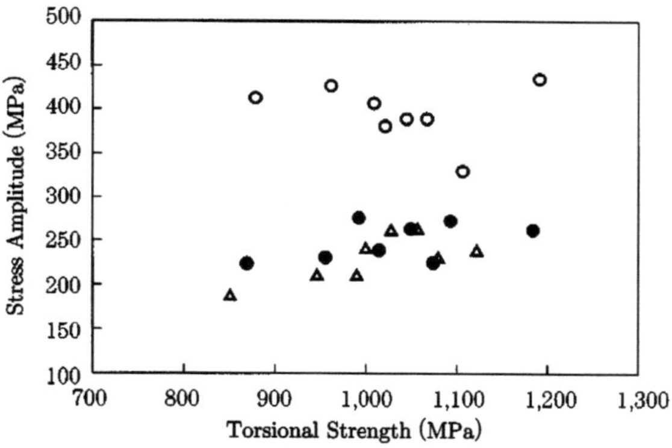

Fig. 2.167. Relation between torsional strength of stainless steel spring wire and fatigue strength of springs made of the same wire [14]

△ as coiling

● low temperature annealing

○ low temperature annealing and shot peening

steel spring wire, there are no clear correlation between tensile strength, and rotary bending fatigue strength and fatigue strength of springs. The research committee of fatigue strength of stainless steel spring wire has taken these issues as one of the most important study issues. Part of its co-research works are shown in Fig. 2.166 and Fig. 2.167 [14]. It can be seen from the Fig. 2.166 and Fig. 2.167 that although the fatigue strength of spring without shot peening improves with the increase of tensile strength and torsional strength, the effect of shot peening seems to be not proportional with the strength. However, the fatigue strength of springs has tendency to improve about 20 MPa by

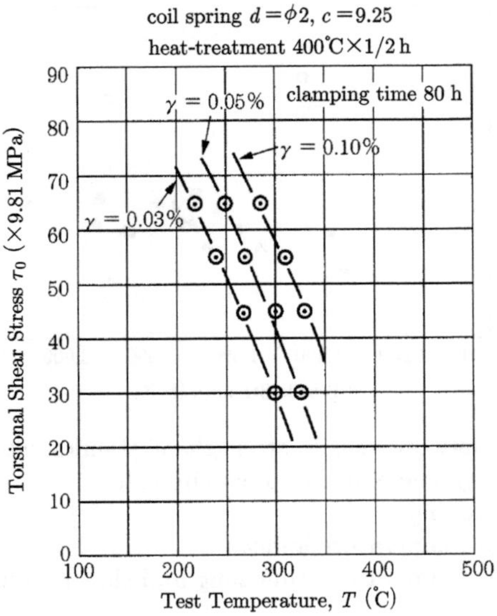

Fig. 2.168. Torsional shear stress-temperature diagram of SUS304-WPB [15]

low temperature annealing (400°C, 30 min) and 150 MPa by shot peening. It was also found that while the fatigue strength of steel spring wire is strongly influenced by the tensile strength, in case of stainless steel spring wire, the fatigue strength is not influenced by the strength so much, but can be strongly influenced by the surface roughness. However, the correlation between tensile strength and fatigue strength show positive when tested in the same surface roughness condition. The committee has been working to make clear the fatigue mechanism of stainless steel spring wire, such as the relation between the fatigue strength and hardness distribution, surface roughness, shot-peening, amount of martensite phase, Poission's raio, residual stress, and crack closing by strain-induced martensite.

JSMA standard No. 12 specifies the method to evaluate permanent set resistance at high temperature. Figure 2.168 and Fig. 2.169 [15] show torsional stress-temperature diagram of SUS304-WPB and SUS631 J1-WPC respectively, and Fig. 2.170 [15] shows a measured example of temperature dependence of tensile strength. In these figures, γ means the residual shear strain,which is defined as shear stress relaxation divided by modulus of rigidity, G. From these diagrams, the temperature which can satisfy the residual shear strain $\gamma \leqq 0.1\%$, is below 300°C for SUS304-WPB ($\tau_0 = 579\,\text{MPa}$, $\tau_0/\sigma_B = 0.39$), and 350°C for SUS631 J1-WPC ($\tau_0 = 667\,\text{MPa}$, $\tau_0/\sigma_B = 0.39$). Therefore, it can be said that stainless steel is superior to SWOSC-V in heat resistance when the service temperature is above 250°C. However, with respect to the permanent set resistance for longer time, the γ of any steel doesn't become constant value even after 3000 hours, and still tends to

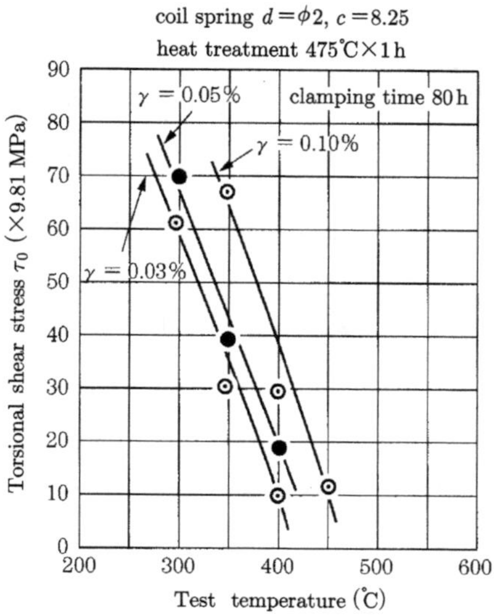

Fig. 2.169. Torsional shear stress-temperature diagram of SUS631-WPC [15]

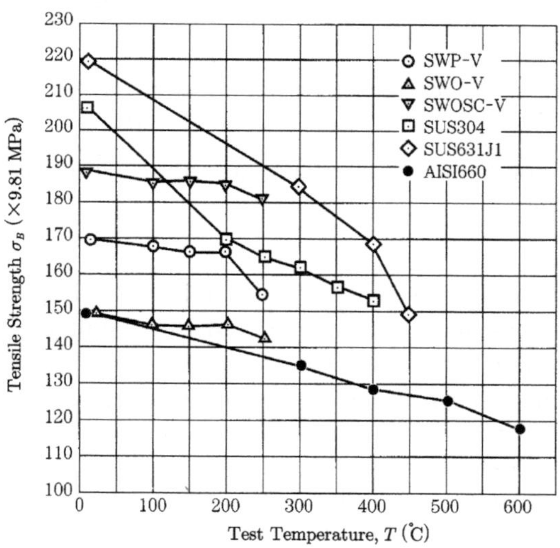

Fig. 2.170. Examples of temperature dependence of tensile strength [15]

increase with time. There is a proposal of constitutive equation [16] where γ is expressed by the variables of temperature, clamping stress, and time.

(vi) Low temperature annealing properties of stainless steel wire

Drawn wire normally undergoes low temperature annealing (stress relieving) or precipitation hardening heat treatment after being formed to the shape of springs. The aim of low temperature annealing is to reduce residual stress accumulated in drawing and coiling process, and to improve mechanical prop-

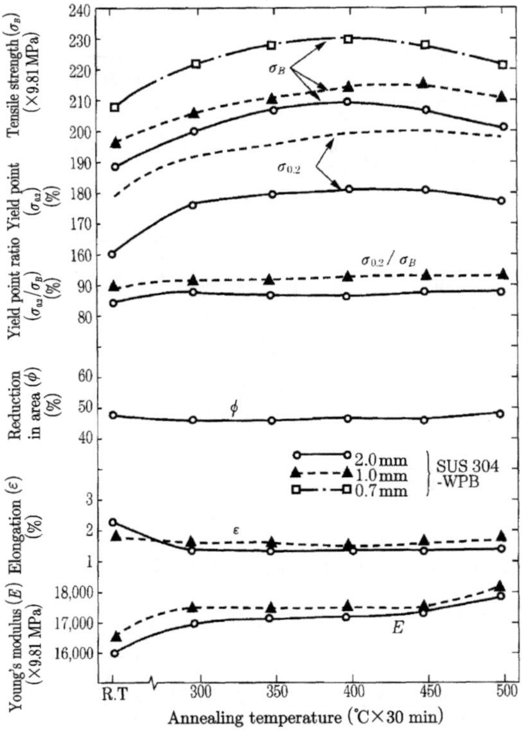

Fig. 2.171. Relation between annealing temperature and mechanical properties of SUS304-WPB [17]

erties by age hardening, especially in order to have the improvement of fatigue strength and permanent set resistance, and the stability of spring shape. In case of precipitation hardening heat treatment, the age hardening plays an important role in strength increase. A typical precipitation hardening condition for stainless steel wire is 400°C, 20–30 minutes for SUS304, and 470°C, 60 minutes for SUS631J1.

SUS304 tensile coil springs that requires initial tension or springs used in the static load can be sometimes annealed at 250–350°C for 15–20 minutes. Figure 2.171 [17] shows the relation between the annealing temperature and tensile properties of SUS304.

(vii) Newly developed stainless steel spring wire

Recently, smaller, lighter, and more precise electronic devices and parts have been in commercial use. Therefore, for the purpose of high strength and multi-function, new stainless steel spring wire have been developed. Table 2.31 summarizes typical examples of brand name and features of newly developed stainless steel spring wires.

Table 2.31. New stainless steel spring wire products of domestic wire makers

required property	item	wire production maker	available diameter range mm	characteristics of products
High tensile strength	Cryo S	Suzuki Metal Industry Co., Ltd.	4.5–1.6	high strength by means of very low temperature drawing high tensile strength regardless of diameter ($\geqq$ 1960 MPa)
	Herculee	Nippon Seisen Co., Ltd.	2.9–0.10	SUS301 stainless steel spring wire high strength equivalent to that of SWPA, B
	Super Dolce	SUMITOMO (SEI) Steel Wire Corp.	2.3–0.5	SUS304 based nitrogen added stainless steel spring wire high strength equivalent to that of SUS631J1-WPC
Corrosion resistance	Dolce	SUMITOMO (SEI) Steel Wire Corp.	4.0–0.50	corrosion resistant better than that of SUS316 tensile strength equivalent to that of SUS304-WPB
	316H1	Nippon Seisen Co., Ltd.	4.0–0.50	corrosion resistant better than that of SUS316 tensile strength equivalent to that of SUS304-WPB
Identification	"Deru" color	Shinko Wire Company, Ltd.	1.4–0.35	take on color by special plating after low temperature annealing treatment 2 series, gold and black type
Electrical conductivity	Elco S	Suzuki Metal Industry Co., Ltd.	0.35–0.20	SUS304/Cu plated steel wire electric conductivity of 15–35% IACS
Precise dimension	Dual S	Shinko Wire Company, Ltd.	1.6–0.35	ferritic/austenitic duplex phase alloy little change in dimension after low temperature annealing

(viii) Remarks in use of stainless steel wire

Generally, steel material for springs shows lower fatigue strength than expected when applied repeated stresses under corrosive gas or liquid environments. This is called corrosion fatigue. Stainless steel spring wire has been used as most common corrosion resistant material. Even though fatigue limit of stainless steel wire at 10^7 cycles is significantly decreased, the decrease rate and the decrement is much lower than that of oil-tempered wire, to show that stainless wire has advantage on fatigue strength in corrosive environments as shown in Fig. 2.172. The corrosion resistance of stainless steel can be mainly attributed to its surface passive layer, and it shows excellent resistance especially for oxidizing acid such as nitric acid. However, this passive layer can be destroyed in the environment which contains chloride ion, to generate micro cell with metal matrix and to have metal ions dissolved into environment. With the progress of this reaction, pitting corrosion takes place on the surface. As shown in Fig. 2.173, it should be noted that the cold working processes of stainless steel wire can cause the increase of martensite, and lattice defects such as dislocations, to give some impact to this pitting corrosion [19].

It should be also noted that stress corrosion cracking (SCC) can easily occur in austenitic stainless steel in the chloride ion environment under tensile stress. Examples of critical chlorine ion concentration-maximum temperature curves above which actual SCC failure in industrial water environment occurs in austenitic stainless steel, are shown in Fig. 2.174. According to the Fig. 2.174, it can be recommended that Mo added stainless steel such as SUS316 should be used in mild environment between room temperature and 100°C.

The shot peening process is considered to be effective to prevent SCC of stainless springs because it can reduce residual tensile stress. However, there is a report [21] saying that shot-peening makes SUS304 stainless steel more susceptable to SCC failure in salt water environment due to accelerated pitting corrosion, see Fig. 2.175. It is concluded that there is some risk of SCC accident in some special environment. From these points, it is important to remove chlorine ion as much as possible, to make periodical cleaning to prevent the condensation of chloride and to decrease the temperature in the environment.

It is well known that the spring diameter of piano wire or hard drawn steel wire decreases and that of stainless steel wire increases by stress relief annealing. This means that as the free height and spring rate of stainless steel wire can decrease by stress relief the coil diameter and hook angle should be formed smaller, estimating the change by stress relief in advance. Although the color of oxide film formed on stainless steel surface when tempered is significantly thinner than normal steel wire, it can be recommendable that the oxide film and Ni coating should be removed to have the surface passivated for corrosion resistance since annealing at too high temperature could spoil corrosion resistance. The condition of passivation of stainless steel spring for corrosion protection can be normally dipping in 20–40% HNO_3 at 50–70° for

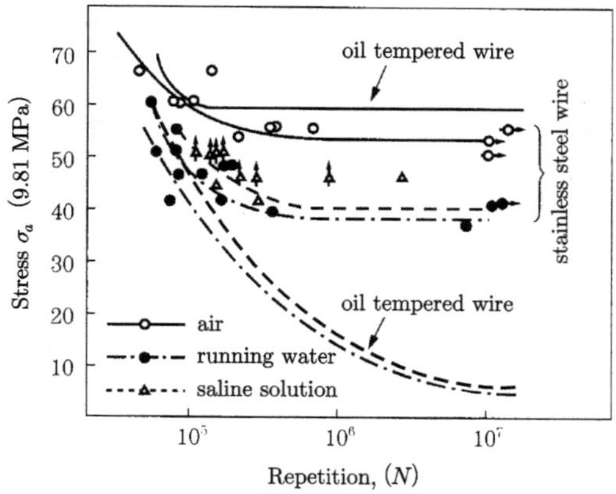

Fig. 2.172. Corrosion fatigue of stainless steel spring wire [18]

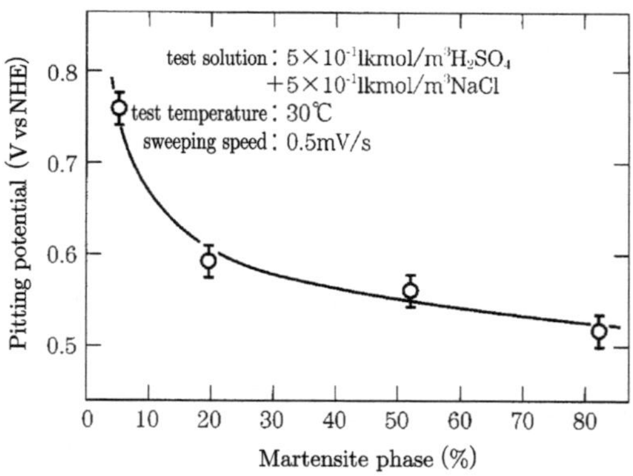

Fig. 2.173. Effect of deformation induced martensite on corrosion pit property of SUS304 [19]

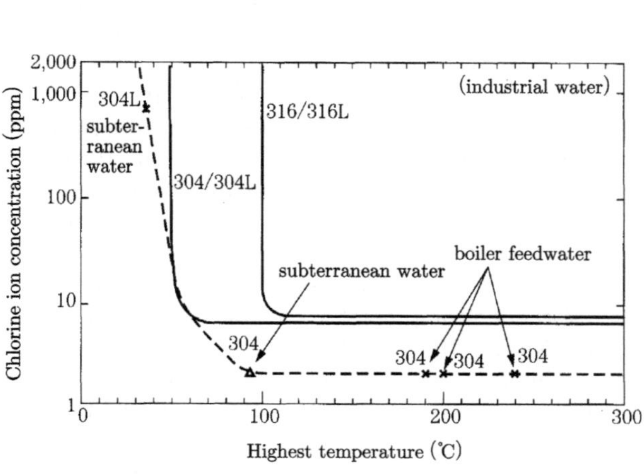

Fig. 2.174. Threshold SCC failure line of 304, 304L, 316, and 316L

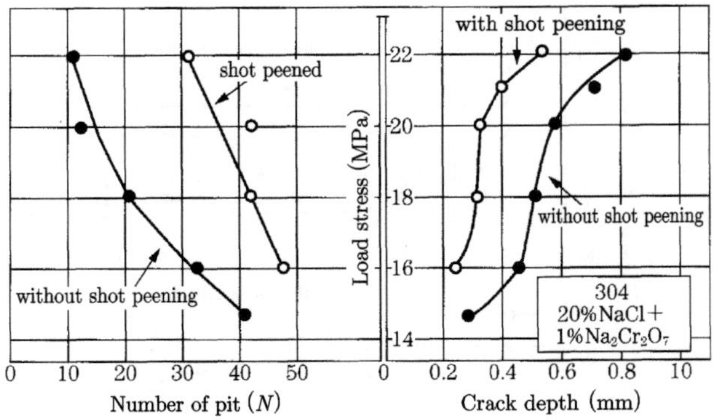

Fig. 2.175. Effect of shot peening on number of corrosion pits and crack depth of SCC testing sample under 20% NaCl + 1% $Na_2Cr_2O_7$ aqueous solution (test period: 50 h)

30–60 mins. or 4% HF + 4% CrO_3 at 60° for 30 mins, cleaning by water and drying.

References

1. Wire and Products Association Wire Product Text (4th ed., Japanese) (1997) p. 548
2. JIS G 4314: Stainless steel wire for springs
3. JIS G 4308: Stainless steel wire rod
4. Wire and Products Association Wire Product Text (4th edition, Japanese) (1997) p. 550
5. Sumitomo Electric Industry: Internal report (1996) p. 3
6. T. Suzuki: Stainless steel handbook (Japanese) (1973) p. 159
7. JSMA and JSSR: Heat treatment manual for small spring (Japanese) (1982) p. 68
8. JSSR spring technical report (Japanese) (1995) p. 10
9. R. Komura: Spring (BANE 3rd edition, Japanese) Maruzen (1982) p. 65
10. JSMA and JSSR: New Spring Technology (Japanese) (1988) p. 162
11. M. Nishihata: 4ND management and technology committee (1994) p. 4
12. JSSR: Fatigue strength of stainless steel wire research committee (1997) p. 2
13. JSSR: Database committee Autumn technical report (1993) p. 45
14. JSSR: Fatigue strength of stainless steel wire research committee (1998) p. 1
15. JSSR: Heat resistance spring material committee data file (1986)
16. JSMA standard No. 12 (1987) p. 10
17. Suzuki Metal Industry: Company catalog (1990) p. 18
18. A. Ohno, H.Koyama, A. Tange: Spring (BANE 3rd edition) Maruzen (1982) p. 364
19. T. Matsuda: Stainless steel handbook (3rd edition, Japanese) (1995) p. 449
20. T. Adachi: Stainless steel handbook (3rd edition, Japanese) (1995) p. 583
21. M. Fujiwara: Stainless steel handbook (3rd edition, Japanese) (1995) p. 459

(4) Maraging steel wire

Maraging steel comprises a special class of high-strength steel which was developed by the International Nickel Co., Ltd. as special material for aerospace applications. The features of maraging steel are derived from low-carbon martensite which involves the carbon content less than 0.03%. The Martensite of spring steel are usually hard and brittle. However, the low-carbon martensite formed in the maraging steel is relatively soft and ductile, about 30 to 35 HRC. The machining and workability of maraging steel are, therefore, good enough to meet with most processings. Maraging steel can be strengthened by the precipitation of intermetallic compounds at temperature of about 480°C. The typical chemical compositions are that the sum of Ni, Co, and Mo is about 30%, adding Al and Ti for precipitation-hardening. The word "maraging" comes from martensite + aging.

Commercial maraging steel is classified into 18% Ni, 20% Ni, 25% Ni type according to its Ni content. 18% Ni type maraging steel is the most popular in commercial applications. Although the strength of 18% Ni maraging steel depends upon Mo, Co and Ti content, the grade of 2050 MPa 0.2% proof stress has been widely used for spring applications. The types and chemical compositions are shown in Table 2.32 [1, 2]. In the case of machine structural uses, solution treated material are worked and age-hardened subsequently to get high-strength. In the case of spring, solution treated wire rod is cold drawn, cold formed to spring shape, and age-hardened to high-strength. According to its high strength, fatigue strength and permanent set resistance are also high. Since the transition temperature of maraging steel is low, it has better resistance to the low-temperature embrittlment. The resistance to the delayed fracture and the stress-corrosion cracking is also relatively high. Figures 2.176–2.178 [3,4] show the relation between mechanical properties (fatigue strengths) and cold drawing reduction, and aging. Figure 2.180 [5] shows the comparison of the delayed fracture properties between maraging steel and low alloy steel. Although the heat-treatment condition of maraging steel can be selected by

Table 2.32. Types and Chemical Compositions of Maraging Steel

Type	0.2% Proof Stress	Chemical Compositions (wt%)							
		C	Si	Mn	Ni	Co	Mo	Al	Ti
18% Ni	1,350 MPa Grade	≦ 0.03	≦ 0.12	≦ 0.12	17.0–19.0	8.0–9.0	3.0–3.5	0.05–0.15	0.15–0.25
	1,700 MPa Grade	≦ 0.03	≦ 0.12	≦ 0.12	17.0–19.0	7.0–8.5	4.6–5.1	0.05–0.15	0.3–0.5
	2,050 MPa Grade	≦ 0.03	≦ 0.12	≦ 0.12	18.0–19.0	8.0–9.5	4.6–5.2	0.05–0.15	0.5–0.8
	2,350 MPa Grade	≦ 0.03	≦ 0.12	≦ 0.12	17.0–19.0	12.0–13.0	3.5–4.5	0.05–0.15	1.2–1.8
20% Ni	1,700 MPa Grade	≦ 0.03	≦ 0.12	≦ 0.12	18.0–20.0	–	–	0.05–0.15	1.3–1.6
25% Ni	1,700 MPa Grade	≦ 0.03	≦ 0.12	≦ 0.12	25.0–26.0	–	–	0.05–0.35	1.3–1.6

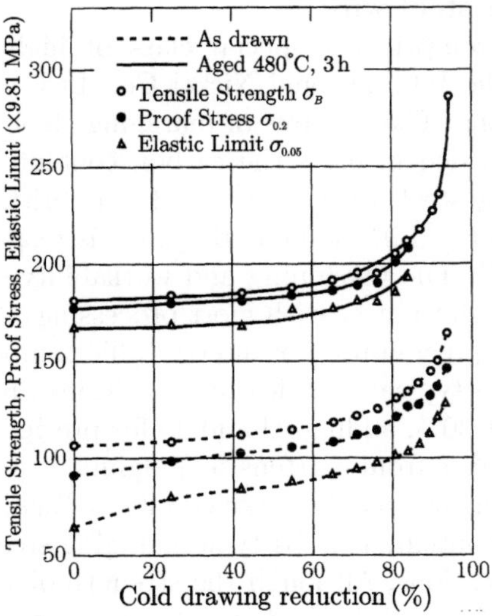

Fig. 2.176. Relation between Tensile Properties and Reduction in cold drawing of 18% Ni Maraging Steel Wire

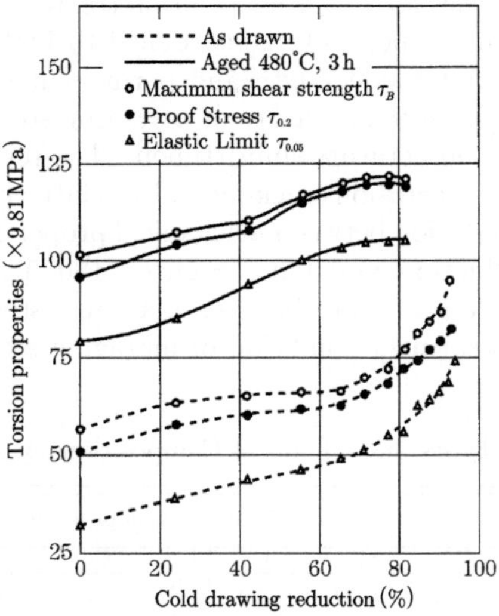

Fig. 2.177. Relation between torsional properties and reduction in cold drawing of 18% Ni maraging steel wire

the grade, size and applications, the typical heating condition can be shown in Fig. 2.179 [6]. Aging is normally done at 480–520°C for 3 to 5 hours.

Corrosion behaviors of maraging steels should be given attentions according to its environmental condition. Maraging steels do not involve Cr element

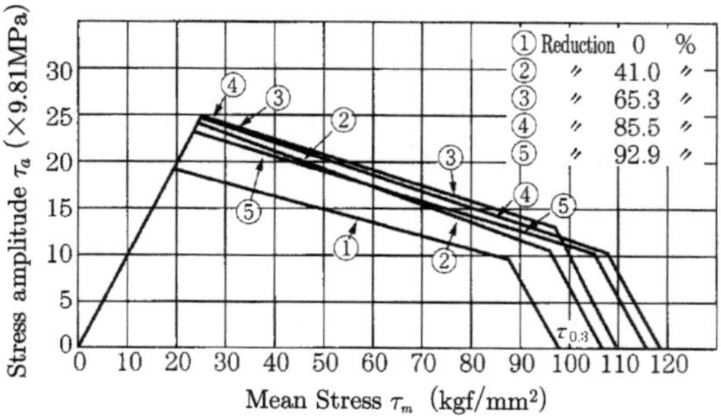

Fig. 2.178. Relation between fatigue property and reduction of 18% Ni maraging steel wire

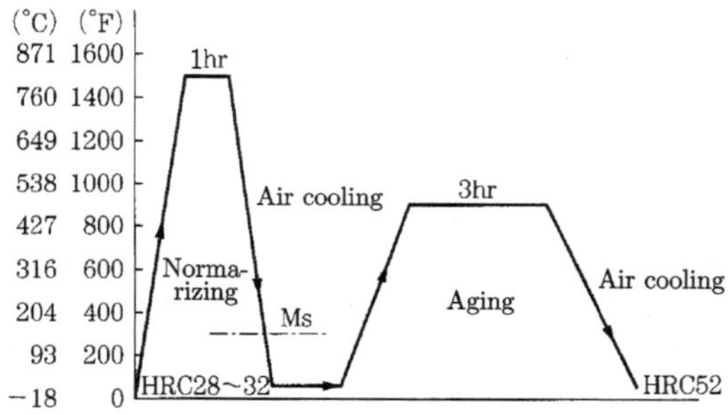

Fig. 2.179. Heat treatment of 18% Ni maraging steel

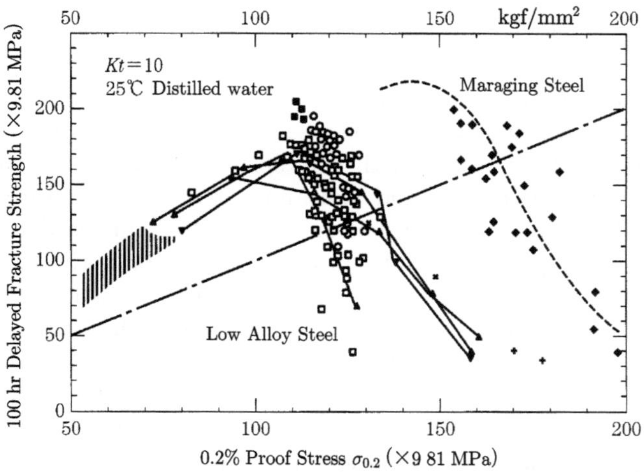

Fig. 2.180. Delayed Fracture Properties of Maraging Steel

which is necessary to produce passivate film at surface layer. In industrial and marine atmospheres, maraging and conventional steels, for example piano wire, have essentially the same corrosion rates. Because maraging steel has extremely high strength, small pitting at surface can cause the early fracture of springs.

References

1. S. Sato, H. Ono: Kinzoku (Japanese) 44, 8 (1974), p. 40
2. Technical report No. 231, Hitachi Metals Ltd. 2
3. M. Osawa, et al.: Journal of Japan Metals Ins. 32 (1) p. 95
4. R. Komura: Spring (Japanese) Maruzen (1982) p. 66
5. Yamamoto, T. Fujita: Climax molybdenum development company (Japan) (1971) p. 283
6. S. Owaku: Bulletin of Japan Inst. of Metals 3, 12 (1964) p. 722

(5) Steel strip for springs

(i) Types of steel strip for springs

Flat spring, snap ring, spring pin, spring washer and other small springs are made of steel strip for springs. As for steel wire for springs, cold-drawing process is used to control the sizes and the cross-sectional shapes of wire. In case of steel strip, repeated cold-rolling is applied to the hot-rolled strip to control a required cross-sectional shape. Cold-rolled strip has usually good size tolerance and good surface quality, compared with hot-rolled strip. The steel material used for springs are carbon steel, alloy steel, and stainless steel. The standards for steel strip for springs are, JIS G 3311 for cold-rolled special steel strip, JIS G 4802 for cold-rolled steel strip for springs, and JIS G 4313 for cold-rolled stainless steel strip for springs.

① Classification by manufacturing process

There are two manufacturing processes for cold-formed springs using steel strip for springs. As the same with wire spring, one is using relatively hard material, forming to springs, and annealing at low temperature. The other is using relatively soft material, forming to springs, and hardening by heat-treatment such as quenching and tempering or aus-tempering. In the former process, although the number of manufacturing process is less, it is not suitable to form a complicate shaped spring due to high hardness material. To the contrary, although the latter process is suitable for forming a complicate shaped spring which should be heat-treated after forming. Therefore, large and continuous facilities for heat-treatment can be required.

② Classification by metallurgical structure

The types of microstructure in steel strip are represented by the following three types.

(i) Annealed steel strip

Annealed steel strip, which consists of carbon steel, alloy steel, martensitic stainless steel or precipitation hardening stainless steel, is cold rolled strip annealed at its final process. Carbon steel, alloy steel and martensitic stainless steel have spheroidizing microstructure by annealing which is soft and ductile, and can be applied to complicate shaped springs. After spring forming, quenching and tempering or aus-tempering can be applied. In case of precipitation hardening stainless steel, precipitation hardening treatment can be applied after spring forming.

(ii) Cold-rolled steel strip

Steel strip as cold-rolled in shipping condition is named cold-rolled steel strip. Carbon steel and low-alloy steel strips are cold-rolled with small reduction after annealing. This skin-rolled strip has slightly higher hardness than annealed steel strip and has sufficient formability to spring shape. Because of higher hardness than annealed steel strip, the burr at blanking cannot be appeared. After spring forming, quenching and tempering or aus-tempering are carried out to have enough strength for springs.

In case of austenitic stainless steel and precipitation hardening stainless steel, heat-treatment such as quenching and tempering or aus-tempering can not be applied. The strength of springs can be applied by work-hardening in cold rolling. Therefore, higher hardness strip rolled by large reduction shows poor formability to form complicate shaped springs. After forming relatively simple shaped spring, low temperature annealing for austenitic stainless steel and aging for precipitation hardening stainless steel can be done to have strength for springs.

(iii) Heat-treated steel strip

As mentioned previously, annealed strip and cold-rolled strip are heat-treated to have enough strength after forming to spring shape. Because of distortions during the heat-treatment process, the process of shape correction can be sometimes required. To avoid these problems, steel strip which is quenched and tempered continuously to cold-rolled steel strip or steel strip which is aus-tempered to have bainitic microstructure, have been in production. The former is called as hardened steel strip and the later is bainitic steel strip. Hardened steel strip consists of carbon steel, low-alloy steel and martensitic stainless steel strip. These strips can be formed to simple-shaped springs at room temperature. However, it is not suitable to be applied to complicate-shaped springs. Having enough strength for springs, the only low temperature annealing is applied after forming. Bainitic steel strip has higher toughness than hardened steel strip, to show better formability to spring-shape. After forming, low-temperature annealing can be applied as the same with hardened steel strip.

③ Classifications by JIS standards

As mentioned previously, steel strip for springs is classified into three types, cold-rolled special steel strip, cold-rolled steel strip for springs and cold-rolled stainless steel strip for springs from the point of JIS standard.

(i) Cold-rolled special steel strip and cold-rolled steel strip

Cold-rolled special steel strip was designated as JIS, Japan Industrial Standard in1952 and has been revised several times and has been widely used. This material standard prescribes the cold-rolled steel strip for special uses with thirty-seven steel grades. Their chemical compositions and typical applications are listed in Table 2.33 [1] and the hardness of strip is shown in Table 2.34 [1]. The applications of cold-rolled special steel strip are not restricted to spring uses. To the contrary, the fact is that the spring applications have been chosen from this standard. The strength of this strip can be given by cold rolling or annealing after cold rolling.

For the steel strip of spring applications, higher strengthened strip such as quenching and tempering or aus-tempering should be required. Considering the application for springs, it has been in demand to have the unique standard for springs, which can make it easier to select a suitable steel grade and dimensions, to assure its quality, and to produce quality springs. Under these circumstances, JSMA, Japan Spring Manufacturers Association has designated the standard of "Steel strips for spring" as JSMA No. 6 in 1964. Based on JSMA No. 6, JIS G 4802 "Steel strips for spring" has also been established in 1973. In this JIS standard, 8 steel grades and 4 types of strength are selected as shown in Table 2.35 [2]. Typical hardness level can be also listed in the Table 2.35. Users can select an appropriate material, referring to those table and specifications. Four types of cold-rolled steel strip for springs consist of A: Annealed, R: Cold-rolled, H: Quenched-and-tempered, and B: Aus-tempered. The Type B is also called as bainitic steel strip since its microstructure is bainite. The value of hardness in Table 2.35 is more severe than that of cold-rolled special steel strip. Although the hardness of Type A is nearly equal to that of cold-rolled special steel strip, the harness of Type R is designated as 230–270 HV, where the occurrence of burr can be minimized.

In the case of Type H and Type B, low temperature annealing is applied after forming springs. Therefore, the hardness of strip can be brought to the hardness of springs. Except Type A, spring makers can select the preferable hardness of strip from wide ranges (types R, H and B) of specifications under the tolerance of ±25 HV.

For cold-rolled steel strip for springs, permissible tolerances in hardness are designated as ±30 HV. Comparing with cold-rolled special steel strip, the specifications of cold-rolled steel strip for springs are different in the shape, size, and tolerances. Dimensional items such as thickness and its tolerances, width and its tolerances, camber and flatness are designated.

Table 2.36 [1,2], shows the thickness and its tolerances of cold-rolled steel strip for springs and cold-rolled special steel strip. Standard thickness in cold-rolled steel strip for springs can be related to the thickness of its products such as, spring washer, spring pin and other springs specified in JIS. The thickness tolerances of cold-rolled steel strip for springs are equal to or smaller than special steel strip and those of ET grade is especially severe.

The comparison of width tolerances is shown in Table 2.37. Several shapes of edge in steel strip are specified. Mill-edge is a edge-shape carried through the cold rolling. Slit-edge is an approximately square edge produced by slitting.

Table 2.33. Chemical compositions and typical applications of special steel strips

mass%

steel	specification	C	Si	Mn	P	S	Cu	Ni	Cr	Ni + Cr	Mo	W	V	typical appl
carbon streel	S 30 CM	0.27–0.33	0.15–0.35	0.60–0.90	≦ 0.030	≦ 0.035	≦ 0.30	≦ 0.20	≦ 0.20	≦ 0.35	–	–	–	retainers
	S 35 CM	0.32–0.38	0.15–0.35	0.60–0.90	≦ 0.030	≦ 0.035	≦ 0.30	≦ 0.20	≦ 0.20	≦ 0.35	–	–	–	office supply friction plate
	S 45 CM	0.42–0.48	0.15–0.35	0.60–0.90	≦ 0.030	≦ 0.035	≦ 0.30	≦ 0.20	≦ 0.20	≦ 0.35	–	–	–	clutches, chains, washers
	S 50 CM	0.47–0.53	0.15–0.35	0.60–0.90	≦ 0.030	≦ 0.035	≦ 0.30	≦ 0.20	≦ 0.20	≦ 0.35	–	–	–	machine parts washers
	S 55 CM	0.52–0.58	0.15–0.35	0.60–0.90	≦ 0.030	≦ 0.035	≦ 0.30	≦ 0.20	≦ 0.20	≦ 0.35	–	–	–	springs
	S 60 CM	0.55–0.65	0.15–0.35	0.60–0.90	≦ 0.030	≦ 0.035	≦ 0.30	≦ 0.20	≦ 0.20	–	–	–	–	chains, saws
	S 65 CM	0.60–0.70	0.15–0.35	0.60–0.90	≦ 0.030	≦ 0.035	≦ 0.30	≦ 0.20	≦ 0.20	–	–	–	–	cluches, springs
	S 70 CM	0.65–0.75	015–0.35	0.60–0.90	≦ 0.030	≦ 0.035	≦ 0.30	≦ 0.20	≦ 0.20	–	–	–	–	saws
	S 75 CM	0.70–0.80	0.15–0.35	0.60–0.90	≦ 0.030	≦ 0.035	≦ 0.30	≦ 0.20	≦ 0.20	–	–	–	–	washer, springs
carbon tool steel	SK 120 M	1.10–1.30	≦ 0.35	≦ 0.50	≦ 0.030	≦ 0.030	≦ 0.25	≦ 0.25	≦ 0.30	–	–	–	–	blades, springs
	SK 105 M	1.00–1.10	≦ 0.35	≦ 0.50	≦ 0.030	≦ 0.030	≦ 0.25	≦ 0.25	≦ 0.30	–	–	–	–	blades, springs
	SK 95 M	0.90–1.00	≦ 0.35	≦ 0.50	≦ 0.030	≦ 0.030	≦ 0.25	≦ 0.25	≦ 0.30	–	–	–	–	gages, springs
	SK 85 M	0.80–0.90	≦ 0.35	≦ 0.50	≦ 0.030	≦ 0.030	≦ 0.25	≦ 0.25	≦ 0.30	–	–	–	–	knives, saws
	SK 75 M	0.70–0.80	≦ 0.35	≦ 0.50	≦ 0.030	≦ 0.030	≦ 0.25	≦ 0.25	≦ 0.30	–	–	–	–	springs, knifes
	SK 65 M	0.60–0.70	≦ 0.35	≦ 0.50	≦ 0.030	≦ 0.030	≦ 0.25	≦ 0.25	≦ 0.30	–	–	–	–	saws, springs

mass%

steel	specification	C	Si	Mn	P	S	Cu	Ni	Cr	Ni + Cr	Mo	W	V	typical appl
alloy tool steel	SKS 2 M	1.00–1.10	≦ 0.35	≦ 0.80	≦ 0.030	≦ 0.030	≦ 0.25	≦ 0.25	0.50–1.00	–	–	1.00–1.50	–	knives, metal-band saws
	SKS 5 M	0.75–0.85	≦ 0.35	≦ 0.50	≦ 0.030	≦ 0.030	≦ 0.25	0.70–1.30	0.20–0.50	–	–	–	–	knives, saws
	SKS 51 M	0.75–0.85	≦ 0.35	≦ 0.50	≦ 0.030	≦ 0.030	≦ 0.25	1.30–2.00	0.20–0.50	–	–	–	–	knives, saws
	SKS 7 M	1.10–1.20	≦ 0.35	≦ 0.50	≦ 0.030	≦ 0.030	≦ 0.25	≦ 0.25	0.20–0.50	–	–	2.00–2.50	–	knives, saws
	SKS 81 M	1.10–1.30	≦ 0.35	≦ 0.50	≦ 0.030	≦ 0.030	≦ 0.25	≦ 0.25	0.20–0.50	–	–	–	–	saws, blades, shutters
	SKS 95 M	0.80–0.90	≦ 0.50	0.80–1.10	≦ 0.030	≦ 0.030	≦ 0.30	≦ 0.25	0.90–1.20	–	–	–	–	chains, springs
chromium steel	SCr 420 M	0.18–0.23	0.15–0.35	0.60–0.85	≦ 0.030	≦ 0.030	≦ 0.30	≦ 0.25	0.90–1.20	–	–	–	–	chains
	SCr 435 M	0.33–0.38	0.15–0.35	0.60–0.85	≦ 0.030	≦ 0.030	≦ 0.30	≦ 0.25	0.90–1.20	–	–	–	–	chains
	SCr 440 M	0.38–0.43	0.15–0.35	0.60–0.85	≦ 0.030	≦ 0.030	≦ 0.30	≦ 0.25	0.90–1.20	–	–	–	–	chains
nickel chromium steel	SNC 415 M	0.12–0.18	0.15–0.35	0.35–0.65	≦ 0.030	≦ 0.030	≦ 0.30	2.00–2.50	0.20–0.50	–	–	–	–	Business machine parts
	SNC 631 M	0.27–0.35	0.15–0.35	0.35–0.65	≦ 0.030	≦ 0.030	≦ 0.30	2.50–3.00	0.60–1.00	–	–	–	–	Business machine parts
	SNC 836 M	0.32–0.40	0.15–0.35	0.35–0.65	≦ 0.030	≦ 0.030	≦ 0.30	3.00–3.50	0.60–1.00	–	–	–	–	Business machine parts
Ni–Cr–Mo steel	SNCM 220 M	0.17–0.23	0.15–0.35	0.60–0.90	≦ 0.030	≦ 0.030	≦ 0.30	0.40–0.70	0.40–0.65	–	–	–	–	chains
	SNCM 415 M	0.12–0.18	0.15–0.35	0.40–0.70	≦ 0.030	≦ 0.030	≦ 0.30	1.60–2.00	0.40–0.65	–	0.15–0.30	–	–	chains
Cr–Mo steel	SCM 415 M	0.13–0.18	0.15–0.35	0.60–0.85	≦ 0.030	≦ 0.030	≦ 0.30	≦ 0.25	0.90–1.20	–	0.15–0.30	–	–	blades
	SCM 430 M	0.28–0.33	0.15–0.35	0.60–0.80	≦ 0.030	≦ 0.030	≦ 0.30	≦ 0.25	0.90–1.20	–	0.15–0.30	–	–	blades
	SCM 435 M	0.33–0.38	0.15–0.35	0.60–0.80	≦ 0.030	≦ 0.030	≦ 0.30	≦ 0.25	0.90–1.20	–	0.15–0.30	–	–	blades
	SCM 440 M	0.38–0.43	0.15–0.35	0.60–0.80	≦ 0.030	≦ 0.030	≦ 0.30	≦ 0.25	0.90–1.20	–	0.15–0.30	–	–	blades
spring steel	SUP 6 M	0.56–0.64	1.50–1.80	0.70–1.00	≦ 0.035	≦ 0.035	≦ 0.30	–	–	–	–	–	–	springs
	SUP 9 M	0.52–0.60	0.15–0.35	0.65–0.95	≦ 0.035	≦ 0.035	≦ 0.30	–	0.65–0.95	–	–	–	–	springs
	SUP 10 M	0.47–0.55	0.15–0.35	0.65–0.95	≦ 0.035	≦ 0.035	≦ 0.30	–	0.80–1.10	–	–	–	0.15–0.25	springs
Mn steel	SMn 438 M	0.35–0.41	0.15–0.35	1.35–1.65	≦ 0.030	≦ 0.030	≦ 0.30	≦ 0.25	≦ 0.35	–	–	–	–	chains
	SMn 443 M	0.40–0.46	0.15–0.35	1.35–1.65	≦ 0.030	≦ 0.030	≦ 0.30	≦ 0.25	≦ 0.35	–	–	–	–	chains

Table 2.34. Hardness of special steel strips HV (JIS G 3311)

Steel	Grade	Annealed	cold-rolled
carbon steel	S 30 CM	≦ 160	160–230
	S 35 CM	≦ 170	170–250
	S 45 CM	≦ 170	170–250
	S 50 CM	≦ 180	180–270
	S 55 CM	≦ 180	180–270
	S 60 CM	≦ 190	190–280
	S 65 CM	≦ 190	190–280
	S 70 CM	≦ 190	190–280
	S 75 CM	≦ 200	200–290
carbon tool steel	SK 120 M	≦ 220	220–310
	SK 105 M	≦ 220	220–310
	SK 95 M	≦ 210	210–300
	SK 85 M	≦ 200	200–290
	SK 75 M	≦ 190	190–280
	SK 65 M	≦ 190	190–280
alloy tool steel	SKS 2 M	≦ 230	230–320
	SKS 5 M	≦ 200	200–290
	SKS 51 M	≦ 200	200–290
	SKS 81 M	≦ 220	220–310
	SKS 7 M	≦ 250	250–340
	SKS 95 M	≦ 200	200–290
Cr-steel	SCr 420 M	≦ 180	180–270
	SCr 435 M	≦ 190	190–270
	SCr 440 M	≦ 200	200–290
Ni–Cr steel	SNC 415 M	≦ 170	170–240
	SNC 631 M	≦ 180	180–240
	SNC 836 M	≦ 190	190–250
Ni–Cr–Mo steel	SNCM 220 M	≦ 180	180–240
	SNCM 415 M	≦ 170	170–240
Cr–Mo steel	SCM 145 M	≦ 170	170–240
	SCM 430 M	≦ 180	180–250
	SCM 435 M	≦ 190	190–270
	SCM 440 M	≦ 200	200–280
spring steel	SUP 6 M	≦ 210	210–310
	SUP 9 M	≦ 200	200–290
	SUP 10 M	≦ 200	200–290
Mn steel	SMn 438 M	≦ 200	200–290
	SMn 443 M	≦ 200	200–290

Round-edge is an prepared edge of a specified contour which is done by rolling, grinding, or cutting. The cross-sectional shapes of edge are shown in Fig. 2.181.

A curvature along the longitudinal direction in both sides is defined as camber. As flat springs are produced continuously using automated pressing machine, the camber of strip can be a serious problem in manufacturing the

Table 2.35. Type and Hardness of Cold-rolled steel strip for springs

HV

Type	Annealing	Cold-Rolling	Quenching-Tempering	Austempering
Grade	A	R	H	B
S50C-CSP	≦ 180	230–270	–	360–440
S55C-CSP	≦ 180	230–270	350–450	360–440
S60C-CSP	≦ 190	230–270	350–500	360–440
S65C-CSP	≦ 190	230–270	–	–
S70C-CSP	≦ 190	230–270	350–550	–
SK5-CSP	≦ 190	230–270	350–600	–
SK4-CSP	≦ 190	230–270	400–600	–
SUP10-CSP	≦ 190	230–270	–	–

* The hardness variation per one slit should be within 30 HV.
The center hardness of R, H and B type should be decided within the range, and its tolerance should be ±25 HV for H and B, and ±20 HV for R.

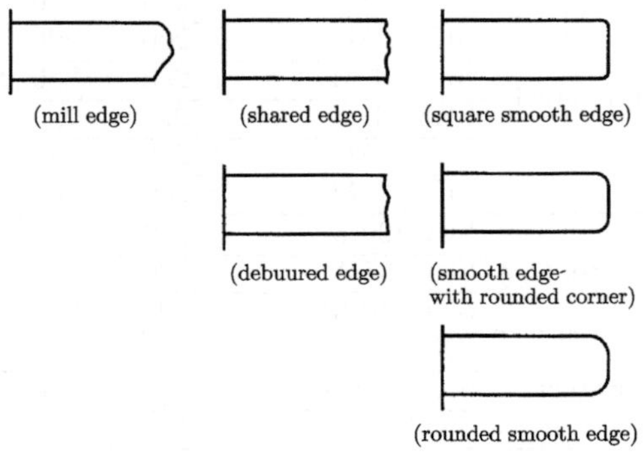

Fig. 2.181. Edge shapes of Steel strip

springs. Table 2.38 [1,2] shows the measuring method of camber and specifications of camber in cold-rolled steel strip for springs and cold-rolled special steel strips. The maximum value of camber in Type H and Type B are smaller than that of Type A and Type R because heat-treated steel strip is straightened by tensile force during heat-treatment.

(ii) Cold-rolled stainless steel strip for springs

Although the feature of stainless steel strip for springs can be said to be its corrosion-resistance, flat spring material used by spring makers and communication equipments makers are shown in Fig. 2.182 [3]. While spring makers use mainly carbon steel strip, stainless steel and non-ferrous metal are used largely by communication equipments makers. Stainless steel strips are widely used for these applications due to its superior corrosion-resistance and reliability, despite of its higher price.

Table 2.36. Thickness tolerance for cold rolled steel strip and cold rolled special steel

unit: mm

Specified Thickness, t	Standard Thickness	Cold – rolled steel strip for spring ① Thickness tolerance			Cold – rolled special steel strip ② Thickness tolerance A		Thickness tolerance B	
		width, w $200 < w$	width, w $200 \leqq w < 500$	width, w $80 < w$ (ET)	width, w $200 < w$	width, w $200 \leqq w < 500$	width, w $200 < w$	width, w $200 \leqq w < 500$
$t < 0.10$		±0.008	–	±0.006	±0.008	–	±0.012	–
$0.10 \leqq t < 0.15$	0.10 0.12	±0.010	–	±0.008	±0.010	–	±0.015	–
$0.15 \leqq t < 0.25$	0.15 0.20	±0.015	±0.020	±0.010	±0.015	±0.020	±0.020	±0.025
$0.25 \leqq t < 0.40$	0.25 0.28 0.30 0.35	±0.020	±0.025	±0.015	±0.020	±0.025	±0.025	±0.035
$0.40 \leqq t < 0.60$	0.40 0.50 0.45 0.55	±0.025	±0.030	±0.020	±0.025	±0.030	±0.035	±0.040
$0.60 \leqq t < 0.90$	0.60 0.80 0.70	±0.030	±0.040	±0.025	±0.030	±0.040	±0.045	±0.055
$0.90 \leqq t < 1.20$	0.90 1.10 1.00	±0.040	±0.050	±0.035	±0.040	±0.050	±0.055	±0.070
$1.20 \leqq t < 1.60$	1.20 1.40	±0.050	±0.060	–	±0.050	±0.060	±0.070	±0.080
$1.60 \leqq t < 2.10$	1.60 2.00 1.80	±0.055	±0.070	–	±0.055	±0.070	±0.075	±0.090
$2.10 \leqq t < 3.00$	2.20 2.80 2.50	±0.065	±0.080	–	±0.065	±0.080	±0.080	±0.090
$3.00 \leqq t < 4.00$	3.00 4.00 3.50	±0.080	±0.090	–	±0.080	±0.090	±0.090	±0.100

Table 2.37. Width tolerance for cold-rolled steel strip and cold-rolled special steel strip

unit: mm

Thickness \ width	Cold-rolled steel strips for spring*			Cold-rolled special steel strip**	
	Width torelance				
	$w < 80$	$80 \leqq w < 200$	$80 \leqq w < 200$	$w < 200$	$200 \leqq w < 500$
$0.25 \leqq t < 0.60$	±0.10	±0.15	±0.25	±0.15	±0.25
$0.60 \leqq t < 1.20$	±0.15	±0.20	±0.30	±0.20	±0.30
$1.20 \leqq t < 4.00$	±0.20	±0.25	±0.40	±0.25	±0.40

* For the strip of thickness less than 0.25 mm, and Type H and B, the tolerance should be decided under the agreement between manufacturers and users.

** For the thickness less than 0.25 and over 4.00 inclusive, and width over 500 inclusive, the tolerance should be decided under the agreement between manufacturers and users.

Table 2.38. Camber Tolerance of Cold-rolled steel strip

(mm)

type \ width	Camber Tolerance for each 1 m				
	$w < 10$	$10 \leqq w < 20$	$20 \leqq w < 40$	$40 \leqq w < 80$	$80 \leqq w$
A and R	—	8	6	3	1
B	—	5	5	2	1
H	2	2	2	2	1

Measeuring method for camber

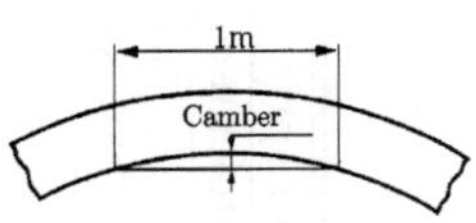

* The maximum camber of cold-rolled special steel strip is 2 mm per any 2 m. The tolerance of the width below 80 mm should be decided between manufactures and users.

Table 2.39. Type and Chemical Compositions of Cold-rolled stainless steel strip for springs

mass %

Type	C	Si	Mn	P	S	Ni	Cr	Others
SUS301-CSP	$\leqq 0.15$	$\leqq 1.00$	$\leqq 2.00$	$\leqq 0.045$	$\leqq 0.030$	6.00–8.00	16.00–18.00	
SUS304-CSP	$\leqq 0.08$	$\leqq 1.00$	$\leqq 2.00$	$\leqq 0.045$	$\leqq 0.030$	8.00–10.50	18.00–20.00	
SUS420J2-CSP	0.26–0.40	$\leqq 1.00$	$\leqq 1.00$	$\leqq 0.040$	$\leqq 0.030$		12.00–14.00	
SUS631-CSP	$\leqq 0.09$	$\leqq 1.00$	$\leqq 1.00$	$\leqq 0.040$	$\leqq 0.030$	6.50–7.75	16.00–18.00	Al 0.75–1.50
SUS632J1-CSP	$\leqq 0.09$	1.00–2.00	$\leqq 1.00$	$\leqq 0.040$	$\leqq 0.030$	6.50–7.75	13.50–15.50	Ti 0.20–0.65 Cu 0.40–1.00

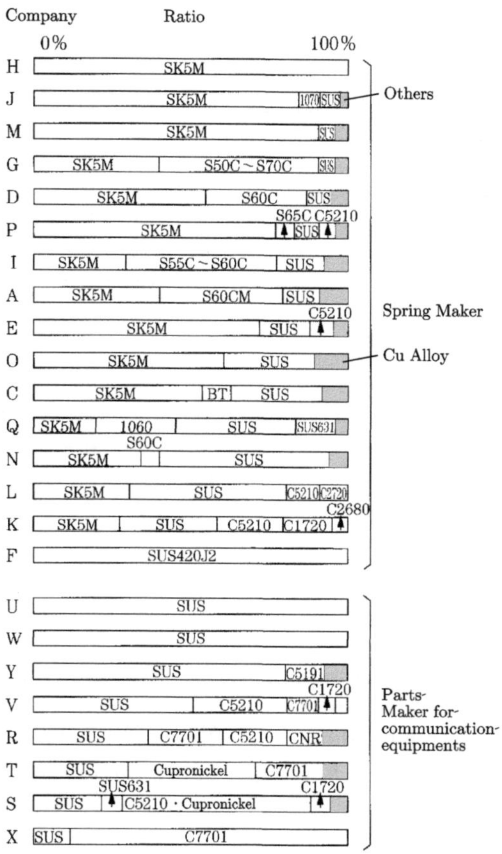

Fig. 2.182. Spring material used by spring markers and part markers for communication equipments

As stainless steel strip for springs in JIS G 4313 five steel grades shown in Table 2.39 [4], are specified. They can be classified roughly into austenitic, martensitic, and precipitation hardening type. Ferritic stainless steel, for example, SUS430, has been also used as spring material recently.

As for the mechanical properties of stainless steel strip for springs, hardness and bendability are specified as shown in Table 2.40 [4]. The bendability can be evaluated by V-block method specified in JIS Z 2248 (Metallic material bending test method). Depending upon the customer's request, tensile test can be conducted instead of hardness.

Since SUS301 and SUS304 are meta-stable austenitic stainless steel, with cold working, its strength can be increased by work-hardening and the occurrence of deformation induced martensite. The degree of cold working is controlled to obtain the desired strength. Five grades of hardness in SUS301 and three grades of hardness in SUS304 are specified, as shown in Table 2.40.

Table 2.40. Hardness and Bendability

Type	Mark	Cold-rolled or Annealed		Precipitation-hardened	
		Hardness HV	Bendability V bend	Mark	Hardness HV
SUS301-CSP	1/2H	≧ 310	≦ 2T	–	–
	3/4H	≧ 370	≦ 2.5T	–	–
	H	≧ 430	–	–	–
	EH	≧ 490	–	–	–
	SHE(1)	≦ 530	–	–	–
SUS304-CSP	1/2H	≧ 250	≦ 2T	–	–
	3/4H	≧ 310	≦ 2.5T	–	–
	H	≧ 370	–	–	–
SUS420J2-CSP	0	≧ 247	–	–	–
SUS631-CSP	0	≧ 200	≦ 0.5T	TH1050 RH950	≧ 345 ≧ 392
	1/2H	≧ 350	≦ 1.5T	CH	≧ 380
	3/4H	≧ 400	–	CH	≧ 450
	H	≧ 450	–	CH	≧ 530
SUS632J1-CSP	1/2H	≧ 350	–	CH	≧ 400
	3/4H	≧ 420	–	CH	≧ 480

T: thickness of strip (1) SEH should be requested within the range of EH by users

SUS301 obtains the same hardness under smaller cold rolling ratio because SUS301 shows higher work-hardening rate compared with that of SUS304. SUS301 also shows better bendability at the same hardness.

SUS631 also called as 17-7PH is precipitation hardening type stainless steel. Using cold-rolled strip designated as Type H, 3/4H or Type 1/2H, precipitation hardening (named as CH) of 475°C, 1 hour is conducted after forming springs. Although the precipitation hardening named as TH1050 or RH950 are also specified using solution treated steel strip, Type O, their complicated processes are not in the practical use. SUS632J1 steel has been added newly in JIS G 4313 since 1996. In the case of SUS632J1, Type 1/2H and Type 3/4H are specified. The precipitation hardening treatment is the same condition as SUS631, or 475°C, 1 hour. SUS420J2 is martensitic stainless steel. Using annealed steel strip designated as type O, which is air-cooled from 750°C or slow-cooled from 800–900°C, springs are quenched and tempered after forming.

The size and its tolerances of cold-rolled stainless steel strip for springs are shown in Table 2.41 and Table 2.42 [4]. The ET and ST grade with severer thickness tolerances are specified as having the allowable maximum value of flatness. The degree of flatness, S is measured as shown in Fig. 2.184. Since cold-rolled stainless steel strip for springs can be produced by the same process as cold-rolled steel strip, Type A and Type R, the maximum value of camber becomes the same value, as shown in Table 2.43 [4]. However, it is possible to be requested up to 1/2 of this maximum value by the agreement between manufacturers and users. In the case of ET or ST, thickness tolerance of cold-

Table 2.41. Thickness torelance of stainless steel spring

Thickness, t \ Width	Thickness tolerance			Thickness tolerance (ET)			Thickness tolerance (ST)		
	$w < 160$	$160 \leqq w < 250$	$250 \leqq w < 600$	$w < 80$	$80 \leqq w < 250$	$250 \leqq w < 600$	$w < 80$	$80 \leqq w < 250$	$250 \leqq w < 600$
$0.10 \leqq t < 0.16$	±0.015	±0.020	±0.020	±0.008	±0.012	±0.015	±0.005	±0.008	±0.010
$0.16 \leqq t < 0.25$	±0.020	±0.025	±0.030	±0.012	±0.015	±0.020	±0.008	±0.010	±0.012
$0.25 \leqq t < 0.40$	±0.025	±0.030	±0.035	±0.015	±0.020	±0.025	±0.010	±0.012	±0.015
$0.40 \leqq t < 0.60$	±0.035	±0.040	±0.040	±0.020	±0.025	±0.030	±0.015	±0.015	±0.020
$0.60 \leqq t < 0.80$	±0.040	±0.045	±0.045	±0.025	±0.030	±0.035	±0.015	±0.018	±0.025
$0.80 \leqq t < 1.00$	±0.040	±0.050	±0.050	±0.025	±0.030	±0.035	±0.015	±0.020	±0.025
$1.00 \leqq t < 1.25$	±0.050	±0.050	±0.050	±0.030	±0.035	±0.040	±0.020	±0.025	±0.030
$1.25 \leqq t < 1.60$	±0.050	±0.060	±0.060	±0.030	±0.035	±0.040	±0.020	±0.025	±0.030

Table 2.42. Width tolerance of stainless steel strip for spring

(mm)

Thickness, t	Width tolerance			
	$w < 80$	$80 \leqq w < 160$	$160 \leqq w < 250$	$250 \leqq w < 600$
$t < 0.60$	±0.10	±0.15	±0.20	±0.25
$0.60 \leqq t < 1.00$	±0.15	±0.20	±0.25	±0.25
$1.00 \leqq t < 1.60$	±0.20	±0.20	±0.30	±0.30

Table 2.43. Camber tolerance of stainless steel strip for spring

(mm)

	Camber tolerance per 1000 mm			
Width	10–20 exclusive	20–40 exclusive	40–80 exclusive	Over 80 inclusive
Max. value	8	6	3	1

rolled stainless steel strip, spring deflection limit in cyclic bending tests can be available under user's requests for reference.

(iii) Tension-annealed stainless steel strip for springs

Although cold-rolled stainless steel strip for springs have excellent corrosion resistances and mechanical properties, it has the tendency of causing non-uniform deformation such as transverse bend during cold forming, which can make it difficult to carry out an uniform process when blanking or bending. To countermeasure with this issue, stainless steel strip annealed under a tensile force has been developed as tension-annealed stainless strip for springs. This stainless steel strip shows high flatness, good spring quality and excellent corrosion resistance. However, it has not been designated in JIS G 4313 yet. The schematic figure of tension-annealing facilities used in commercial productions is shown in Fig. 2.183 [5]. The facilities consist of heating furnace, coilers, bridle rolls and tension detecting devices. The steel strip is continuously heated under a constant tensile force. As the strip is treated at higher heating temperature, higher strength and spring deflection limit can be obtained. The improvement of spring deflection limit and flatness is remarkable under a short time of tension-annealing treatment. Figure 2.184 shows the

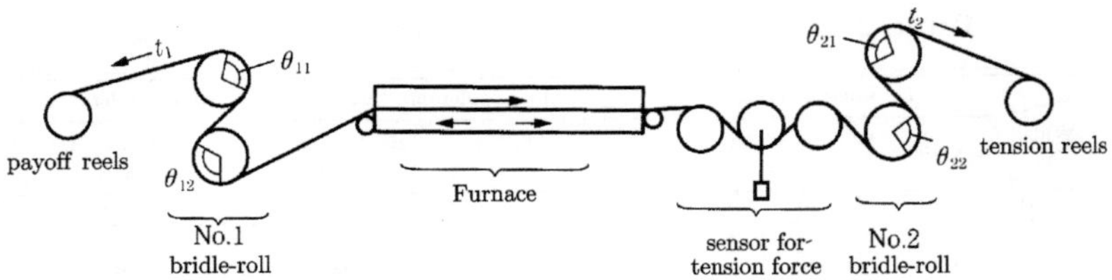

Fig. 2.183. Schematic figure in tention-annealing

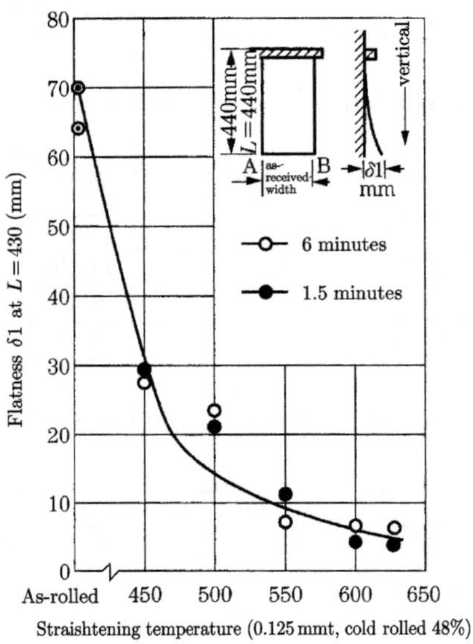

Fig. 2.184. Effect of time and temperarure on flatness of stainless steel strip

effect of straightening temperature on flatness of strip. However, higher temperature brings thicker oxide film on the surface of strip, to deteriorate the corrosion resistance and solder-ability. To solve those problems, acid pickling with mixture of fluoric acid and nitric acid is usually carried out after tension-annealing treatment. Tension-annealed stainless steel strip for springs are applied for the spring of telephones, relays, switches and floppy disc drives.

(iv) Solder clad stainless steel strip

Stainless steel strip is considered as one of the materials which are difficult for soldering. If necessary, manual soldering operation is done after forming the spring. Recently stainless steel strip on which the solder plating be done, has been developed to be applied to springs for telephone. During the processes of tension-annealing, one side or both side of strip surface can be plated with immersing in a molten bath of solder. The thickness of solder are about 2–10 μm. The tensile strength and elongation of solder clad stainless steel strip are the same as those of non-clad stainless steel strip. Because of good adhesion to strip, no peeling-off in blanking or bending can be seen. These materials

Table 2.44. Properties in solder clad stainless steel strip

	Tensiles strength	Elongation	maximum working stress for spring	fatigue strength	HV hardness	bendability	Young's modulus
	MPa	%	MPa	MPa		(90°)	GPa
0	≧ 490	≧ 40	–	–	≦ 200	–	–
1/2	≧ 830	≧ 6	≧ 390	≧ 390	≧ 250	0.2R	≧ 172
3/4	≧ 930	≧ 3	≧ 490	≧ 440	≧ 300	0.2R	≧ 172
H	≧ 980	–	≧ 635	≧ 490	≧ 350	0.4R	≧ 172

(Sampling: Rollong direction)

are applied to the connector and its small parts. Table 2.44 [6] shows the properties of solder clad stainless steel strip.

(v) High strength steel strip for springs

The Type R strip as specified in JIS G 4802 is used for springs heat-treated after forming. Therefore the hardness of strip is low for the formability of springs. However, like spiral springs which are not complicate shapes, it can be possible to increase the hardness of steel strip. As examples of those applications, retractor springs for the automotive seat-belt and spiral springs for an extension cord of vacuum cleaners. This high strength steel strip can be manufactured by two different processes. One is cold rolling patented wide steel strip, thereafter, low-temperature annealing and slitting it. Another is cold-rolling music wire or high carbon steel wire to strip, and annealing at low temperature. When heat resistance required, Si–Cr wire can be used.

SUS301type stainless steel strip is also used for retractor springs for automotive seat belts.

(ii) Steel grade for springs

The steel grade used for spiral springs, coil springs, disc springs and washers is designated as JIS G 3311 and JIS G 4802. The chemical compositions have been shown in the Table 2.33 and Table 2.34. As for stainless steel strip, martensitic stainless steel, austenitic stainless steel and precipitation hardening stainless steel are designated as JIS G 4313, as shown in the Table 2.40.

In addition to the JIS standards, ferritic stainless steel such as SUS430 is applied to springs for specific purposes.

Carbon steel strip with the C contents of 0.3 to 1.2% is designated as cold-rolled special steel strip. However, for spring applications, carbon steel with the C contents of 0.5 to 1.0%, has been used. It can be realized that relatively lower C steel is applied for complicate-shaped springs and higher C steel is applied for springs required higher strength. As cold-rolled steel strip for springs, steels with 0.5 to 1.0% carbon contents are designated. Heat-treatment of springs after forming falls into two methods-quenching and tempering and aus-tempering. In case of quenching, oil-quenching is normally conducted. Although the quenching hardness increases with increasing the carbon content, the hardness saturates around 0.6–0.7% carbon. Table 2.45 [7] shows the re-

Table 2.45. Quenching temperature and As-quenched hardness

Steel Grade	Quenching Temp. (°C)	As-quenched Hardness (HV)	Steel Grade	Quenching Temp. (°C)	As-quenched Hardness (HV)
S50CM	830–860	720–770	S75CM	790–820	800-850
S55CM	820–850	750–800	SK4M	780–810	830–880
S60CM	810–840	760–810	SK5M	780–810	820–870
S70CM	790–820	790–840	SUP6M	820–850	780–830

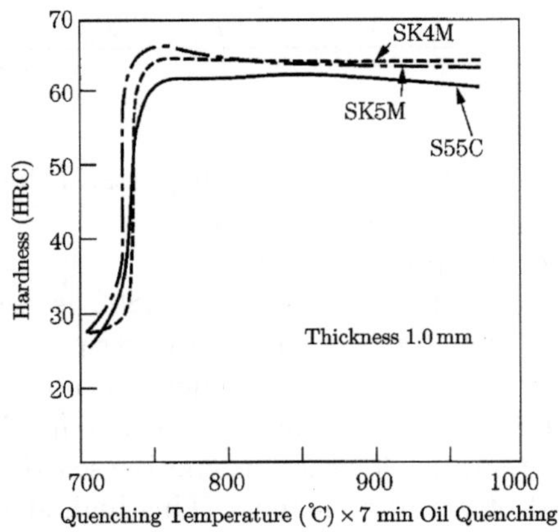

Fig. 2.185. Quenching Hardness Curves of S55C, SK4M, SK5M

lationship between quenching temperature and hardness in typical steels. To improve the hardenability of steel, Cr is occasionally added within permissible ranges. However, when the thickness of strip is thin, it is not required to change the chemical contents with controlling quenching temperature, atmosphere of furnace and cooling condition. Even if the quenching is inadequate, the same hardness can be attained by low-temperature tempering. However, lower elongation, proof stress and fatigue strength can be realized, compared with those of fully-quenched material. Therefore, the quality for springs is not sufficient. Since the quenching hardness depends upon the carbon content, the completeness of quenching can be predictable by quenching hardness. Figure 2.185 and Fig. 2.186 show the quenching hardness curves and tempering hardness curves of typical steels.

In the case of aus-tempering, formed springs are fully austenitized at high temperature, then dipped into a salt-bath at 300–500°C. Springs are kept at the temperature until bainitic transformation is finished sufficiently. Tempering is not required in aus-tempering. Figure 2.187 [8] shows the relation between hardness and isothermal transformation temperature in aus-tempering. It is obvious that the hardness can be different under the same aus-tempering condition, depending upon the carbon content. For material of heat-treated cold-rolled steel strip for springs (Type H) and bainite steel strip (Type B) JIS SK5 and carbon steel with 0.5–0.65% carbon can be mostly used respectively.

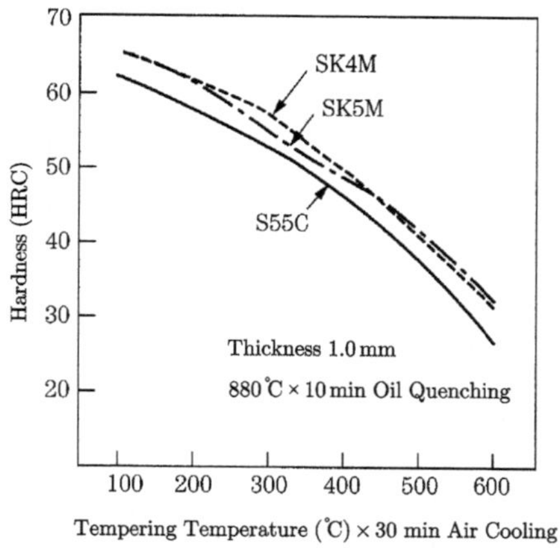

Fig. 2.186. Tempering Hardness Curves of S55C, SK4M, SK5M

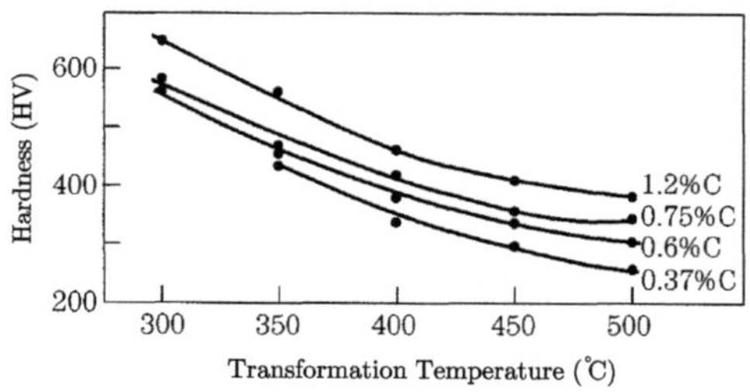

Fig. 2.187. Effect of C content on the relation between isothermal transformation temperature and Hardness

Stainless steel strip for springs shown in Table 2.39, has good corrosion resistance and heat resistance because of high Cr content. Therefore, the applications can be extended by using these properties.

Since JIS SUS301 and SUS304 are austenitic stainless steel, they show marked work-hardening during cold working and hardening caused by martensitic transformation simultaneously. Therefore, the hardening during cold working is remarkable. Figure 2.188 [3] shows the difference of work-hardening between stainless steel and carbon steel. SUS301 shows higher strength at the same reduction, compared with SUS304. SUS301 also shows larger elongation at the same strength. Although it can be said that SUS301 is more suitable for spring material, SUS301 is more sensitive to the fluctuation of cold rolling condition, and required to have the strict control of cold rolling condition. As for SUS304, the change of quality with the fluctuation of cold rolling condition is small. Due to large cold reduction ratio in the applications to springs, it should be noted that the anisotropy of mechanical properties for both steel strips is large. Figure 2.189 shows the relation between the Young's modulus and spring deflection limit, and cold rolling reductions, as for SUS304.

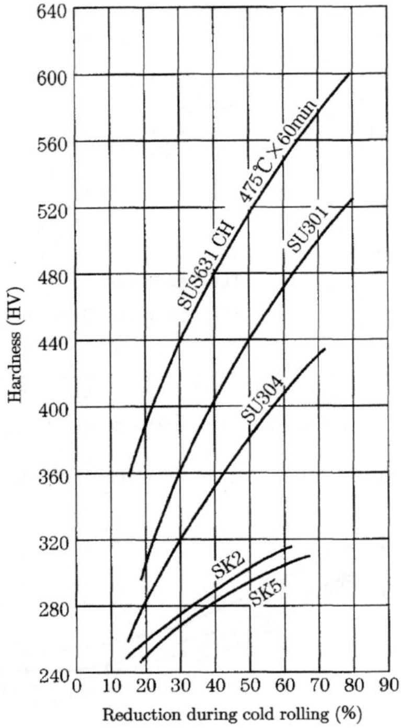

Fig. 2.188. Hardness of cold-rolled

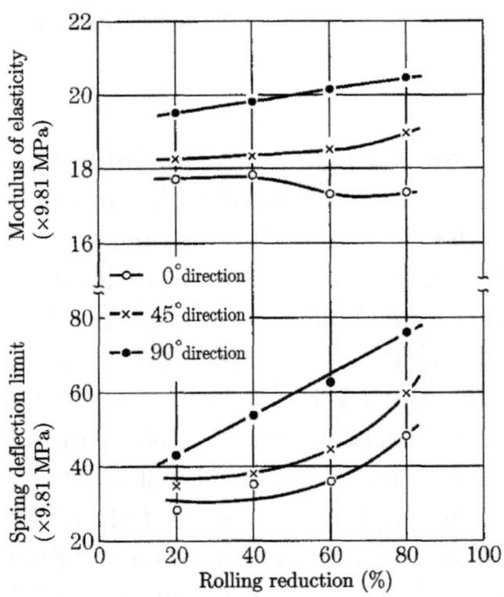

Fig. 2.189. Anisotropy of stainless steel strip for spring (SUS304)

It can be realized that the impact of cold working on the anisotropy of mechanical properties is large. This stainless steel strip can be applied to springs as the cold-rolled condition. However, as shown previously, applying tension-annealing, the flatness and spring properties can be improved remarkably.

Figure 2.190 [3] shows the tempering hardness curves of SUS420J2 martensitic stainless steel. When required high hardness, higher temperature quenching is applied and when required high toughness, lower temperature quenching is applied. Table 2.46 [4] shows the relation between the hardness and the heat-treatment condition of SUS420J2 which is described in JIS as reference values.

Precipitation hardening type stainless steel applied to springs is mainly SUS631. This material shows higher spring properties, superior fatigue strength and smaller anisotropy in mechanical properties, compared with the other stainless steel. As solution treated condition, it has high toughness to make it possible to form a complicate shaped spring. Conducting heat-treatment after spring forming, superior spring properties can be obtained. Table 2.47 [3] and Table 2.48 [10] show the heat-treatment conditions and mechanical properties. SUS632J1 stainless steel which is added Si, Cu, Ti to the basic composition of 15Cr-7Ni, shows the higher hardness increase by precipitation treatment than SUS631. Therefore, more complicate shaped springs can be made under

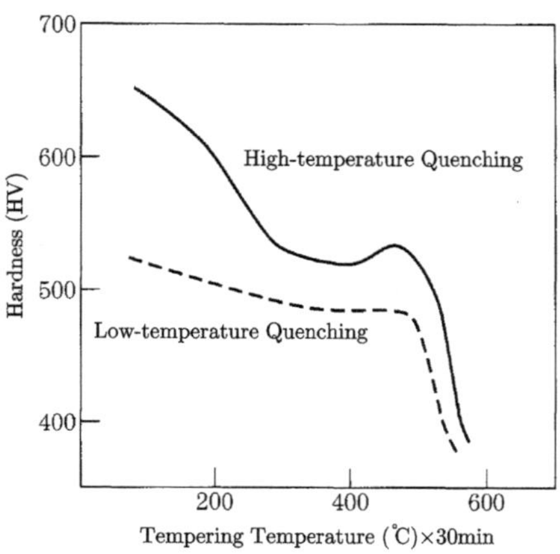

Fig. 2.190. Tempering hardness curves of SUS420J2

Table 2.46. Hardness and Heat treatment of SUS420J2

Heat treatment condition		Hardness (HV)
Quenching	Tempering	
900	300–400	410–460
950	300–400	460–510
1000	300–400	490–540
1050	300–400	510–570

Table 2.47. Heat treatment of SUS631 stainless steel

Designation	Symbol	Condition
Solution treatment	S	1038 ± 10°C Rapid cooling
Intermediate treatment	T	S → Forming, 760°C × 90 min Air cooling
	R	S → Forming, 950°C × 15 min Air cooling −73°C × 8 h
	C	S → Cold rolling
Precipitation hardering treatment	TH 1050	T → 560 ± 5°C × 90 min Rapid cooling
	TH 950	T → 510 ± 5°C × 30–60 min Air cooling
	RH 950	R → 510 ± 5°C × 30–60 min Air cooling
	CH 900	C → 480 ± 5°C × 30 min Air cooling

Table 2.48. Mechanical properties of SUS631 stainless steel sheet

Item \ Condition	S	T	R	TH1050	TH950	RH950	C	CH900
Tensile strength (MPa)	893	991	1207	1373	1481	1589	1481	1717
Proof stress (MPa)	324	853	785	1236	1275	1501	1275	1648
Elongation (%)	39	9	10	6	8	5	2	1
Hardness (HR)	B85	C31	C35	C43	C45	C46	C43	C49
0.2% Compressive proof Stress Longitudinal (MPa)	–	–	–	1393	1403	1648	1020	1511
Transverse	–	–	–	1393	1403	1648	1472	1766

solution-treated condition and higher strength after precipitation-hardening treatment than SUS631 can be obtained.

(iii) Manufacturing process of steel strip for springs

Steel strip for springs can be manufactured by cold-rolling and annealing repeatedly, using hot-rolled strip produced by hot-rolling billet or slab. A typical manufacturing flow of steel strip for springs is shown in Fig. 2.191 [3]. Hot-rolled strip is de-scaled by acid-pickling or shot-blasting before cold rolling. Hard and brittle oxide scale must be removed completely. The existence of oxide scale can cause the scratch of roll and poor strip surface appearance. Hot-rolled steel strip which is cooled rapidly from high temperature after hot rolling has relatively high strength to show poor formability. To improve the formability, annealing (carbon steel and martensitic stainless steel) or solution-treatment (austenitic stainless steel) can be carried out. Annealing is carried out for a long time heating at the temperature just over Ac_1 transformation temperature, approximately 700°C, to have uniform spheroidizing microstructure. Since the surface flaw and decarburization at surface are especially harmful for the quality of carbon steel springs, it is important to control the atmosphere of furnace.

At the following process of cold rolling, as steel strip for springs tends to show heavy work-hardening during cold rolling, it is difficult to take a

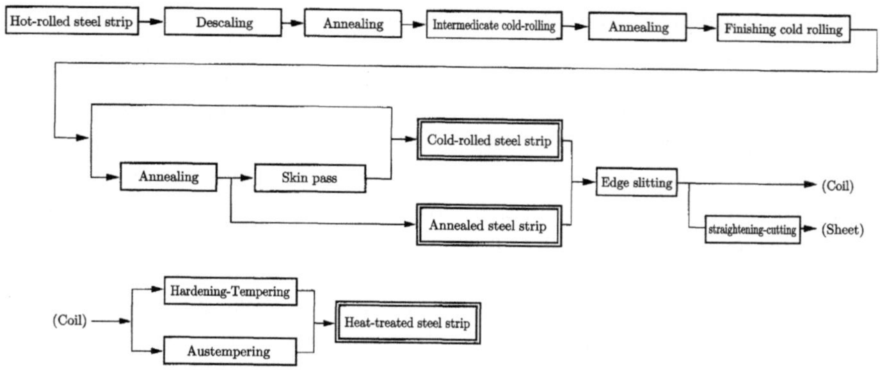

Fig. 2.191. Manufacturing Processes of steel strip for spring

large reduction at one time. Therefore, cold rolling and intermediate annealing can be repeated several times to have the final thickness of steel. At the stage of finishing cold rolling, material with certain mechanical properties can be obtained by selecting the final reduction ratio, considering the relation between the cold reduction ratio and hardness is almost constant. For the finishing cold rolling process, a rolling mill with four-high roll-stands can be employed to satisfy severe tolerances. A rolling mill with twenty-high roll strands (Sendzimir mill) [11] can be applied for steel strip with specially severe tolerances. A high crown control mill for more accurate shape control, has been developed and in commercial applications [12]. Active thickness controlling technology during cold rolling has been also in commercial uses.

Skin pass rolling which is light cold rolling with a few percent reduction, is applied by a rolling mill with two-high roll stands, to improve flatness and punching quality. Then, steel strip slit into a desired width can be supplied as steel strip for springs or as raw material for heat-treated steel strip. Besides coil material of steel strip, sheet material straightened and cut to a certain length is available commercially.

The hardened steel strip and bainitic steel strip which possess required mechanical properties can be manufactured by using cold-rolled steel strip or annealed steel strip as raw material. Hardened steel strip is normally oil quenched and tempered by a continuous furnace which has heat resistant tubes (muffle) and is filled with protective gas to prevent decarburization and oxidization of strip surface. It should be noted that in order to prevent the distortion or deformation of strip during the heat treatment the practices with quenching bed and straightening equipments have been carried out by each manufacturer. Bainitic steel strip is manufactured through the similar heat treatment line. Instead of oil quenching and tempering, aus-temper is conducted. Based on commercial requirements, the surface treatment such as coloring or polishing can be applied.

(iv) Properties of steel strip for spring

The properties required for steel strip can be said to be mechanical properties, fatigue strength, formability, and so on. However, the formability for springs can be also said to be a common requirement. Figure 2.192 [13] shows the results of manufacturer's survey for the features of flat springs. It can be seen that the shape and load are important factors in flat spring. This may be due to the fact that flat springs have highly three-dimensional degree of freedom in design and normally show complicated shapes.

When estimating the formability of springs, it is effective to apply the plastic constant of material, F-value and n-value. The plastic constant can be also used for designing dies to estimate the amount of spring-back. Generally, a stress-strain curve of material can be estimated by an approximate equation, $\sigma = F\varepsilon^n$, where σ: true stress, F: plastic coefficient, ε: true strain, and n: work hardening exponent. The F-value and n-value are plastic constant. It is said that for most metallic materials, that n-value is 1/2–1/10 and F/E is 1/100 to 1/500 (E: Young's modulus). Table 2.49 [13] shows examples of F-value and n-value.

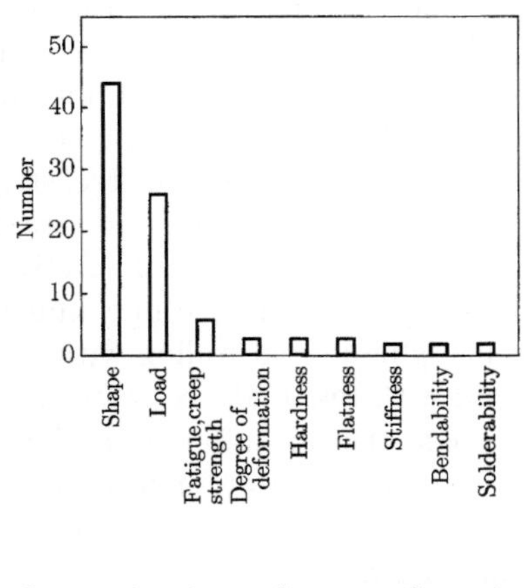

Fig. 2.192. Inspection items for manufacturing flat spring

Table 2.49. Plastic constants in steel strip for springs

		$t = 0.5$ mm		$t = 0.25$ mm	
		F (MPa)	n	F (MPa)	n
SK5	HV400	1644	0.089	1884	0.116
SK5	HV460	1854	0.076	2084	0.086
Bainitic steel	HV400	1977	0.093	1886	0.086
Bainitic steel	HV460	2536	0.120	2544	0.121
SUS301	HV400	1954	0.132	1729	0.125
SUS304	HV360	2252	0.148	1899	0.107

F: plastic cofficient
n: work hardening exponent

Table 2.50. Minimum bend radius at 90° V bending test of steel strip

thickness	0.5 mm				1.0 mm			
bending direction	Transverse to rolling direction		longitudinal to rolling direction		Transverse		longitudinal	
TYPE / hardness (HV)	bainitic	SK5 Q–T	bainitic	SK5 Q–T	bainitic	SK5 Q–T	bainitic	SK5 Q–T
320	0.6R	1.3R	1.0R	1.6R	1.1R	2.5R	1.5R	3.0R
360	1.0	1.6	1.3	2.0	2.3	3.5	3.2	4.0
400	1.3	1.6	1.5	2.5	2.8	4.0	3.3	4.5
440	1.3	2.0	1.8	2.5	3.5	5.0	4.4	5.0
SK5 As-annealed					180° Bending to solid is OK			

When manufacturing springs with steel strip, the plastic deformation is mainly bending and shearing. The shearing formability depends upon the design of dies. The bendability of steel strip can be estimated by the minimum bending radius at bending test. Softer material normally shows a smaller bending radius. As for annealed steel strip, it becomes possible to bend to the minimum bending radius close to zero. Table 2.50 [14] shows the minimum bending radius of bainitic steel strip and hardened steel strip. The bendability does not only depend upon steel type and grade, but also the rolling direction. Therefore, it should be required to consider the rolling direction of steel strip in designing springs.

As normal springs, it will be required for flat springs to have high fatigue strength and permanent set resistance. There is not so many fatigue strength data of steel strip compared with that of hot-wound spring steel and steel wire. As one reason, the fatigue testing of steel strip is difficult and there are many scattering in fatigue testing results owing to a variety of testing methods and equipments. Recent improvements on testing methods and equipments have made it possible to have many research works related with the fatigue strength of steel strip. It is said that as the factors which can influence the fatigue strength of flat springs are hardness, toughness, microstructure, non-metallic inclusions, surface flaws and decarburizations, surface roughness, edge shape after blanking and so on.

When the strip type A and R used for springs, the heat-treatment can be applied to the springs after spring forming. The relation between tempering temperature and hardness in JIS SK5 which is a typical steel grade used for Type A and Type R, is shown in Fig. 2.193 [15]. It is obvious that the hardness and tensile strength decrease and the elongation increase linearly with increasing the tempering temperature.

① Difference between hardened steel strip and bainitic steel strip

In the case of manufacturing springs by using heat-treated steel strip, the material can be selected from two steel strip types, hardened steel strip and bainitic steel strip. The microstructure of type H is tempered-martensite,

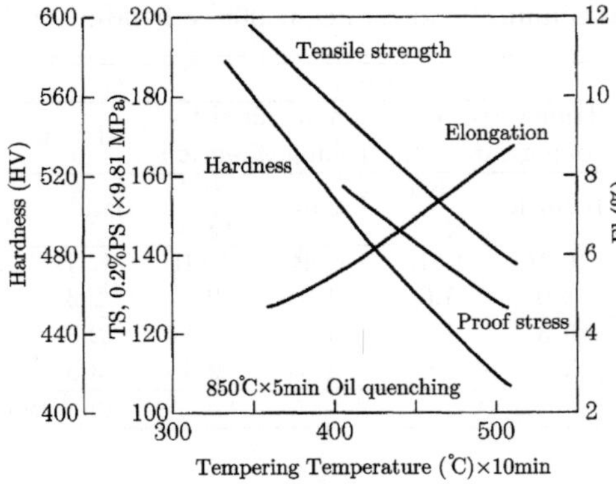

Fig. 2.193. Relation between Mechanical properties and tempering temperature in SK5M

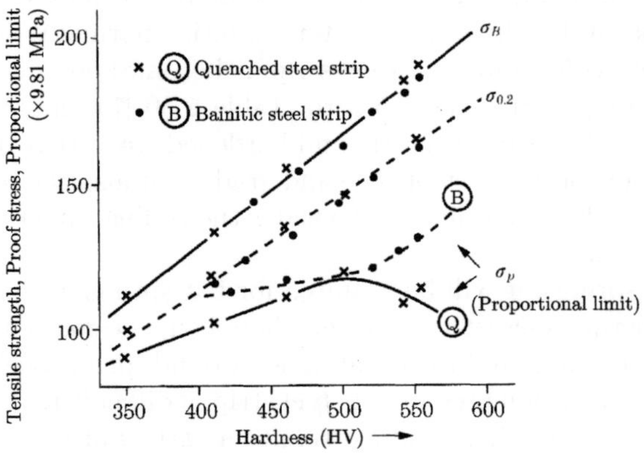

Fig. 2.194. Tensile properties of quenched and tempered steel strip and bainitic steel strip

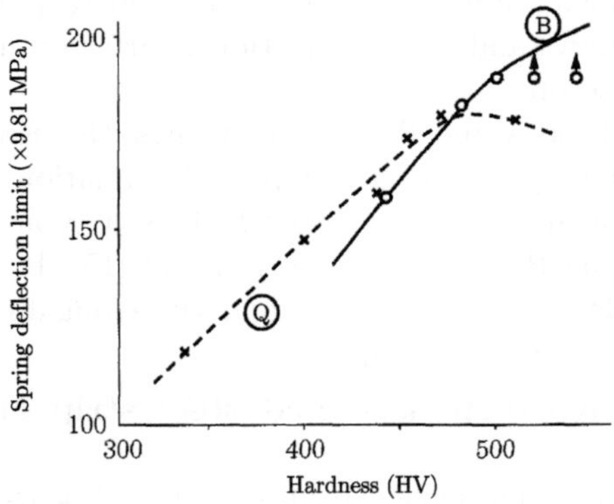

Fig. 2.195. Spring deflection limit of quenched and tempered steel strip (Q) and bainitic steel strip (B)

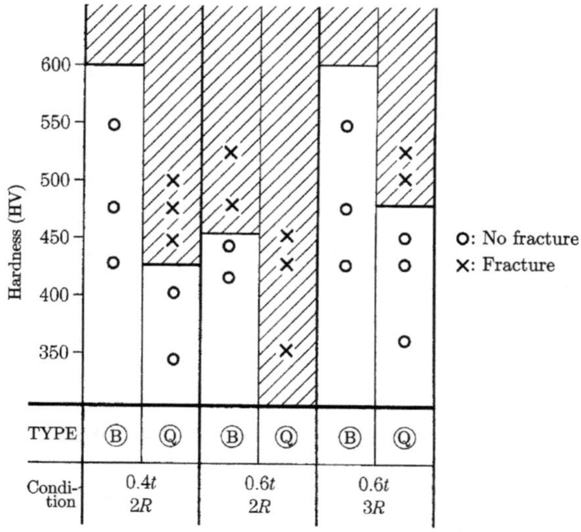

Fig. 2.196. Bendability of quenched—tempered steel (Q) and austempered steel (B)

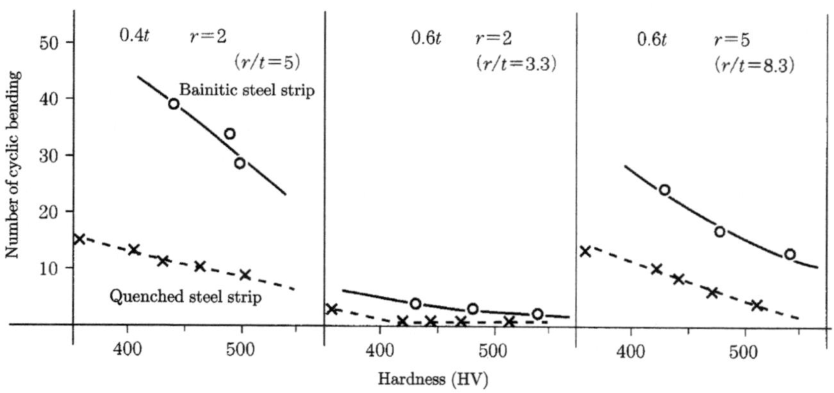

Fig. 2.197. Cyclic bendability of quenched tempered steel strip and austempered steel strip

and that of type B is bainitic microstructure. Compared two materials at the same hardness, bainitic steel shows many advantages such as good bendability, good permanent set resistance, and small distortions at heat treatment. Figure 2.194–Fig. 2.197 and Table 2.51 [16] show the comparison between hardened steel strip and bainitic steel strip. The bainitic steel strip also shows low sensitivity to the hydrogen embrittlement at electrogalvanizing process [17].

Table 2.51. Permanent set test results

Shape	Type	Hardness (HV)		Sagging (mm)	
		$\overline{X}$	R	$\overline{X}$	R
U-Shape	Oil-quenching–Tempering	398.2	20	0.190	0.30
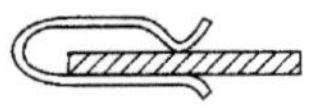	Aus-tempering	433.5	20	0.068	0.05
		490.4	20	0.002	0.03
bent-shape	Oil-quenching–Tempering	402.0	20	0.040	0.19
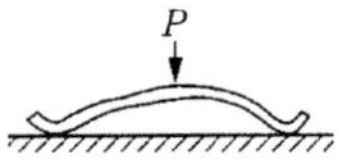	Aus-tempering	435.2	20	0.011	0.04
		494.4	20	0.003	0.03

② Difference in performance between hardened steel strip and bainitic steel strip

Figure 2.198 [13] shows the spring deflection limit, Kb, of quenched and tempered steel strip (SK5M) and bainitic steel strip (S60CM). Harder material shows higher spring deflection limit. As for bainitic steel strip, the spring deflection limit can be followed with the regression equation of, Kb $= 0.22\,\text{HV}+5$ (Scattering is within ± 100 MPa), where HV: Vicker's hardness.

However, the values of quenched and tempered steel strip are not always consistent with the equation. It should be noted that the spring deflection limit largely depends upon steel grade, processing and heat-treatment conditions. Fatigue properties of hardened steel strip and bainitic steel strip are shown in Fig. 2.199 [13] and Fig. 2.200 [16] respectively. The tested results show some difference between the testing machines. Mechanical testing machine shows larger scattering than the electromagnetic machine. For flat springs, the scattering in testing results tends to occur by some reasons originated

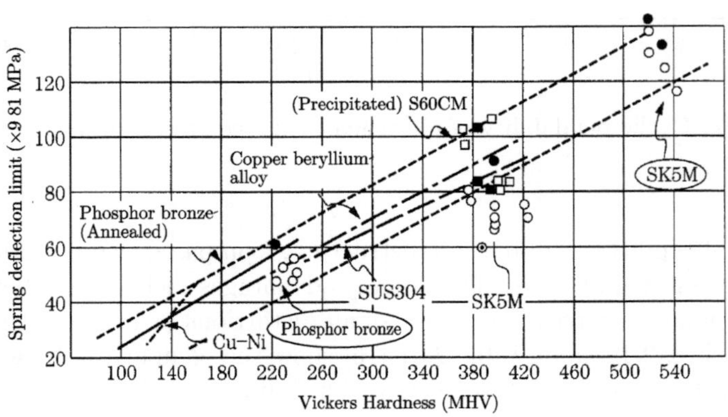

Fig. 2.198. Hardness and spring deflection limit of quenched-tempered steel strip for springs (SK5M is the old name of SK85M, see Table 2.33 and 2.34)

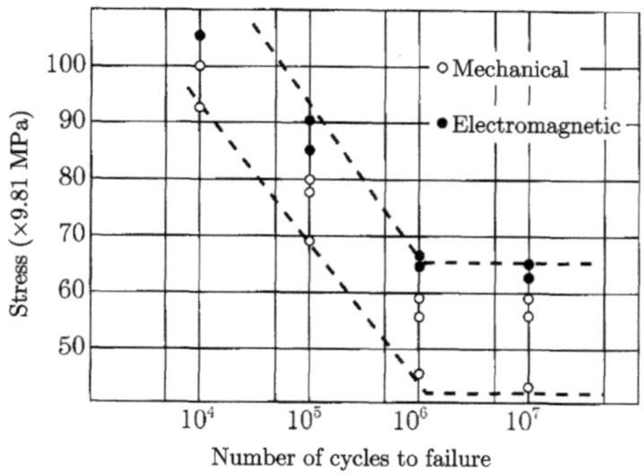

Fig. 2.199. Fatigue life of quenched and tempered steel strip, SK5M

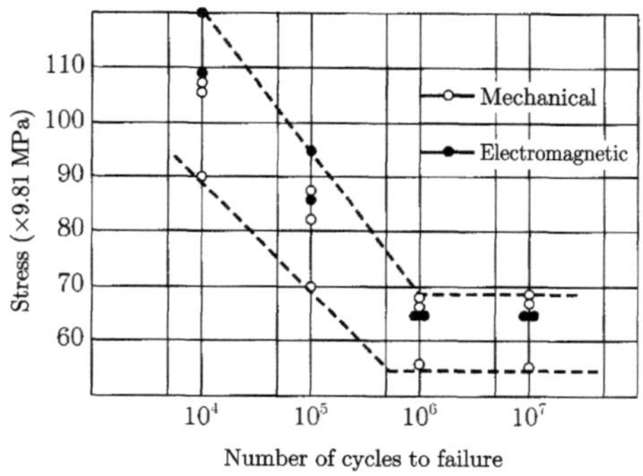

Fig. 2.200. Fatigue life of aus-tempered (bainitic) steel strip, S60CM

from the shape of flat springs. It should be noted to finish the test specimen such as the finishing condition of edge with less scatter. It is well known that the fatigue strength decreases by the existence of fine non-metallic inclusions in the case of flat springs even if the material satisfies the specifications of chemical compositions. The effect of sulfur contents on the fatigue strength of SK5 flat springs is shown in Fig. 2.201 [18].

In the case of springs with heat-treated steel strip, residual stress after bending, or even after blanking can deteriorate the spring deflection limit and fatigue strength. Figure 2.202 [19] shows the relation between annealing temperature and, hardness and spring deflection limit in blanked material. To eliminate this influence, it should be required to conduct the low-temperature annealing of 200°C, 30 minutes at least.

The relation between temperature and stress relaxation (creep strain) of quenched and tempered steel strip applied the stress of 294 to 720 MPa, is shown in Fig. 2.203 [13]. This result does not change when the thickness changes. It can be said that there are no difference of creep deformation (permanent set) when applied a constant stress as far as the stress is below the 50% of spring deflection limit.

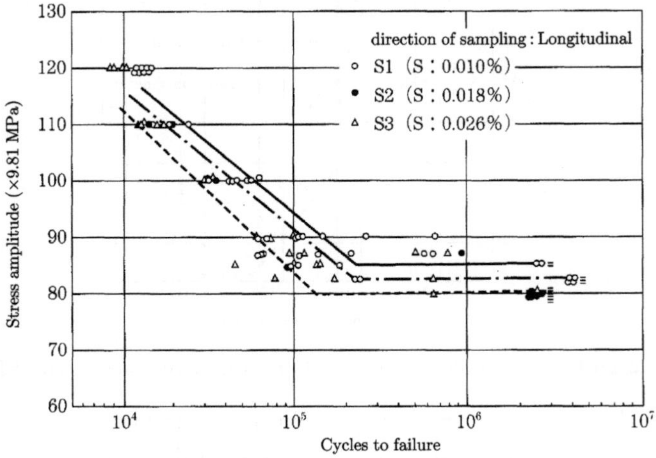

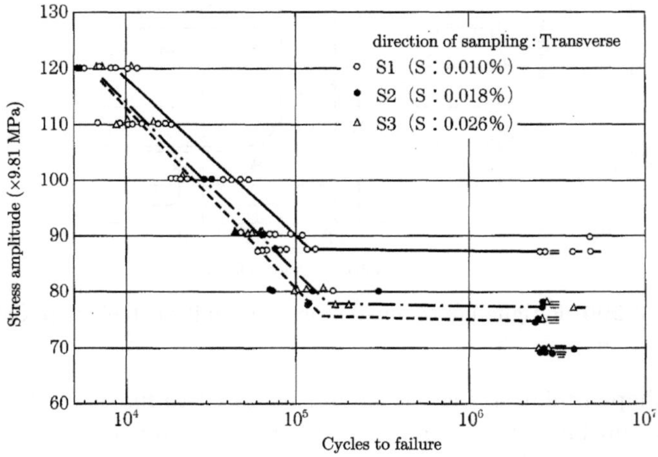

Fig. 2.201. Effect of S content on fatigue properties of quenched and tempered steel strip

③ Stainless steel strip for springs

As the same with stainless steel wire, stainless steel strip for springs is widely used for the applications where high corrosion resistance and high spring performance are required. As for springs applied to business equipments and communication equipments due to its economical reason, the stainless steel strip has been widely used instead of copper alloy except the applications where high conductivity or non-magnetic property is required. Figure 2.204 [13] shows the relation between hardness and spring deflection limit of stainless steel strip, SUS304-CSP. Figure 2.205 [13] also shows examples of the fatigue strength.

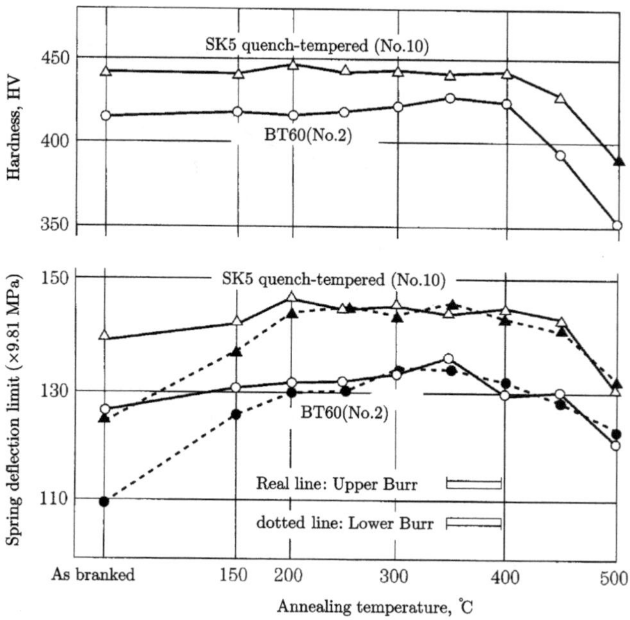

Fig. 2.202. Relation between annealing temperature, spring deflection limit for spring and hardness

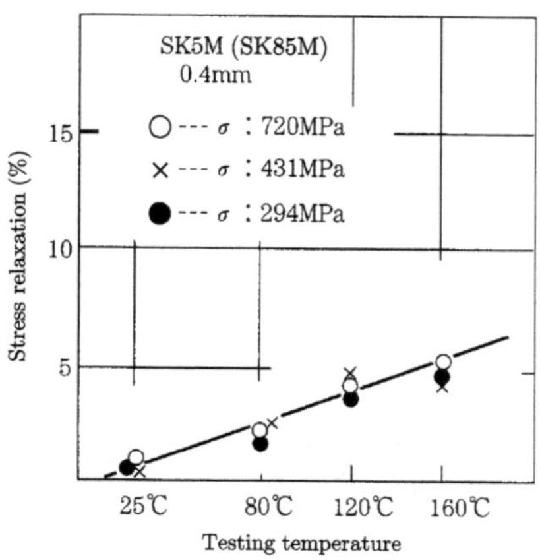

Fig. 2.203. Creep deformation and testing temperature (Holding time: 100 hr)

For springs of austenitic stainless steel, low-temperature annealing is applied after forming. If the annealing is conducted at the temperature higher than 400°C, the surface oxidization occurs to have the surface color changed and stained. This oxidization is said to deteriorate corrosion resistance. Table 2.52 [20] shows the relation between corrosion resistance and annealing conditions for SUS301-CSP. Figure 2.206 [20] also shows the relation between

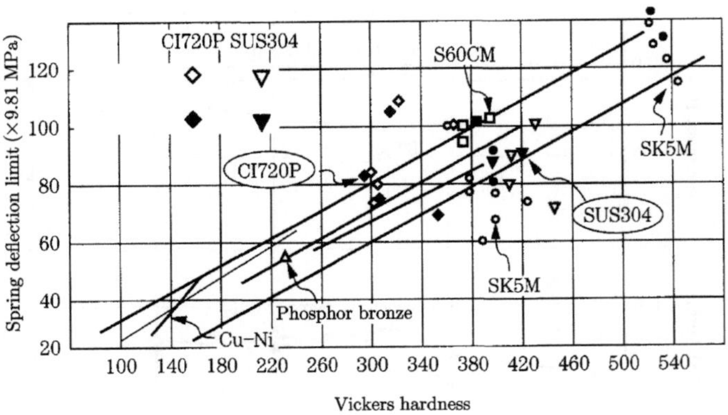

Fig. 2.204. Hardness and spring deflection limit of stainless steel strip

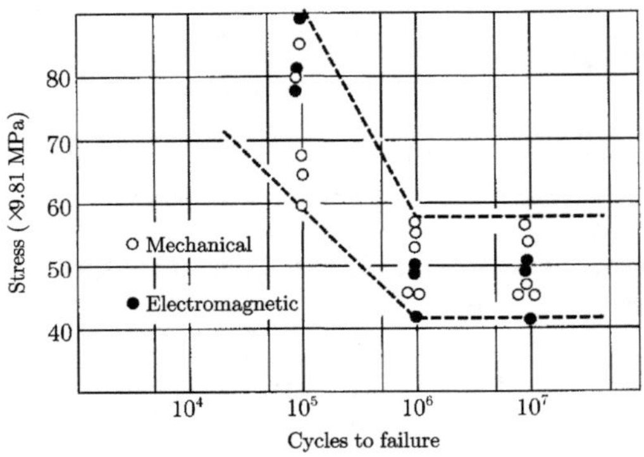

Fig. 2.205. Fatigue life of stainless steel strip, SUS304

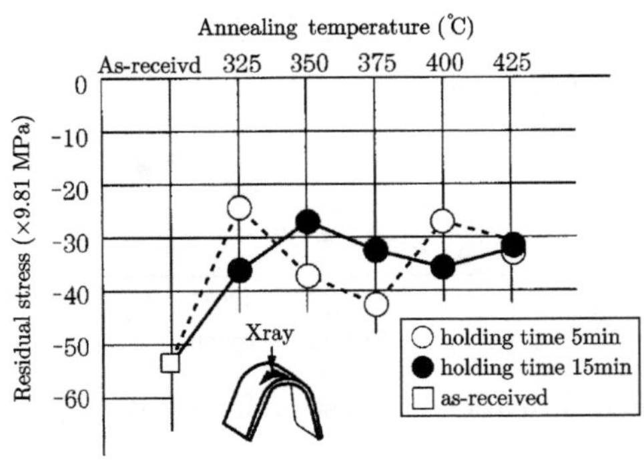

Fig. 2.206. Annealing condition and residual stress of stainless steel strip for springs

annealing temperature and residual stress. Since the difference of residual stress between 325°C and 425°C is small and there are not any difference of permanent set and fatigue strength as well, it is said that the low temperature

Table 2.52. Rust generation and annealing condition of stainless steel strip for springs

temperature (℃)	holdingtime (min)	time to rusting (0, 24, 120, 240)
325	5	● 240 ⇨ (No rust)
	15	● 240 ⇨
350	5	● 240 ⇨
	15	● 240 ⇨
375	5	● 240
	15	● 240
400	5	● 24
	15	● 24
425	5	● 24
	15	● 24

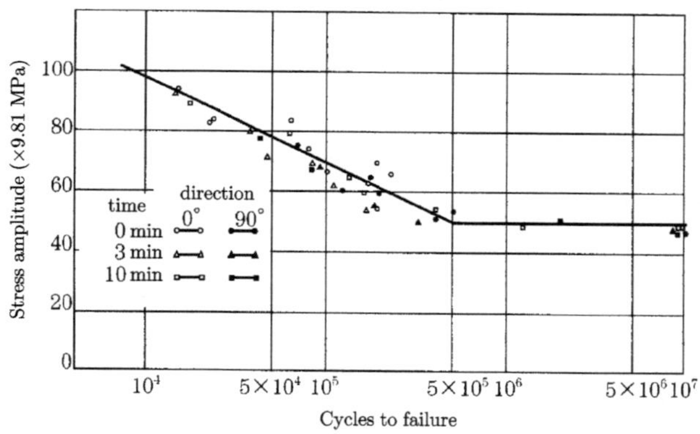

Fig. 2.207. Effect of acid pickling time on fatigue properties of stainless steel strip

annealing between 325°C and 375°C is appropriate in the sense of corrosion resistance.

As described previously, tension-annealed stainless steel strip has been widely used for electronic and communication equipments. Figure 2.184 shows the relation between tension annealing temperature and flatness for 0.125 mm thickness and 48% reduction stainless steel strip. Spring properties such as tensile strength and spring deflection limit, can be also improved with higher tension annealing temperature. Processing too high temperature may cause thicker oxide film on the surface to deteriorate the solder-ability and corrosion resistance. To measure this problem, acid-pickling in hydrofluoric acid-nitric acid mixture can be conducted after tension annealing. As shown in Fig. 2.207 [5], this acid-pickling does not affect the fatigue property.

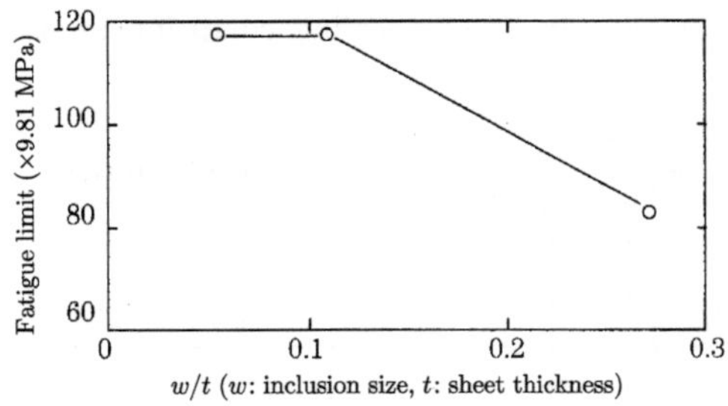

Fig. 2.208. Effect of non-metallic inclusion on fatigue strength of stainless

Non-metallic inclusions can deteriorate the fatigue strength of stainless steel strip, as the same as carbon and low alloy steel strip. Figure 2.208 [21] shows the effect of the size of non-metallic inclusion on the fatigue strength of tension annealed stainless steel strip. Even for a small inclusion around 5 μm diameter, it can be said that there is a great impact on the fatigue strength of thin steel strip around 20 μm thickness.

References

1. JIS G 3311 Cold-rolled special steel strips
2. JIS G 4802 Cold-rolled steel strips for spring
3. Gunji Kudo: Bane 3rd ed. (in Japanese) Maruzen, (1982) p. 71
4. JIS G 4313 Cold-rolled stainless steel strips
5. Mikio Nishihata, Yasuo Kumakura: Transactions of JSSR (1975) p. 14
6. Mikio Nishihata, Shigeaki Otake, Keiichi Komura: Recent Technologies for Spring, JSSR, (1988) p. 165
7. Sub-committee of cold-rolled special steel strip: Tokushu-kou, 27, 6 (1978) p. 11
8. Japan Society for precision engineering and Japan Society for technology of plasticity: Workability of bainitic steel strip and its application to precision equipment (in Japanese), Symposium text (1974) p. 11
9. Mikio Nishihata, Shigeaki Ohtake, Takuo Okumura: Transactions of JSSR, 25 (1980) p. 19
10. Sakamoto: J. Japan Inst. Metals 3 (1964), p. 649
11. Electric furnace steel forum ed.: Handbook of special steel (in Japanese), Rikougakusha (1969) p. 10
12. Daido steel co. ltd: Denki Seiko (Japanese) 51 (1980) p. 231
13. Committee on strength of flat spring: Transactions for JSSR, 31 (1986) p. 76
14. Sub-committee of cold-rolled special steel strip: Special Steel (Japanese) 26 (1977) p. 27
15. Daido steel Co. Ltd: Denki Seiko (Japanese) 51 (1980) p. 231
16. Toshiyuki Nakamura, Hideo Tosino, Mitsuyoshi Kato: JSSR Spring technical report (1973) p. 1
17. Hiroshi Morita, Masatoshi Fujimura: JSSR Autumn technical report (1974) p. 9
18. Kiyosu Omori, Yasuyuki Kono, Tomoaki Teramoto: JSSR Autumn Technical. Report (1977) p. 13

19. Yasuo Sakaba: JSSR Autumn Technical Report (1982) p. 5
20. Kimiaki Nakanishi, Seiji Nonoyama: JSSR Autumn Technical Report (1990) p. 27
21. Jun Nakatsuka, Wataru Murata, Hidehiko Sumitomo, Kiyoshi Sawatani: JSSR Spring Technical Report (1988) p. 13

2.2 Heat Resisting Material for Springs

2.2.1 Introduction

Although there is no clear definition of heat resisting springs, heat resisting springs in practical applications is to be used in wide temperature ranges from room temperature to extremely high temperature. Depending upon the applications and environments, various materials such as carbon steel, alloy steel and stainless steel, have been chozen as a heat resisting spring. Figure 2.209 [1] shows examples of spring material applications for the elevated temperature. In this chapter, alloy tool steel, Fe-base superalloy, Ni-base superalloy and Co-base superalloy are introduced and mainly focused on the properties and manufacturing processes. As for references, various applications for heat-resisting spring are summarized in Table 2.53 [1]

2.2.2 Types of Heat Resisting Spring Material

(1) Alloy tool steel

Hot-work tool steel (SKD series in JIS) contains relatively large amounts of Cr, Mo, W, V and shows high strength at high temperature. They can be used as materials for heat resisting springs. Chemical compositions and heat-treated hardness of these alloy tool steels are shown in Table 2.54 and 2.55. As shown in Fig. 2.210, which indicates tempering-hardness curves, high alloyed hot-work steels SKD and SKH series in JIS show typical secondary-hardening behavior where fine secondary carbides are precipitated in the matrix when tempered at 500–550°C. Since the fine carbide is stable under the elevated temperature lower than the tempering temperature, these hot-work tool steel can show good properties for heat resisting spring material. These hot-work tool steels are basically hyper-eutectoid steel. Therefore, large number of primary carbides and secondary carbides exist in the final heat-treated condition. Although it is important to control the size and the morphology of residual carbides, the manufacturing processes of hot-work tool steel are basically the same as those of spring steel and low-alloy steel.

(2) Fe-base superalloy

Austenitic stainless steel has face centered cubic lattice structure (fcc) which can be stable up to a certain temperature as high as above 1000°C and contains high Ni and Cr, to show high oxidization-resistance and corrosion-resistance under the elevated temperature. Fe-base superalloy can be said to

Table 2.53. High temperature resistant spring materials and their applications

Materials \ Application	Hydraulic	Brake	Clutch	Seal	Shock absorber	Steel Mill	Fuel-injector	Turbine	Engine	Torque converter	Exhaust system	Machine	Stationery	Copy machine	Binding	Others
Carbon steel																
SAE 1070				○												
〃 1090				○												
Carbon tool steel																
SK 5	○															
Special steel strip																
SK 5 M															○	
S 65 CM															○	
Hard drawn steel wire																
SWB				○					○							
SWC									○							○
Piano wire																
SWPA	○○		○		○				○			○		○		○
SWPB	○				○											
SWPV	○	○							○○○○○○○							
OT-wire (carbon steel)																
SWO-V			○			○			○○○○							
OT-wire (alloy steel)																
SWOCV-V		○	○○○						○○							
SWOSC-V	○○○○○	○○	○○○○○○				○		○○○○○○○○○○○							
SWOSM-B	○○															
Spring steel																
SUP									○			○				
SUP 6	○															
SUP 9	○○○○															
SUP 10	○															○○
Alloy tool steel																
SKD 4						○										
SKD 6																○
SKD 61															○	
Stainless steel																
SUS 301-CSP				○							○	○	○○	○		
〃 302-WPB													○			
〃 304-WP	○○○○		○	○○○○○○				○				○	○○○		○	○
〃 304-CSP				○○												○
〃 316																○
〃 316-WPA	○			○○○												
NAS 305				○												
PH stainless steel																
SUS 631 J 1-WPC	○							○			○○					
〃 631-CSP						○					○					○
15-7 PH				○												
Maraging steel												○				
18 Ni-Maraging						○										○
Ni-base super-alloy																
Hastelloy B				○												
〃 C				○												
Monel				○												
Inconel																○
〃 718						○										
〃 X-750	○			○○○		○					○				○	
Ni-Co super-alloy																
Refractaloy 26						○		○								

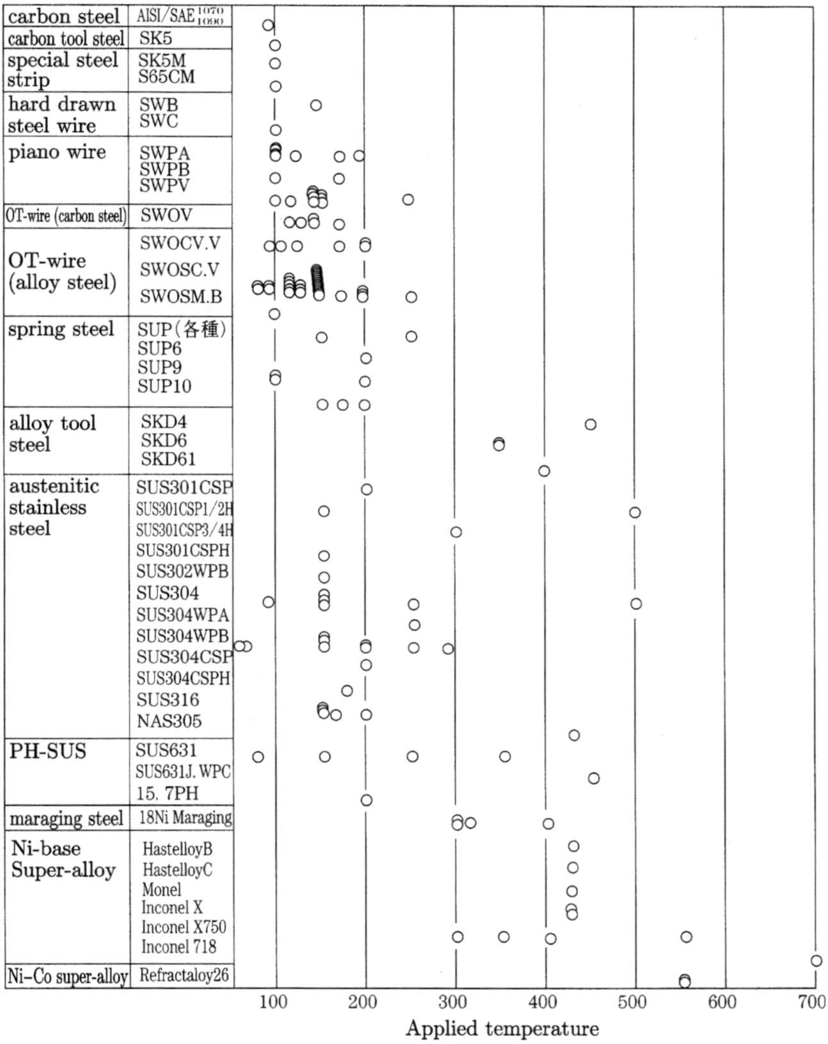

Fig. 2.209. Examples of spring materials for high temperature

be material which is added some chemical elements to austenitic stainless steel for improving the strength at high temperature. As heat-resistant alloy, this alloy can be positioned between martensitic stainless steel and Ni or Co-base superalloy. Table 2.56 shows the chemical compositions of typical Fe-base superalloys. Fe-base superalloy is strengthened by solid-solution hardening or precipitation-hardening. Interstitial elements such as C and N, and substitutional elements such as Mo and W are added for solid-solution hardening. As for precipitation hardening, carbides and inter-metallic compounds precipitate in metal matrix to strengthen Fe-base superalloy. The carbide such as

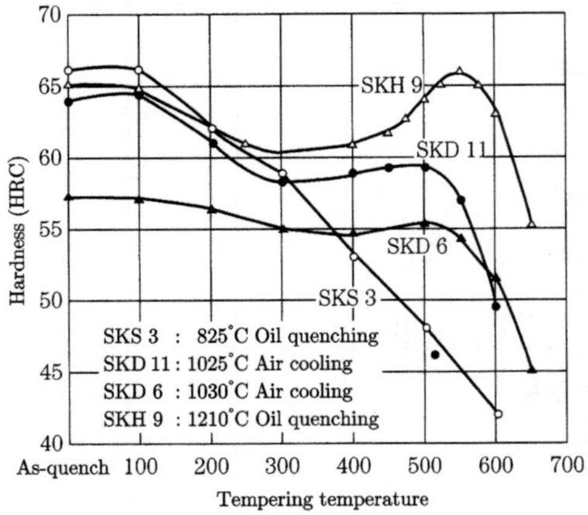

Fig. 2.210. Relations between tempering temperature and hardness of SKS3, SKD11, SKD 6, and SKH 9

Table 2.54. Chemical composition of hot-die steel

	Chemical compositions %										Applications
	C	Si	Mn	P	S	Ni	Cr	Mo	W	V	Co	
SKD4	0.25–0.35	≦ 0.40	≦ 0.60	≦ 0.030	≦ 0.020	—	2.00–3.00	—	5.00–6.00	0.30–0.50	—	
SKD5	0.25–0.35	0.10–0.40	0.15–0.45	≦ 0.030	≦ 0.020	—	2.50–3.20	—	8.50–9.50	0.30–0.50	—	Die, Tools, Shearing brade
SKD6	0.32–0.42	0.80–1.20	≦ 0.50	≦ 0.030	≦ 0.020	—	4.50–5.50	1.00–1.50	—	0.30–0.50	—	
SKD61	0.35–0.42	0.80–1.20	0.25–0.50	≦ 0.030	≦ 0.020	—	4.80–5.50	1.00–1.50	—	0.80–1.15	—	
SKD62	0.32–0.40	0.80–1.20	0.20–0.50	≦ 0.030	≦ 0.020	—	4.75–5.50	1.00–1.60	1.00–1.60	0.20–0.50	—	Die, Extrusion
SKD7	0.28–0.38	0.10–0.40	0.15–0.45	≦ 0.030	≦ 0.020	—	2.70–3.20	2.50–3.00	—	0.40–0.70	—	Die, Extrusion
SKD8	0.35–0.45	0.15–0.50	0.20–0.50	≦ 0.030	≦ 0.020	—	4.00–4.70	0.30–0.50	3.80–4.50	1.70–2.10	4.00–4.50	Die
SKT3	0.50–0.60	≦ 0.35	≦ 0.60	≦ 0.030	≦ 0.020	0.25–0.60	0.90–1.20	0.30–0.50	—	—	—	
SKT4	0.50–0.60	0.10–0.40	0.10–0.40	≦ 0.030	≦ 0.020	1.50–1.80	0.80–1.20	0.35–0.55	—	0.05–0.15	—	Forging die, Pressing die, Extrusion die
SKT6	0.40–0.50	0.10–0.40	0.20–0.50	≦ 0.030	≦ 0.020	3.80–4.30	1.20–1.50	0.15–0.35	—	—	—	

$Cr_{23}C_6$ precipitates at rather low rate in small quantity, to have relatively small hardness increase. This type of precipitation hardening alloy is said to be low precipitation type. Working this type of material below the recrystallization temperature, higher strength due to the acceleration of aging can be obtained, compared with the conventional hot-working and aging. A Hot-Cold-Work (HCW) treatment applied to 19-9DL and Timken 16-25-6 alloy shown in Table 2.56 is a typical example.

Table 2.55. Quenched-tempered hardness of hot die steel

TYPE	Designation	Heat-treatment condition, °C		Hardness
		Quenching	Tempering	HRC
Hot-die steel	SKD 4	1080 Oil quenching	600 Air cooling	≧ 42
	SKD 5	1150 Oil quenching	600 Air cooling	≧ 48
	SKD 6	1050 Air cooling	550 Air cooling	≧ 48
	SKD 61	1020 Air cooling	550 Air cooling	≧ 50
	SKD 62	1020 Air cooling	550 Air cooling	≧ 48
	SKD 7	1040 Air cooling	550 Air cooling	≧ 46
	SKD 8	1120 Oil quenching	600 Air cooling	≧ 48
	SKD 3	850 Oil quenching	500 Air cooling	≧ 42
	SKD 4	850 Oil quenching	500 Air cooling	≧ 42
	SKD 6	850 Oil quenching	180 Air cooling	≧ 52

Table 2.56. Chemical compositions of Fe-base superalloys

	Alloy	C	Si	Mn	Ni	Cr	Fe	Mo	Co	Al	Ti	Nb+Ta
Low precipitation hardening	NCF 800	0.04	0.35	0.75	32.5	21.0	bal.	—	—	0.4	0.4	—
	NCF 800H	0.08	0.35	0.75	32.5	21.0	bal.	—	—	0.4	0.4	—
	Incoloy 801	0.05	0.35	0.75	32.0	20.5	bal.	—	—	—	1.1	—
	Incoloy 802	0.35	0.35	0.75	32.5	21.5	bal.	—	—	—		—
	Incoloy 807	0.05	0.40	0.50	40.0	20.5	bal.	5.0W	8.0	0.2	0.3	—
	NCF 825	0.03	0.35	0.75	42.0	21.5	bal.	3.0	—	—	0.8	—
	N 155	0.15	0.5	1.5	20.0	21.0	bal.	3.0	20.0	—	—	1.0
	19-9DL	0.30	0.50	1.0	9.0	19.0	bal.	1.4	1.4W	—	—	0.5
	Timken 16-25-6	0.10	1.0	2.0	25.0	16.0	bal.	6.0	—	—	—	—
High precipitation hardening	SUH 660 (A 286)	0.05	0.5	1.4	26.0	15.0	bal.	1.3	—	0.2	2.0	—
	Discaloy	0.05	0.8	0.9	26.0	13.5	bal.	2.7	—	0.1	1.7	—
	V-57	0.05	0.5	0.3	27.0	14.8	bal.	1.3	—	0.3	3.0	—
	Refractaloy 26	0.03	0.8	1.0	38.0	18.0	bal.	1.2	20.0	0.2	2.7	—
	Alloy 901	0.05	0.1	0.1	42.5	12.5	bal.	5.7	—	0.2	2.8	—
	Incoloy 903	—	—	—	38.0	—	bal.	—	—	0.9	1.4	3.0
	Incoloy 907	—	0.15	—	38.0	—	bal.	—	13.0	0.03	1.5	4.7
	Incoloy 909	—	0.4	—	38.0	—	bal.	—	13.0	0.03	1.5	4.7
	Pyromet CTX-3	0.05	0.2	—	38.3	0.2	bal.	—	13.6	0.1	1.6	4.9

Superalloy using inter-metallic compounds, γ' (Ni_3Al) as precipitates shows a large amount of hardening under the normal aging, to be said as high precipitation type. Although this γ' phase, Ni_3Al can be used in Ni-base superalloy, in Fe-base superalloy the precipitation speed and amount can be increased by the substitution of aluminum to titanium.

(3) Nickel-base superalloy

Nickel-base superalloys are classified into solution-hardening type and γ' (gamma prime) precipitated type. As for manufacturing processes, they are also classified into wrought alloy and cast alloy.

Since addition of a large amount of Ti and Al makes the alloy less hot-workable, casting superalloy, directional-solidification and single crystal type, have been interested in gas turbine application. However, it is normally difficult to form a spring shape of the cast-Ni-base superalloy. Therefore, in this Chapter, wrought alloy is mainly described.

The chemical compositions of typical Ni-base superalloys are listed in Table 2.57 [2]. The solid-solution hardened type alloy of wrought Ni-base superalloy has high strength at the elevated temperature because of the solution hardening with heavy metals such as W and Mo. Although the strength of solid-solution type alloy is relatively low compared with γ' precipitation strengthened type alloy, the solid-solution strengthened superalloy has been formed to tube or sheet metals by using its good workability, drawability and weldability.

Precipitation strengthened alloy has much higher strength at the elevated temperature. Because of including a large amount of active metallic elements such as Al and Ti, the vacuum melting process becomes indispensable in alloy making processes.

As for γ' precipitation strengthened alloy, Inconel 718 is the most popular alloy in commercial applications. This alloy is hardened by the precipitation of

Table 2.57. Chemical compositions of Ni-base superalloys

Type	Alloy	Ni	Co	Fe	Cr	Mo	W	Al	Ti	Nb	Ta	C	B	Zr	Hf	others
solid-solution alloy	Inconel600	76.0		8.0	15.5							0.08				
	Nimonic75	76.0		3.0	19.5				0.4			0.10				
	Inconel690	60.0		9.5	30.0							0.03				
	Inconel601	60.5		14.1	23.0			1.4				0.05				
	Cabot214	75.0		2.5	16.0			4.5								Y: 0.01
	Hastelloy X	47.0	1.5	18.5	22.0	9.0	0.6					0.10				
	Inconel625	61.0		2.5	21.5	9.0		0.2	0.2	3.6		0.05				
	Hastelloy S	67.0		1.0	15.5	14.5		0.3					0.009			La: 0.05
	Inconel617	54.0	12.5		22.0	9.0		1.0	0.3			0.07				
	Haynes230	57.0			22.0	2.0	14.4	0.3				0.10				La: 0.02
γ' precipitation-hardening alloy	Inconel706	41.5		40.0	16.0			0.2	1.8	2.9		0.03				
	Inconel718	52.5		18.5	19.0	3.0		0.5	0.9	5.1		0.04				
	InconelX-750	73.0		7.0	15.5			0.7	2.5	1.0		0.04				
	Inconel751	72.5		7.0	15.5			1.2	2.5	1.0		0.04				
	Nimonic80A	76.0			19.5			1.4	2.4			0.06	0.003	0.06		
	Nimonic90	59.0	16.5		19.5			1.5	2.5			0.07	0.003	0.06		
	Nimonic105	53.0	20.0		15.0	5.0		4.7	1.2			0.13	0.005	0.10		
	Nimonic115	60.0	13.2		14.3			4.9	3.7			0.15	0.160	0.04		
	Nimonic263	51.0	20.0		20.0	5.9		0.5	2.1			0.06	0.001	0.02		
	M 252	55.0	10.0		20.0	10.0		1.0	2.6			0.15	0.005			
	Waspaloy	58.0	13.5		19.5	4.3		1.3	3.0			0.08	0.006			
	Rene 41	55.0	11.0		19.0	10.0		1.5	3.1			0.09	0.005			
	Udimet500	54.0	18.5		18.0	4.0		2.9	2.9			0.08	0.006	0.05		
	Udimet520	57.0	12.0		19.0	6.0	1.0	2.0	3.0			0.05	0.005			
	Udimet700	55.0	17.0		15.0	5.0		4.0	3.5			0.06	0.030			
	Udimet710	55.0	15.0		18.0	3.0	1.5	2.5	5.0			0.07	0.020			
	Udimet720	55.0	14.7		17.9	3.0	1.3		5.0			0.03	0.033	0.03		

largely body-centered tetragonal Ni_3Nb (γ'' gamma double prime) and have high strength at the intermediate temperature region. Ni_3Nb phase is stable up to the temperature of 700°C, however, over 700°C, it transforms to stable phase which does not contribute to the precipitation hardening. Therefore, Inconel 718 loses rapidly its strength in the temperature range from 650 to 800°C. Inconel 718 has been widely applied to gas turbine parts for aircraft and power generator and also to fasteners and springs.

Inconel X-750 is γ' precipitation hardened alloy which is added Al, Ti and Nb to Inconel 600. Due to its large amount of Ni, the heat resistance of Inconel X-750 is stable up to relatively high temperatures. The Inconel X-750 has been also applied to turbine blades of gas-turbine for power generator, and bolts and springs for high temperatures. Inconel 751, which is added higher Al to Inconel X-750, has the stability of γ' phase to keep high temperature strength up to 870°C. This alloy is mainly applied to the engine valve of automobiles. Nimonic 80A, adding Al and Ti to Nimonic 75, is γ' precipitation strengthened alloy. The properties of Nimonic 80A are similar to those of Inconel X-750 and Inconel 751. The γ' phase of Nimonic 80A is more stable due to higher Al amount with no Nb. Ni-base superalloy with Co shown in Table 2.57 can be applicable to higher temperature conditions compared with those without Co.

(4) Co-base superalloys

Table 2.58 shows Co-base superalloy and its chemical compositions. Cobalt has the hcp structure below 417°C and the fcc structure at the higher temperature. For higher temperature applications, improving phase-stability in fcc structure, makes it possible to have high temperature strength, oxidization resistance and corrosion resistance. Austenite stabilizing elements in Co alloy are Ni, Fe, Mn, Ti, Zr, Cu, B, C and so on. MP35N and MP159 alloy which are subject to thermo-mechanical treatment have a large amount of Fe and Ni in order to improve the cold workability. Co-base superalloy which has high temperature strength, contains Ni of 10% or more. Since Co-base superalloy

Table 2.58. Chemical compositions of Co-base superalloys

Alloy	Chemical composition (%)									
	C	Fe	Ni	Cr	Mo	W	Nb	Ta	Co	Others
AiResist 213	0.18			19		4.7		6.5	66	3.5Al, 0.2Zr, 0.1Y
Elgiloy	0.1	bal.	15	20	7				40	2Mn, 0.04B
Haynes 188 (UNSR30188, HA188)	0.1	< 3	22	22		14			39	0.07La, Si, Mn
L605 (HA25, UNSR30188, AISI670)	0.1		10	20		15			53	Si, Mn
Hastelloy 150		bal.		28					50.5	0.07Si
MAR-M918	0.05		20	20				7.5	52.5	0.1Zr
MP 35N (UNSR30035)			35	20	10				35	
MP 159		9	25.5	19	7		0.6		36	0.2Al, 3Ti
Nivco	0.03		22.5						bal.	0.22Al, 1.8Ti
Stellite 6B	1.1	< 3	< 3	30	< 1.5	4.5			bal.	2Mn, 2Si
S-816	0.4	3	20	20	4	4	4		bal.	1.2Mn
V-36	0.3	2.5	20	25	4	2	2		bal.	1Mn
VM 103	0.5	< 0.1		3		25			bal.	

cannot be strengthened by the precipitation hardening in contrast to Ni-base superalloy, it can be characterized by a solid-solution strengthened fcc matrix in which a small quantity of carbides is distributed. The element for solution hardening in Co-base alloy are W, Mo, Ta, Nb, Ni and Cr. The adding of W is the most effective. However, while Mar-M918 replaced W to Ta has been found to improve oxidization resistance without decreasing the strength, the superalloy added W and Ta has been in commercial use.

The carbides observed in Co-base superalloy are MC, M_7C_3, $M_{23}C_6$, and M_6C. The type of carbide which precipitates in matrix can be decided by the ratio of Cr to C and the quantity of Mo, Nb, Ta, Zr and Ti. Oxidization resistance of and corrosion resistance of Co-base superalloy mainly depends upon the Cr content in the alloy. Therefore, Co-base superalloy has a large quantity of Cr compared with Ni-base superalloy. Addition of La, Y and Th can raise the adhesion of surface protecting film to improve oxidization and corrosion resistance. Because the solubility of alloy elements in Co-base superalloy is small in comparison with that of Ni-base superalloy, it becomes important to make alloy composition design to avoid the precipitation of harmful intermetallic compounds.

2.2.3 Manufacturing Process of Heat-Resisting Spring Material

The manufacturing process of superalloy is basically the same as that of stainless steel especially after hot forming processes. However, since the quality requirements for superalloy are severe and complicated chemical compositions are involved, a sophisticated melting process should be applied in manufacturing superalloy. Fe–Ni–Cr alloy and Ni–Cr–Fe alloy such as Incoloy 800 and Inconel 600, can be melted by AOD (Argon–Oxygen–Decarburization) process which is the same melting method as stainless steel. However, most superalloys have been manufactured with re-melting ingots by ESR (Electro-Slag-Remelting) or VAR (Vacuum-Arc-Remelting). Especially, the manufacturing of superalloy containing a large amount of reactive elements such as Al and Ti should be made by a sophisticated melting process such as Vacuum-Induction-Melting and Vacuum-Arc-Remelting.

Figure 2.211 shows a schematic illustration of Vacuum–Arc–Remelting furnace. The following effect can be expected on re-melting under the vacuum pressure of 0.1–1 Pa [3].

① Reducing the dissolved gas, O_2 and N_2
② Decreasing volatile impurity elements.
③ Improving the cleanliness of metal matrix by eliminating non-metallic inclusion, oxide and nitride.
④ Reducing segregations of alloy elements using directional solidification.

ESR process also can be said to aim the similar effect to the VAR process.

The ingot obtained by melting-process is formed to bar or wire required for spring material. Since superalloy shows poor hot-formability especially at

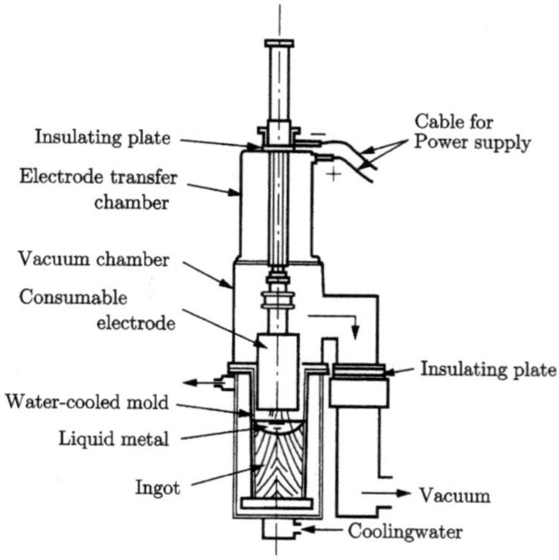

Fig. 2.211. Vacuum arc re-melting furnace

Table 2.59. Forging temperature ranges and forgeabilities for several superalloys

Alloy	Forging temperature °C	Forgeabilities*
N 155 (SUH 661)	1150– 950	1
L 605	1180– 955	3
S 816	1180– 950	3
Incoloy 800 (NCF 2)	1230–1010	1
Incoloy 801	1230–1010	1
Hastelloy B	1190– 980	4
Hastelloy C	1230–1010	3
Hastelloy X	1200– 980	3
Inconel 600 (NCF 1)	1200– 950	1
A 286 (SUH 660)	1180– 950	1
René 41	1180– 950	4
Udimet 500	1200–1000	4
Inconel X-750 (NCF 3)	1200– 980	2
Inconel 700	1175–1000	4
Inconel 718	1120– 930	3
Nimonic 75	1200– 950	2
Nimonic 80A	1150– 950	2
Nimonic 90	1150– 950	3

*Higher number shows more difficult forging.

high deformation rate, it is easy to crack at blooming mill. Therefore, hot forging process instead of hot rolling can be normally applied. Table 2.59 shows forging temperature ranges and the forgeability for several superalloys. The superalloys which possess higher strength at high temperature tend to show smaller forging temperature ranges. Hot extrusion process can be applied for the superalloys which contain higher Al and Ti.

In case of hot rolling, high reduction pass is difficult. In order to keep deformation resistance lower, rolling is normally applied with a slow speed. The considerations regarding to product quality; such as dimension, tolerances, and surface roughness, can be similar to those of stainless steel.

2.2.4 Characteristics of Superalloy

As for Fe-base alloy and Ni-base superalloy, physical properties are shown in Table 2.60 and 2.61 respectively, and their mechanical properties [4] are shown in Tables 2.62, 2.63 and 2.64. Ni-base superalloys normally have larger specific gravity and smaller coefficient of thermal expansion, compared with Fe-base superalloys.

Proof stress and creep strength at elevated temperatures for typical superalloys are shown in Fig. 2.212 and Fig. 2.213. γ' strengthening type superalloys have high strength at relatively low temperature ranges, compared with solid-solution strengthening type. However, the strength of γ' strengthening superalloys drops rapidly at higher temperature ranges. Comparing superalloys with a same strengthening mechanism, Co-base and Ni-base superalloys show much higher strength in comparison with Fe-base superalloys. For the

Table 2.60. Physical properties of Fe-base superalloy

Alloy	specific gravity	specific heat (J/kg · °C)	Thermal conductivity (W/m · °C)		Coefficient of thermal expansion (10^{-6}/°C)	
			Temp (°C)		Temp. ranges (°C)	
NCF 800	7.95	460	800	24.7	20–800	18.0
NCF 800H	7.95	460	800	24.7	20–800	18.0
Incoloy 801	7.97	452	800	24.5	20–800	18.4
Incoloy 802	7.83	425	870	24.2	20–870	18.2
Incoloy 807	8.32				20–870	17.6
NCF 825	8.14	440	870	24.1	20–540	14.0
N 155	8.20	399	650	27.8	20–700	17.3
19-2 DL	7.93	383	650	28.5	20–650	17.9
Timken 16-25-6	8.06	402	600	34.8	20–760	19.4
SUH 660 (A 286)	7.94	383	650	32.7	20–760	18.6
Discaloy	7.97	433	650	30.6	20–760	17.6
V-57	7.96				20–760	19.4
Refractaloy 26	8.12	414			20–650	15.1
Alloy 901	8.21				20–540	15.3
Incoloy 903	8.25	435	540	20.9	20–540	8.6
Incoloy 907	8.33	431			20–540	7.7
Incoloy 909	8.30	422	600	21.8	20–540	7.7

Table 2.61. Physical properties of wrought Ni-base superalloys

Alloy	Specific gravity	Melting temp. °C	Specific heat (J/kg · °C)			Thermal conductivity (W/m · °C)			Coefficient of thermal expansion		Electrical resistivity ($10^{-8}\Omega \cdot m$)
			21°C	538°C	871°C	21°C	538°C	871°C	538°C	871°C	
Inconel 600	8.41	1,355–1,415	445	555	625	14.8	22.8	28.8	15.1	16.4	103
Nimonic 75	8.37		460						14.7	17.0	
Inconel 690	8.14	1,345–1,375				13.3	22.8	27.8			115
Inconel 601	8.05	1,300–1,370	450	590	680	11.3	20.0	25.7	15.3	17.1	119
Cabot 214											
Hastelloy X	8.21	1,260–1,355	485		700	9.1	19.6	26.0	15.1	16.2	118
Inconel 625	8.44	1,290–1,350	410	535	620	9.8	17.5	22.8	14.0	15.8	129
Hastelloy S	8.76	1,335–1,380	405	495	595		20.0	26.1	13.3	14.9	
Inconel 617	8.36	1,330–1,375	420	550	630	13.6	21.5	26.7	13.9	15.7	122
Haynes 230	8.8	1,300–1,370	397	473	595	8.9	18.4	24.4	14.0	15.2	125
Inconel 706	8.08	1,335–1,370	445	580	670	12.6	21.2		15.7		
Inconel 718	8.22	1,260–1,335	430	560	645	11.4	19.6	24.9	14.4		125
Inconel X-750	8.25	1,395–1,425	430	545	715	12.0	18.9	23.6	14.6	16.8	122
Inconel 751	8.16	1,395–1,425							14.9	16.9	124
Nimonic 80A	8.19	1,360–1,390	460			8.7	15.9	22.5	13.9	15.5	124
Nimonic 90	8.00	1,335–1,360	460	585	670	9.8	17.0		13.9	16.2	118
Nimonic 105	7.85		420	545	670	10.8	18.6	24.0	13.9	16.0	131
Nimonic 115	8.36		460			10.7	17.6	22.6	13.3	16.4	139
Nimonic 263	8.25	1,315–1,370	460			11.7	20.4	26.2	13.7	16.2	115
M 252	8.19	1,330–1,355				11.8	20.4	26.2	13.0	15.3	
Waspaloy	8.25	1,315–1,370				10.7	18.1	24.1	14.0	16.0	124
René 41	8.02	1,300–1,395		545	725	9.0	18.0	23.1	13.5	15.6	130.8
Udimet 500	8.21	1,260–1,405				11.1	18.3	24.5	14.0	16.1	120.3
Udimet 520	7.91	1,205–1,400									
Udimet 700	8.08			575	590	19.6	20.6	27.7	13.9	16.1	
Udimet 710											
Astroloy	7.91								13.9	16.2	
René 95						8.7	17.4				

Table 2.62. Creep-rupture strength of Fe-base superalloys (MPa)

Alloy	649°C		760°C	
	100 h	1,000 h	100 h	1,000 h
NCF 800	220	165	82	66
NCF 800H	230	180	95	70
Incoloy 801	228	180		
Incoloy 802	275	170	145	110
Incoloy 807	248	179	145	105
NCF 825	—	—	—	—
N 155	359	295	166	140
19-9 DL	303	255	138	69
Timken 16-25-6	345	248	125	75
SUH 660 (A 286)	421	315	173	105
Discaloy	359	275	125	60
V-57	530	485	241	
Refractaloy 26	530	434	275	159
Alloy 901	620	525	—	205
Incoloy 903	586	510	—	—
Incoloy 907	—		—	—
Incoloy 909	—	345	—	—

Table 2.63. Mechanical properties of wrought Ni-base superalloys

Alloy	Heat treatment	Elastic constant (GPa)					1,000 hr creep-rupture strength (MPa)			
		21°C	540°C	650°C	760°C	870°C	650°C	760°C	870°C	980°C
Inconel600 bar	1,120°C × 2 hAC	214	184	176	168	157			30	15
Nimonic75 bar	1,050°C × 1 hAC	221	186	176	170	156	170	50	5	
Inconel690	1,040°C × 1 hAC	210	180	171	162	153	100	50		
Inconel601 plate	1,150°C × 2 hAC	207	175	166	155	141	195	60	30	15
Cabot214	1,120°C								30	15
HastelloyX plate	1,175°C × 1 hAC	197	161	154	146	137	215	105	40	15
Inconel625 bar	1,150°C × 1 hWQ	208	179	170	161	148	370	160	50	20
HastelloyS bar	1,065°AC	212	182	174	166			90	25	
Inconel617 bar	1,175°AC	210	176	168	160	150	360	165	60	30
Haynes230	1,230°AC	211	184	177	171	164		125	55	15
Inconel706 bar	980°C × 1 hAC + 845°C × 3 hAC + 720°C × 8 hFC + 620°C × 8 hAC	210	179	170			580			
Inconel718 bar	980°C × 1 hAC + 720°C × 8 hFC + 620°C × 8 hAC	200	171	163			595	195		
InconelX-750 bar	1,150°C × 2 hAC + 845°C × 24 hAC + 750°C × 20 hAC	214	184	176	166	153	470		50	
Inconel751	1,150°C × 4 hAC + 845°C × 2 hFC + 705°C × 20 hAC	213					465		70	
Nimonic80A bar	1,080°C × 8 hAC + 705°C × 16 hAC	219	188	179	170	157	420	160		
Nimonic90 bar	1,180°C × 8 hAC + 705°C × 16 hAC	226	190	181	170	158	455	205	60	
Nimonic105 bar	1,150°C × 4 hAC + 1060°C × 16 hAC + 850°C × 16 hAC	223	186	178	168	155		330	130	30
Nimonic115 bar	1,190°C × 1.5 hAC + 1,100°C × 6 hAC	224	188	181	173	164		420	185	70
Nimonic263 plate	1,150°C × 0.2 hWQ + 800°C × 8 hAC	222	190	181	171	158				
M252 bar	1,040°C × 4 hAC + 760°C × 16 hAC	213	184	177	168	158	565	270	95	
Waspaloy bar	1,080°C × 4 hAC + 845°C × 24 hAC + 760°C × 16 hAC						615	290	110	
René41 bar	1,065°C × 4 hAC + 760°C × 16 hAC						705	345	115	
Udimet500 bar	1,080°C × 4 hAC + 845°C × 24 hAC + 760°C × 16 hAC	222	191	183	173	161	760	325	125	
Udimet520 bar	1,105°C × 4 hAC + 845°C × 24 hAC + 760°C × 16 hAC						585	345	150	
Udimet700 bar	1,175°C × 4 hAC + 1,080°C × 4 hAC + 845°C × 24 hAC + 760°C × 16 hAC	224	194	186	177	167	705	425	200	55
Udimet710 bar	1,175°C × 4 hAC + 1,080°C × 4 hAC + 850°C × 24 hAC + 760°C × 16 hAC	222					870	460	200	70
Udimet720 bar	1,115°C × 2 hAC + 1,080°C × 4 hOQ + 650°C × 24 hAC + 760°C × 8 hAC						670			
Astoloy bar	1,095°C × 4 hOQ + 870°C × 8 hAC + 980°C × 4 hAC + 650°C × 24 hAC + 760°C × 8 hAC						770	425	170	55
René95	900°C × 24 h + 1,105°C × 1 hOQ + 730°C × 64 hAC	222	191	183	173	161	860			

Table 2.64. Typical properties of Co-base superalloys

	Tensile strength (0.2% proof stress) [MPa]				
	Room temp.	540°C	760°C	980°C	Features and applications
AiResist213	1,120 (625)		485 (385)		sheets, tubings
Elgiloy	690 (480)				
Haynes188 (UNS R30188, HA188)	960 (485)	740 (305)	635 (290)	210 (140)	high oxidization resistance up to 1100°C, gas turbine combustors
L605 (HA25, UNS R30188, AISI670)	1,005 (460)	820 (250)	455 (260)		
Hastelloy 150	925 (315)				Ta added instead of W
MAR-M918	895 (895)				fastner (bolt) used up to 650°C
MP35N (UNS R30035)	2,025 (1,620)				
MP159	1,895 (1,825)	1,560 (1,495)			
Nivco					
Stellite6B	1,101 (635)				steam turbine blades, high abrasion and corrosion resistance for erosion-shield
S-816	880 (700)	825 (430)	430 (380)		600–1150°C heat-resistance
V-36	2,070 (1,100)	1,520 (900)			guide vanes after-burners, Springs
VM103	1,100 (690)		690 (480)	350 (275)	NASA alloy excellent erosion resistance

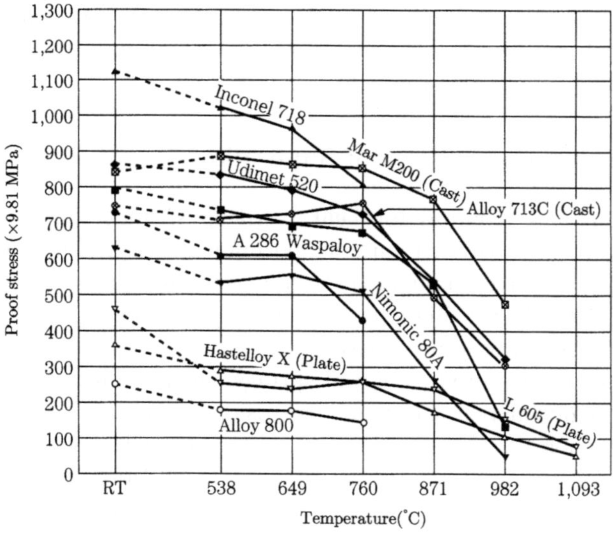

Fig. 2.212. Proof stress of typical superalloys

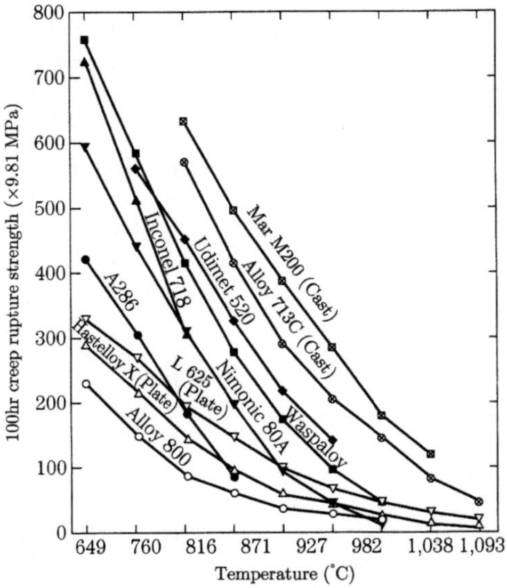

Fig. 2.213. 100 hr creep rupture strength of typical superalloys

applications of high temperature resistant springs, strain-hardening by cold working can obtain higher strength. However, softening in practical uses, can occur, depending upon the applied temperature.

2.2.5 Spring Characteristics of Superalloy

The characteristics required for heat resisting springs are to keep having enough fatigue strength without any permanent set at high temperatures. Figure 2.214 shows stress relaxation properties of heat resisting springs made of Inconel X-750. At the high temperature applications, the permanent set and stress relaxation become a critical issue.

The properties required to actual springs, such as fatigue strength, creep strength, oxidization resistance and corrosion resistance, depend upon the environment where springs can be used. Therefore, it is essential to select the appropriate spring material considering those conditions. Especially, at high temperature, it should be noted that residual stress can be decreased and microstructure can change to have the material softened and embrittled under long time exposures at high temperature. Although heat resisting material contains a large amount of alloy elements to prevent softening and to have the precipitation hardening stabilized for a long period, it can be required to take the durable years of springs in use into consideration.

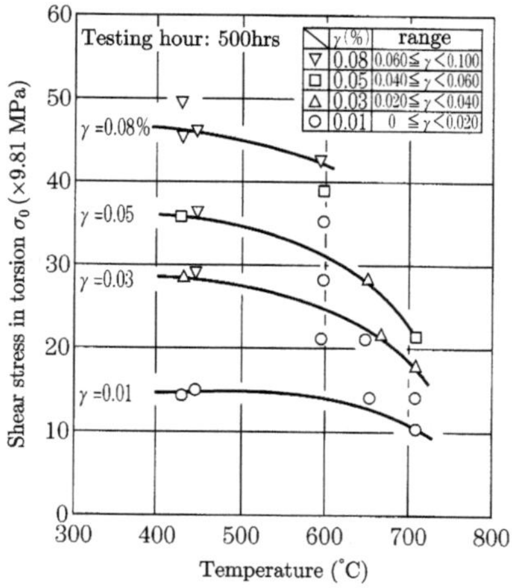

Fig. 2.214. Shear stress-temperature diagram in torsion of Inconel X-750 (γ: residual shear strain)

References

1. Heat resisting material committee of JSSR, Data Book Elevated-temperature strength of heat resisting materials for springs (in Japanese), Technical Report of JSSR (1986) p. 4
2. N.S. Stoloff: Metals Handbook, 10th ed. vol. 1 (1990) p. 950
3. A. Hauff: Proc. of 9th int. vacuum metallurgy conf. on special melting, San Diego (USA), (1988) p. 411
4. C.T. Sims, N.S. Stoloff and W.C. Hagel: SUPERALLOYS II (1987), John Wiley & Sons.
5. Inconel Alloy X-750, Huntington Alloy Handbook (1979)

2.3 Constant Modulus and High Modulus Spring Material

2.3.1 Constant Modulus Spring Material

By the French physicist, Guillaume, Fe–36%Ni alloy of which thermal expansion is extremely small around room temperature, was invented in 1896. This alloy was named the Invar alloy based on the French word “invariable” that is also the same meaning in English. Although the temperature dependency of thermal expansion coefficient of this alloy was small, the temperature dependency of elastic coefficient was large. After that, the research by Guillaume was continued, then the alloy of which the temperature dependency of elastic coefficient was small, namely the Elinvar alloy was developed in 1919. This alloy was called the constant elastic alloy or the elastic constant steel, but origin of the word came from a French word of "élasticité invariable" having the meaning of invariable elasticity.

As for the usual metal and alloy, when the temperature rises, the distance between atoms of metal crystal becomes large and the restoring force becomes weak, so the elastic coefficient decreases. The temperature coefficient is approximately 10^{-4}/°C. However, regarding the precise equipments which utilizes elastic characteristic, such as aircraft instrument parts, spiral springs for clock, springs for seismometers, precise dynamometers for calibration and weighing appliances, some problems may cause when the elastic coefficient changes with the surrounding temperature.

For example, a mechanical type watch using a spiral spring made of the conventional material has 16 seconds delay per day as temperature becomes high by 1°C. This means that time lag and advancing can change with the temperature change of seasons. As for the applications for precise equipments, the temperature coefficient to elastic modulus can be desirable to be 10^{-5}–10^{-7}/°C. The elinvar alloy can be used for this kind of applications. Table 2.65 shows the comparison of temperature dependency to elastic coefficient between Ni-Span C which is a kind of Elinvar alloy, and various alloys [2,3].

The alloy which shows this kind of characteristics was developed in large numbers. They are called elinvar type alloy and can be classified into Fe–Ni type alloy and Fe–Co type alloy from the chemical compositions.

(1) Fe–Ni type elinvar alloy

Figure 2.215 shows the effect of Ni and Cr contents on the temperature coefficient of modulus of rigidity (or thermoelastic coefficient) for Fe-Ni-Cr

Table 2.65. Temperature coefficient of various alloys [2,3]

Material	Average temperature coefficient of rigidity modulus (−50–50°C) $\times 10^{-5}$/°C	Modulus of elasticity GPa	Change of temperature coefficient −46–66°C %
Ni-Span C	0	186	0
0.9% Carbon steel	−25.2	206	2.9
17% Cr Stainless steel	−25.4	196	2.9
3.5% Nickel steel	−25.9	206	3.0
Monel	−28.4	172	3.3
Beryllium copper	−33.5	127	3.9
18% Nickel silver	−33.5	132	3.9
Phosphorus bronze	−36.4	108	4.2
18-8 Stainless steel	−39.1	181	4.5
70/30 Brass	−37.1	98	4.3
Duralmin	−56.0	69	6.5
36% Ni Invar	−48.2	144	5.4

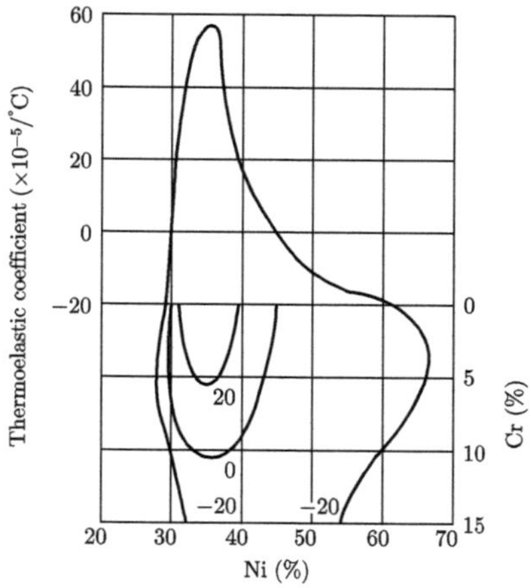

Fig. 2.215. Temperature coefficient of Rigid ratio for Fe–Ni–Cr based alloy g [1]

and Fe-Ni alloys. When no Cr is contained in the alloys, the temperature coefficient has its maximum value and zero value at the Ni content of 34% and both at 30% and 43%, respectively. The temperature coefficient of Fe-Ni-Cr alloy is shown in the lower part of Fig. 2.215. The ternary alloy's chemical compositions that show the zero thermoelastic coefficient are numerous and are along the line zero in the Fig. 2.215. However, these ternary alloys with zero thermoelastic coefficient have rather low modulus of elasticity and tensile strength. To improve these drawbacks, C and W are alloyed further, which is called the Elinvar alloy. In Fe-Ni base alloys, Meterinvar is alloyed with Mo and W, Ni-SpanC and Durinvar are alloyed with precipitation hardening elements, Ti and Al, and Nivarox is alloyed with W, Cr, Mo and Be. Table 2.66 shows the chemical compositions of constant modulus alloys and their various characteristics [2,4,5,7].

Among these alloys, Fe–Ni–Cr–Ti–Al type alloy such as Ni-Span C, YniC and sumi Span-3 have been widely used. Nivarox shows weak magnetism and is influenced little with the magnetic field. Therefore, it has been used widely as a spiral spring of a watch. However, the quantity consumed has been decreasing from the time before due to the spread of the quartz (crystal oscillation) and the digital watch.

Fe–Ni–Cr–Ti–Al type alloy represented by Ni-Span C is precipitation hardening type alloy, that is applied cold working after 950–1,000°C solution heat treatment, and followed by precipitation hardening process. The aging for precipitation hardening can be conducted at such a condition as 600°C × 48 hrs. to 730°C × 3 hrs. for post solution heat treatment or as 600°C × 4 hrs. to 730°C × 1 hr. for post 50% cold working, depending upon the size of material to be heat treated [6]. During the aging for precipitation hardening processing, some part of Ti and Cr form carbides with the carbon which is included as the impurity, so the characteristics can change due to the reduction of effective

Table 2.66. Chemical compositions and characteristics of constant elastic materials

Classification		Commodity name	Chemical compositions (mass %) C	Si	Mn	Ni	Co	Cr	W	Mo	Cu	V	Al	Ti	Be	Cited reference	Coefficient of thermal expansion (×10⁻⁶)	Temperature coefficient of elastic ratio (×10⁻⁵/°C)	Tensile strength MPa
Fe–Ni base	Fe–Ni–Cr	Elinvar	0.8	1~2	1~2	36		12	1~3	—	—	—	—	—	—	1), 2)	0.8	±0.37	735
	Fe–Ni–Cr–Co	Super Elinvar	—	—	0.7	36~39	5~10	5~10	—	—	—	—	—	—	—	1), 2)	—	—	—
	Fe–Ni–Cr–Mo	Iso–elastic	—	—	—	36	—	7~8	—	0.5	—	—	—	—	—	1), 2)	2.2	−1.1~+0.8	1177
			0.1	0.5	0.6	36	—	7.2	—	0.5	0.2	—	—	—	—	3)	7.2	−3.6~+2.7	
	Fe–Ni–Cr–W	Metelinvar	0.6	—	2	40	—	6	3	1.5	—	—	—	—	—	1), 2)	—	—	1432
	Fe–Ni–Cr–Ti	Elinvar Extra	—	—	—	43	—	5	—	—	—	—	—	2.8	—	2)	—	—	—
			0.6	0.5	0.5	42	—	5.5	—	—	—	—	0.6	2.5	—	3)	7.5	0	
			0.04	0.5	0.6	43	0.35	5	—	—	—	—	0.3	2.75	—	1)	—	—	
			1		1	35	—	5	2	—	—	—	—	—	—	1)	8	—	
		sumi span–3	—	—	—	42	—	5.5	—	—	—	—	—	2~3	—	4)	—	±1.0	
	Fe–Ni–Cr–Ti–Al	Ni–Span C	—	—	—	41~43	—	4.9~5.5	—	—	—	—	≦0.1	2.1~2.5	—	2)	2.5	±0.6	612~1373
			≦0.06	≦0.1	≦0.8	42	—	5.2	—	—	—	—	—	2.3	—	3)	—	—	
			0.02	0.33	0.52	42.27	—	5.32	—	—	0.05	—	0.46	2.46	—	1)	8.1	±1.8	
		YniC	0.03			41~43	—	5.1~5.5	—	—	—	—	0.5~0.6	2.2~3.0	—	1)	—	—	
	Fe–Ni–Mo	Vibralloy	≦0.02	—	0.1~0.65	38~43	≦0.5	—	—	15.7	—	—	—	—	—	2)	—	—	—
			0.3	—	2.2	34.8	1.2	—	—	—	—	—	—	—	—	2)	—	—	—
			—	—	—	40	—	—	—	10	—	—	—	—	—	3)	—	—	
			—	—	—	39	—	—	—	9	—	—	—	—	—	1)	—	0.8	
	Fe–Ni–Al–Ti	Durinvar	0.1	—	2	40~42	—	—	—	6	—	—	1.9~2.0	2.1	—	2)	—	±1.0	—
	Fe–Ni–Be–W	Nivarox	—	0.1	0.8	36	—	—	8	—	—	—	—	—	1	2)	—	—	—
	Fe–Ni–Be–Mo		—	0.1	0.8	36	—	—	—	7	—	—	—	—	1	2)	—	—	—
			—	—	—	30	—	—	—	9	—	—	—	—	1	1)	—	—	
			—	—	—	40	—	—	—	9	—	—	—	—	1	1)	—	—	
	Fe–Ni–Be–Cr–Ti		—	—	—	36	—	8	—	—	—	—	—	1	1	1)	7.5	±2.5	
			0.2	1	1	38	—	8			—	—	—	1	1	1)	—	—	
	Fe–Ni–Be–Ti		—	—	—	36	—	—	—	—	—	—	—	1	1	1)	—	—	
Fe–Co base	Fe–Co–Cr	Co–elinvar	—	—	—	—	60	15	—	—	—	—	—	—	—	2)	14	−0.9	—
			—	—	—	16.7	34.6	10.8	—	—	—	—	—	—	—	2)	7.7	0.8	—
			—	—	—	15~20	25~30	8~12	—	—	—	—	—	—	—	3)	—	±2.0	
			—	—	—	—	60	10	—	—	—	—	—	—	—	1)	5.1	−0.2	
			—	—	—	9.1	43.6	12.7	—	—	—	—	—	—	—	1)	7.4	0	
			—	—	—	23.1	27.7	10	—	—	—	—	—	—	—	1)	8.1	−0.3	
	Fe–Co–V	Velinvar	—	—	—	—	60	—	—	—	—	10	—	—	—	1), 2)	7.4	−0.7	—
			—	—	—	20	37.6	—	21.5	—	—	10	—	—	—	2)	7.8	−0.5	—
			—	—	—	30	20	—	—	—	—	10	—	—	—	1)	11.6	−0.7	
	Fe–Co–W	Tungelinvar	—	—	—	—	50	—	19	—	—	—	—	—	—	2)	7.4	−0.7	—
			—	—	—	—	50	—	21.5	—	—	—	—	—	—	1)	7.4	−0.7	
			—	—	—	10	39	—	19	—	—	—	—	—	—	1)	7.8	−4.3	
	Fe–Co–Mo	Moelinvar	—	—	—	—	50	—	—	17.5	—	—	—	—	—	2)	9.6	−0.2	—
			—	—	—	30	10	—	—	15	—	—	—	—	—	2)	9.8	−0.4	—
	Fe–Co–Ni–Mo–Cr	Elcolloy	—	—	—	15~18	35~40	5	4~5	5	—	—	—	—	—	3)	—	—	
			—	—	—	16	36	5	4	4	—	—	—	—	—	1)	9.0	+0.5	
			—	—	—	15	35	5		5	—	—	—	—	—	1)	5.0	−0.2	

1) Metal manual Revision 3edition 2) Steel material manual 1967 3) Metal 1971 November 1st number 4) Metal 1970 July 15th number

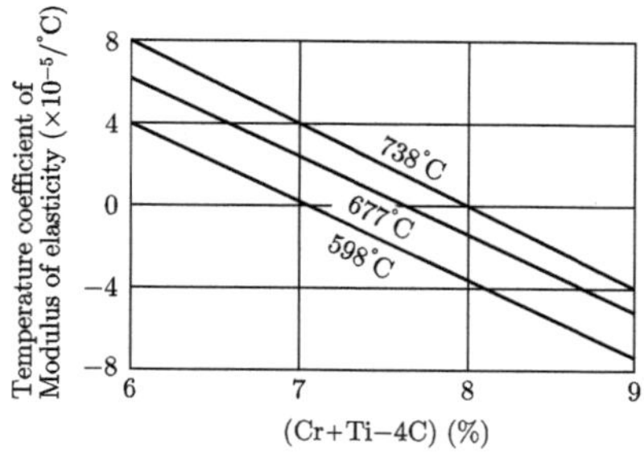

Fig. 2.216. The influence of C, Ti, Cr amount of Ni-spanC and Precipitation hardening processing temperature for Thermoelasticity factor [2]

content of Ti and Cr, depending upon the precipitation hardening condition. Therefore, in order to obtain the intended constant modulus characteristics, the precipitation hardening condition should be determined taking carbon contents (or effective content of Ti and Cr) into consideration. Figure 2.216 shows the influence of precipitation hardening temperature and C, Ti, and Cr contents on Ni-SpanC alloy's thermoelastic coefficient (temperature coefficient of modulus of elasticity) [2].

(2) Fe–Co base elinvar type alloy

Although Fe–Ni based alloy has been in the center of research works, in Japan, Fe–Co based alloy has become popular after the constant elastic alloy (Co-elinvar) had been found by Masumoto and Saito of Institute of Materials Research, Tohoku University. Figure 2.217 shows the effect of Co content on thermoelastic coefficient for Fe-Co, Fe-9%V-Co, and Fe-10%Cr-Co alloy [1]. As for the Fe-Co base alloy without Cr and V, there is a gradual negative dependency, however, with adding Cr or V, a typical characteristic of Elinvar type alloy can appear. The similar phenomenon occurs not only with Cr and V but also with Mo and W. Figure 2.218 shows the chemical compositional curves on which the thermoelastic coefficient becomes zero for Fe-Co-Cr, Fe-Co-Mo and Fe-Co-V alloy. In case of Fe-Co-W alloy, the thermoelastic coefficient does not reach to the plus range. However, when Ni is added, it becomes possible to improve the characteristic and to save the Co content. Co-elinvar is named for Fe-Co-Cr alloy, Moelinvar is for Fe-Co-Mo, Velinvar is for Fe-Co-V, and Tugelinvar is for Fe-Co-W. Since Nivarox of Fe-Ni alloy is added Be, it is said that wide scatterings of the product quality may occur due to the oxidization of Be [7]. Since Co-elinvar is stable in the sense of productivity, and shows small scatterings between the production lots, it had been used for more than 90% of spiral spring (hair spring) material of watches in Japan [7].

The Elinvar type alloy is designed in metal physics standpoint to make magnetostriction[1] large enough to compensate the normal temperature coefficient at room temperature. One characteristic of this kind of alloy is easy

[1] Magnetostriction: Strain when it is subjected to a magnetic field.

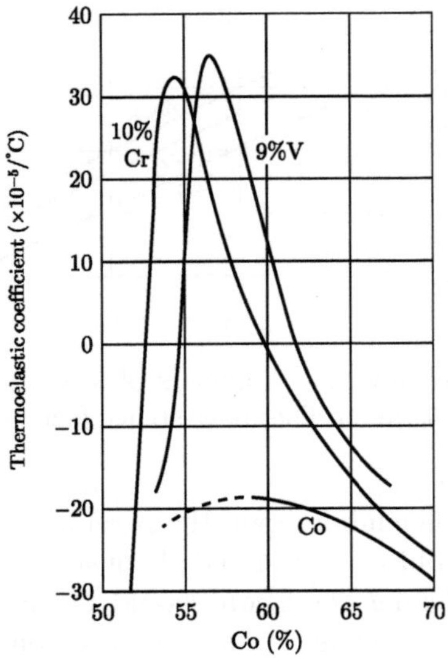

Fig. 2.217. Effect of Co content on thermoelastic coefficient for Fe–Co, Fe–Co–10%Cr, Fe–Co–9%V [1]

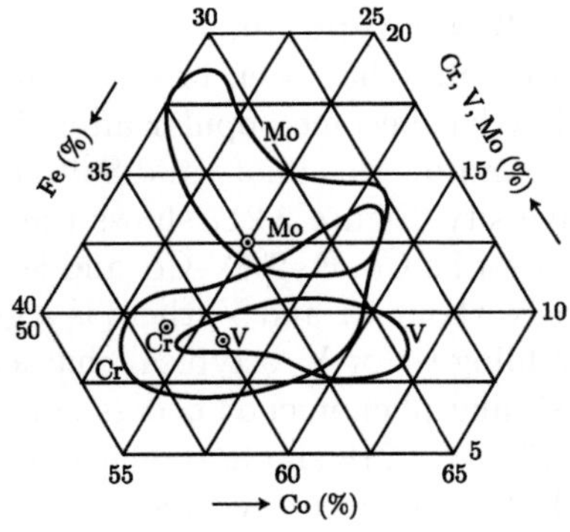

Fig. 2.218. Chemical compositional curves for Fe–Co–Cr, Fe–Co–V, Fe–Co–Mo alloys with thermoelastic coefficient $g = 0$, Outside is $g < 0$, ◉ mark shows positive max.value [1].

to be influenced by external magnetic condition. When the external magnetic field becomes large, thermoelastic coefficient gradually increases. At the magnetic saturation condition, it shows the same temperature dependency as the paramagnetic material that is magnetized slightly in magnetic field direction, and no longer shows the original characteristic of Elinvar type alloy. In addition, the Elinvar type alloy magnetized once shows the temperature dependency of the elastic coefficient which is similar to the general metal, if having residual magnetization state after removing the magnetic field. Therefore, it is necessary to sufficiently pay attention to the external magnetic field for using the Elinvar type alloy.

2.3.2 High Modulus Spring Material

High modulus spring material is Co base alloy containing 35% to 45% Co or Co–Ni base alloy. It is somewhat confusing with constant modulus spring material because containing much Co. However, the applications are quite different each other. A typical example of high modulus spring material is a power spiral spring (main spring) for a mechanical type watch that has been recently replaced at large by the electronic digital type watch and the quartz watch. The carbon steel was mainly used as material of this power spiral spring before the World War II. Since the carbon steel with quenching and tempering, could show higher elastic limit in comparison with the other metals, it was suitable for the raw material of the spiral spring, therefore, still used for the power spiral spring at present.

The spiral spring has a characteristic of increasing the duration of generating force, as its length becomes long, and in case of a clock, it is necessary to be operated for a long time with the torque as weak as possible, so the shape of a clock spring should be thin and long. In addition, when using a spring for a long time, it occasionally cause the breakage. Generally, a thicker spiral spring may have a possibility of early fracture. As a characteristic of the power spring material of clocks, high elastic modulus, large σ/E (σ: Elastic limit, E: Young's modulus), high fatigue characteristic, good corrosion resistance, low notch sensitivity, and good forming ability can be required [5].

After the World War II, many new alloys were developed for the watch spring, and one of them was Elgiloy of Co base alloy. Elgin National Watch Co. in Illinois, U.S.A., received the information concerning the complaint that watch springs broke during the maneuvers by the return soldiers of the World War II, then requested to Battelle Institute to develop ultimate spring material. Then, in 1949, Elgiloy of Co–Cr–Ni–Fe–Mo base alloy was born. Besides Elgiloy, alloys such as Dynavar, Diaflex, Citizen alloy, SPRON100, SPRON510, Nivaflex, Phynox, Vimetal, and NASV604PH were developed. The chemical compositions of these high elastic spring materials and characteristics are shown in Table 2.67 [2, 3, 8, 9]. The Modulus of elasticity of these materials is 206 to 226 Gpa, which is equivalent to or more than the quenched and tempered steel, and show good corrosion resistance with low notch sensitivity and nonmagnetism. A heavy working in cold condition and precipitation hardening heat treatment can be used for these alloys in order to obtain high spring characteristic. But in case of strip shaped material, it is necessary to pay attention to the strong anisotropy in bending. Although the precipitation hardening condition can be different for each alloy, it can be chosen at 450 to 600°C for 2 to 5 hours. The degree of cold working and mechanical properties of Elgiloy sheet and wire are shown in Fig. 2.219 and Fig. 2.220 [10], and the rotating bending fatigue test results for SPRON100

Table 2.67. Chemical compositions and charaeteristies of high modulus spring material

Material name	C	Si	Mn	Co	Cr	Ni	W	Mo	Ti	Al	Be	Fe	Linear expansion coefficient (× 10^{-6})	Modulus of elastlcity (GPa)	Thermoelastic coefficient (× 10^{-5}/°C)	Hardness (HV)	Tensile strength (MPa)	Elastic limit (MPa)	σ^2/E
Elgiloy	0.15	—	2	40	20	15	—	7	—	—	0.04	Bal.	12.7	211	−39.6	700	2450	1667	1.35
Phynox	0.15	—	2	38	20	17	—	7	—	—	—	Bal.	12.7	206	−40.0	650	2400	1667	1.38
Nivaflex	0.02	0.5	1	45	12	26	4	4	1	—	0.3	Bal.	—	221	—	710	2250	1736	1.40
Dynavar	0.2	—	1.6	42.5	20	13	2.8	2	—	—	0.04	Bal.	—	—	—	—	—	—	—
NAS V 604 PH	0.10–0.15	≦ 0.5	≦ 1.5	40 ≦	20.5–22.5	15.5–17.5	—	5.8–6.8	—	≦ 0.35	—	Bal.	—	—	—	—	—	—	—
Vimetal	—	—	0.5	34	14	—	—	6	Cb:2.5, Ta: 0.5			Bal.	—	—	—	—	—	—	—
SPRON 100	≦ 0.03	0.9	0.8	39	12	16.5	4	3.9	0.6	0.08	—	Bal.	13	216	—	560	1960	—	—
SPRON 510	≦ 0.04	≦ 0.1	≦ 0.5	Bal.	20	32.5	—	10	0.5	—	—	1.6	13	226	—	670	2450	—	—
Diaflex	—	—	1	40	15	20	4	4	1	—	—	Bal.	12.6	226	—	670	2450	1765	1.41
Citizen alloy	0.1	—	1.5	41	21	16	—	6.5	—	—	—	Bal.	—	203	—	720	2350	—	—

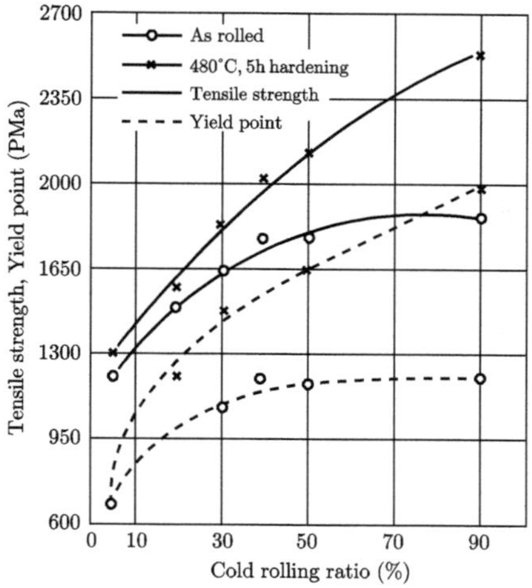

Fig. 2.219. Characteristics of Elgiloy sheet as-rolled and as-age-hardened [10]

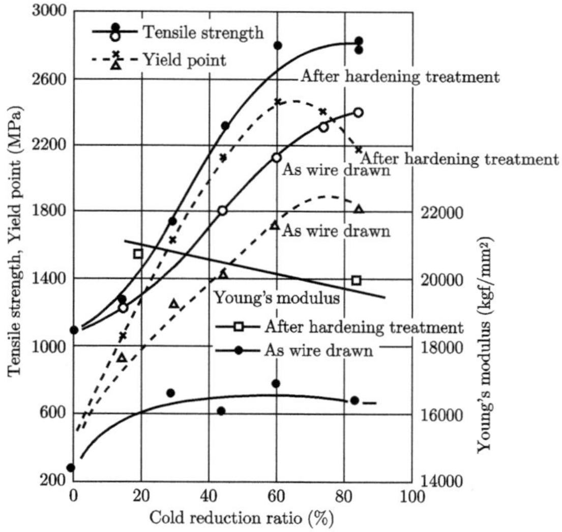

Fig. 2.220. Characteristics of Elgiloy wire as-drawn and as-age-hardened [10]

and SPRON510 wire are shown in Fig. 2.221 and Fig. 2.222 [9]. Because of good characteristics as spring material, Eligiloy was used for power springs of clocks and belt for printers in the early time [11]. Additionally, this material has been also used for aerospace, medical, and petrochemical fields. As for examples, the applications are extended to springs for various meters, metal diaphragms of valves, medical tweezers, and brace wire for orthodontics and so on [12,13].

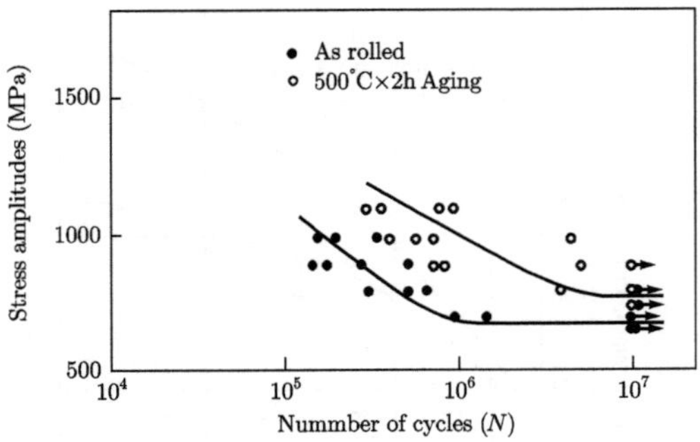

Fig. 2.221. S-N Curve of SPRON 100, 60% cold worked material [9]

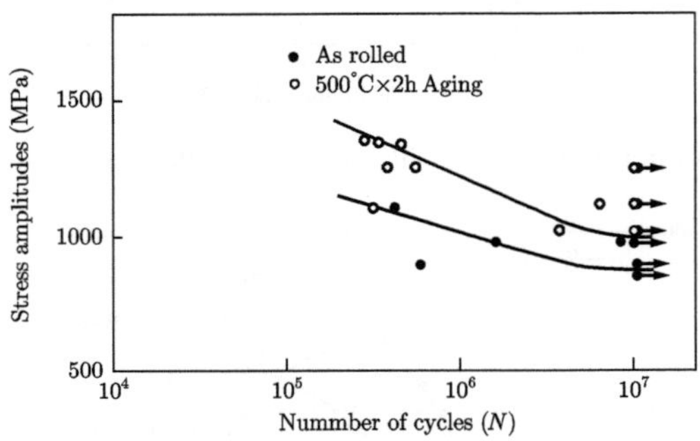

Fig. 2.222. S-N Curve of SPRON 510, 60% cold worked material (wire diameter 0.40 mm) [9]

References

1. Y. Shirakawa: The Iron and Steel Institute of Japan Steel manual (Japanese) (1962) p. 1460
2. H. Saito: The Japan Institute of Metals and The Iron and Steel Institute of Japan Steel material manual (Japanese) (1967) p. 832
3. H. Saito: The Japan Institute of Metals, High elastic material and Constant elastic material (Japanese) (1962) p. 1
4. H. Saito: The Japan Institute of Metals, Metal manual revised 3rd edition, Maruzen (1971) p. 1049
5. Y. Shirakawa: Metal (Japanese) Oct. 15th (1971) p. 54
6. L. Flants: AMS International, Metals Handbook 10th Edition Vol. 2 (1990) p. 895
7. H. Saito: Metal, July 15th (1970), p. 39
8. T. Oya and M. Hasegawa: Stainless steel manual (Japanese) Nikkan Kogyo Shimbun Ltd. (1973) p. 1325
9. SII Micro Parts Ltd.: SPRON 100 Technical data, SPRON 510 Technical data
10. M. Nishihata: Spring 3rd edition (Japanese), Maruzen, (1982), p. 85, p. 93
11. M. Nishihata: Metallic material for precise equipments (Japanese) Nikkan Kogyo Shimbun Ltd. (1985) p. 127
12. SII Micro Parts Ltd. Catalog
13. Nippon Gakki Co., Ltd. Catalog

2.4 Copper Base Alloy Spring Material

2.4.1 General Remarks

(1) Features

Copper has features of good corrosion resistance, good formability, good electrical and thermal conductivity, and non-magnetism. Copper has been used for ornaments and machine parts by bronze (with Sn), brass (with Zn), nickel silver (with Ni and Zn) since ancient times. However, in modern times after electricity being invented, it has been used as material of wiring and switching parts of electric circuit.

Although copper alloy is more expensive than steel, it is mostly used for the purpose of reducing the resistance or heat generation along the electrical circuit.

On the other hand, since Young's modulus of copper is as low as approximately 60% of steel, it is possible to make a pliable spring that has more displacement compared with steel at the same shape and strength.

(2) Various standards

In Japan, copper alloy is standardized in JIS, as shown in Table 2.68. JIS H3130 (sheet and strip of phosphor bronze for spring, nickel silver and copper beryllium) includes the standardized copper alloy for springs. However, other alloys have been used extensively as spring material. The expression of alloy number can be shown in Table 2.69 [1].

In the U.S., five digits number can be registered by applying to CDA (Copper Development Association). Only chemical compositions and physical properties are standardized, without specifying the properties for the grade [2]. ASTM (American Society for Testing and Materials) defines the mechanical properties such as the grade of tensile strength and hardness to CDA Number [3]. DIN (Deutsche Industrial Norme) is relatively popular in Europe. As world standard, ISO has been proposed and has been included in JIS.

Table 2.68. JIS standard of Copper alloy

JIS No.	Contents
H3100	Copper and Copper alloy, sheets, plates and strips
H3110	Phosphor bronze. Nickel silver, sheets, plates and strips
H3130	Copper beryllium alloy, Copper-titanium alloy, Phospher bronze, Copper-nickel-tin alloy and Nickel-silver sheets, plates and strips for springs
H3250	Copper and Copper alloy rods and bars
H3260	Copper and Copper alloy wire
H3270	Phosphop bronze, Nickel silver, Copper beryllium, bars, strips, and wire

Table 2.69. Expression of JIS Alloy's Number [1]

Alloy No.	Principal component
C1xxx	Cu-high Cu base alloy
C2xxx	Cu–Zn base alloy
C3xxx	Cu–Zn–Pb base alloy
C4xxx	Cu–Zn–Sn base alloy
C5xxx	Cu–Sn base alloy·Cu–Sn–P base alloy
C6xxx	Cu–Al base alloy·Cu–Si base alloy·Special Cu–Zn base alloy
C7xxx	Cu–Ni base alloy·Cu–Ni–Zn base alloy

(3) Classifications of Copper alloy for springs

Although there are various classifications for copper alloy [4], standing on spring manufacturer and user side in this book, they are classified as shown in Table 2.70.

Strength of each alloy can be determined by the combination of the following conditions.

1) solid solution strengthening with alloy elements
2) annealing condition
3) cold working ratio by rolling, drawing and so on.
4) low temperature stress relief (annealing)
5) age hardening

Chemical compositions of typical copper alloys for springs are shown in Table 2.71.

(i) Brass (Low cost spring material)

Brass is used for the spring with low cost, because its cost becomes lower as the Zn (zinc) element increases. The strength level is low like 400 MPa to 600 MPa. However, the electrical conductivity is 28%[2],therefore, Brass is popular in various applications where some amount of current flows, such as plugs, switches, and terminals of lighting equipments, and also battery terminals of toy and connector for the circuit of lighting equipments of automobiles. Usually, spring manufacturers form the rolled and drawn material to springs. However, since Brass has the issue of stress corrosion cracking and large permanent set at high temperature, many improved copper alloys have been proposed by Copper alloy manufacturers.

(ii) Phosphor bronze and nickel silver (general-purpose spring material)

Although the strength and price of phosphor bronze increase with adding Sn (tin), it has been widely used as electric conductive spring material such as various switches and relays for industry and home electronic equipments

[2] Electrical conductivity: The capacity of material to conduct electric current expressed as a percentage of that of the International Annealed Copper Standard measured at 20°C (1.7241 $\mu\Omega$). The unit is IACS.

Table 2.70. Classification of copper alloy for springs

JIS standard	Classification	Strengthening method	Selection criterion	Main applications
Available	① Brass	Work hardening Stress relieving (low temperature annealing)	Cost-conscious	Electrical outlet Terminal for lightings Harness termianl for automotive equipments Contact point Spring for toys
	② Phosphor bronze · Nickel silver	Work hardening Stress relieving (low temperature annealing)	Versatility	Connecter, Switch, Relay for electric and electronic parts
	③ Copper beryllium	Work hardening Precipitation hardening	Lighter, more compact High reliability	Connecter, Switch for electronic parts Connecter, Socket for industry Contact point Spring, Relay for communication equipment
Not available	④ Other copper alloys	Work hardening Low temperature stress relieving (low temperature annealing) Precipitation hardening Fine particle dispersion Spinodal decomposition	Dedicated design Improved characteristic Cost reduction	Lead frame Harness connector for automobile Connecter, switch, and relay for electric and electronic parts

Table 2.71. Chemical compositions of typical copper alloy for springs [5]

Alloy name	Code	Principal component (%)	Shape
Brass	C2600	Cu–Zn(30)	Sheet, Strip, Bar, Wire
	C2680	Cu–Zn(35)	Sheet, Strip
	C2700	Cu–Zn(35)	Bar, Wire
	C2720	Cu–Zn(37)	Sheet, Strip, Wire
	C2800	Cu–Zn(40)	Bar, Wire
	C2801	Cu–Zn(40)	Sheet, Strip
Phosphor bronze	C5111	Cu–Sn(4)–P(0.1)	Sheet, Strip, Bar, Wire
Nickel silver	C5102	Cu–Sn(5)–P(0.1)	Sheet, Strip, Bar, Wire
	C5191	Cu–Sn(6)–P(0.1)	Sheet, Strip, Bar, Wire
	C5212	Cu–Sn(8)–P(0.1)	Sheet, Strip, Bar, Wire
	C7351	Cu–Ni(18)–Zn(10)	Sheet, Strip, Bar, Wire
	C7451	Cu–Zn(25)–Ni(10)	Sheet, Strip, Wire
	C7521	Cu–Zn(18)–Ni(18)	Sheet, Strip, Bar, Wire
	C7541	Cu–Zn(24)–Ni(14)	Sheet, Strip, Bar, Wire
Phosphor bronze	C1700	Cu–Be(1.7)–Co(0.25)	Sheet, Strip
for spring	C1720	Cu–Be(1.9)–Co(0.25)	Sheet, Strip, Bar, Wire
Nickel silver	C5210	Cu–Sn(8)–P(0.2)	Sheet, Strip
Copper beryllium	C7701	Cu–Zn(26)–Ni(18)	Sheet, Strip, Bar, Wire

because of a good balance of various characteristics. Regarding spring use, the alloy containing 8% tin is standardized in JIS. The material to which low temperature annealing (stress relief) is applied after rolling is supplied to spring manufacturers.

Nickel silver, containing Zn and Ni, shows silver white color and good corrosion resistance. Although its electric conductivity is as low as 5%, the strength is 450 MPa to 850 MPa. It has been used as the standard material for wire spring of the crossbar switch for a long time. At present, it is used for the parts that need corrosion resistance, such as transistor cases and moving contact of rheostats.

As for the spring uses, the alloy containing 26% Zn and 18% Ni is standardized in JIS. The material can be used after low temperature stress relief like Phosphor Bronze.

(iii) Copper beryllium (high grade spring material)

Copper beryllium is the conductive spring material having good high temperature characteristics. When the 2% maximum of beryllium is added, it shows the tensile strength of 1,500 MPa and electric conductivity of 22% minimum. However, because of high cost, it is applied to small-sized parts and the parts in the system where a longtime reliability is required, which cannot be designed with phosphorus bronze.

In the U.S., copper beryllium has been used for the connector of military use, and has spread to switches, relays, and connectors for industries and consumer applications. In Japan, it has become the indispensable material to meet with the requirement of lighter and more compact electronic equipments in consumer uses. Generally, the rolled material is formed and heat treated by spring manufacturers. However, the consumption of mill hardened material which does not need any heat-treatment, has been increasing.

(iv) Other copper alloys (exclusive or developed alloy)

Although the above-mentioned typical spring material can meet with most applications for electric conductive springs, as for lead frames of semiconductor packages and harness connectors of automobiles, dedicated developments to meet required characteristics have been made aggressively, because of extremely large quantity requirements for both parts. It can be also seen that materials with characteristics between phosphor bronze and copper beryllium have been developed.

There are various kinds of alloys such as Cu–Fe base, Cu–Sn base, Cu–Ni–Si base, Cu–Ni–Sn base, Cu–Ti base and so on. Their tensile strength is 400 to 800 MPa, and electrical conductivity is 30 to 80% IACS. As for Cu–Ti, its tensile strength is 700 to 1,100 MPa, and electrical conductivity is 10 to 16% IACS. These materials can be formed by spring manufacturers after low temperature stress relief of cold rolled material or after precipitation hardening treatment. There is a case where precipitation hardening treatment is carried out after spring forming for Cu-Ti alloy strip.

(4) Selection method for copper alloy spring material

(i) Tensile strength and electrical conductivity

Generally, tensile strength and electrical conductivity are incompatible characteristics. When the tensile strength is increased with adding elements to Cu, the electrical conductivity becomes low. If the tensile strength is increased with heavy cold forming like rolling, electrical conductivity slightly falls. In case of strengthening with the precipitation hardening, the tensile strength and electrical conductivity rises simultaneously. If tensile strength and price is the same, the degree of freedom for designing becomes high for alloy which has higher electrical conductivity. Therefore, developments of new material have been continued by material manufacturers. The relation between tensile strength and electrical conductivity of typical copper alloy is shown in Fig. 2.223 [6].

(ii) Method of choosing temper designation (tensile strength and forming ability)

The temper designations are specified on the grade of strength or hardness, which can be determined by the combination of plastic working degree and heat treatment. This is summarized in JIS, shown in Table 2.72 [7]. The temper designation, H can be obtained by cold working with approximately 37% reduction in area for sheet, strip and approximately 60% for wire. In the U.S., A is sometimes used. A is equivalent to O (zero) in JIS which is the softest condition after annealing or solution heat treatment. M means that material is shipped after mill hardening at plant. T shows the characteristic after heat treatment at spring manufacturers after shipping.

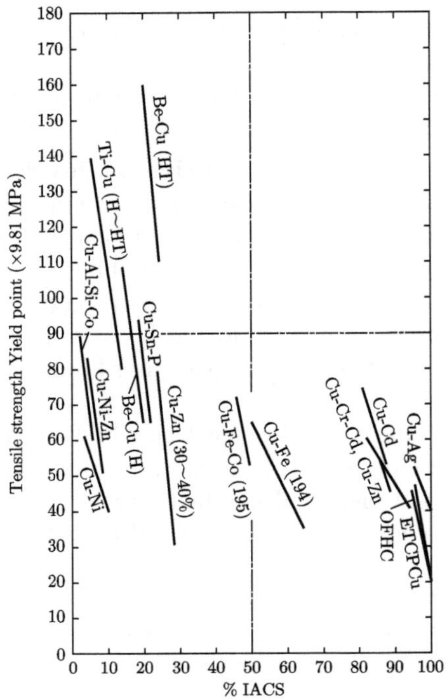

Fig. 2.223. Tensile strength and electric conductivity of various copper alloys [6]

Table 2.72. Strength grade specified by JIS

Strength grade	Definition
F	As fabricated. No description of mechanical properties. F: Fabrication
O	Fully recrystallized or annealed. Tensile strength (TS) is the lowest.
OL	Annealed or lightly worked. TS is the same as O.
1/8H	Work hardened to the TS between O and 1/4H.
1/4H	Work hardened to the TS between 1/8H and 1/2H.
1/2H	Work hardened to the TS between 1/4H and 3/4H.
3/4H	Work hardened to the TS between 1/2H and H.
H	Work hardened to the TS between 3/4H and EH.
EH	Work hardened to the TS between H and SH. EH: Extra hard
SH	Heat-treated to the highest TS. SH: Spring hard
SR	Stress relieved. SR: Stress relief

The materials classified by high temper designation like EH and SH have high tensile strength. However, the forming ability of bending and drawing of sheet and strip is poor due to low elongation. Therefore, it is necessary to decide whether the first priority is taken to finished spring characteristics or spring formability. When both characteristics are needed, the practice of using the one class higher alloy can be recommended. In addition, anisotropy of high temper designation material is large because of heavy plastic working. Especially when bending along the rolling direction, special attention for crack prevention is required. The bending test method is specified in JIS Z 2248 such as the right angle bend, 60 degree W bend, and 180 degree bend.

(iii) Availability and international commonality

The copper alloy with the standard thickness size specified in JIS can be comparatively easily obtained. While electronic parts manufacturers have transferred their productions to overseas, the need of local procurements of material has been increased.

A crosscheck between CDA, ASTM and ISO can be required.

(5) Remarks for spring design

(i) Young's modulus

Young's modulus is the index which influences directly on the spring load. The value obtained by tensile tests, have been usually used. Young's modulus of copper alloy varies a little by additional element shown in Fig. 2.224 [8]. There is a tendency that calculation results correspond to experimental results when a deflection factor is used for a spring. The deflection factor can be obtained with the displacement at the loading point, by using the spring deflection limit tester specified in the 6.4 section of JIS H3130.

(ii) 0.2% proof stress and spring deflection limit value

0.2% proof stress, the stress when the 0.2% permanent strain occurs in the tensile test, is generally used as an index of the spring design. On the other hand, the spring deflection limit (Kb value), which is a surface bending stress

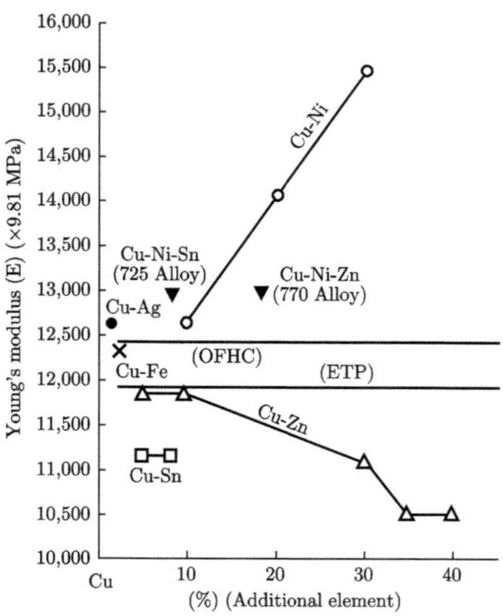

Fig. 2.224. Additional element and change of Young's modulus for Copper [8]

when a constant permanent deformation occurs at the loading point in deflecting and unloading a plate, is standardized in JIS H3130 (Japan) and DIN 1777 (Germany). The measuring method of Kb value, which falls into repeatedly deflecting method and moment method, is designated in JIS H3130. The moment method can be suitable for measuring material with high strength and large deflection material such as copper beryllium. Figure 2.225 [9] shows how to apply the load in the moment method.

It depends upon the applications that either 0.2% proof stress or Kb value should be used. For example, it should be safe to use 0.2% proof stress for the parts with complicated bending like connectors or when stress can be applied on the sheared surface with stamping. In the U.S., the research and development of connectors has been popular, so in ASTM, only 2% proof stress is standardized. In case of no bending and no plating process like some switches or relays, it becomes possible to draw out the capability of material by using Kb value. Especially, this tendency is more prominent in hearily worked material with high quality classification. Since the Kb value is a bending characteristic, the conditions around surface layer become important. The Kb value of material with heavy plating sometimes shows an extremely low value. JIS designates the applicable thickness range for Kb value to be between 0.2 mm and 1.6 mm. When stress relief (low temperature annealing) is applied to work-hardened material, the Kb value can be improved without decreasing elongation, to be able to produce springs with more complicate shape and high strength, refer to Fig. 2.226 [10]. The actual movement of complicatedly shaped springs has become possible to be analyzed by FEM recently. Therefore, it may be difficult to design with the point datum like 0.2% proof stress and Kb value. In this case, stress-strain curve and Poisson's ratio can be required.

(iii) Fatigue characteristics

The thickness range of electrical conductive spring material is from 0.1 mm to 0.4 mm in general. The testing method of this kind of thin sheet has not

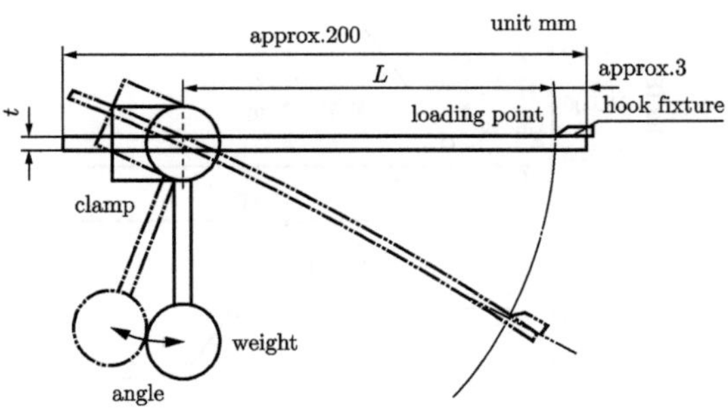

t : thickness of specimen
L : distance between fixed end of specimen and loading point $L=\sqrt{3000t}$

Fig. 2.225. Installation of specimen and loading method [9]

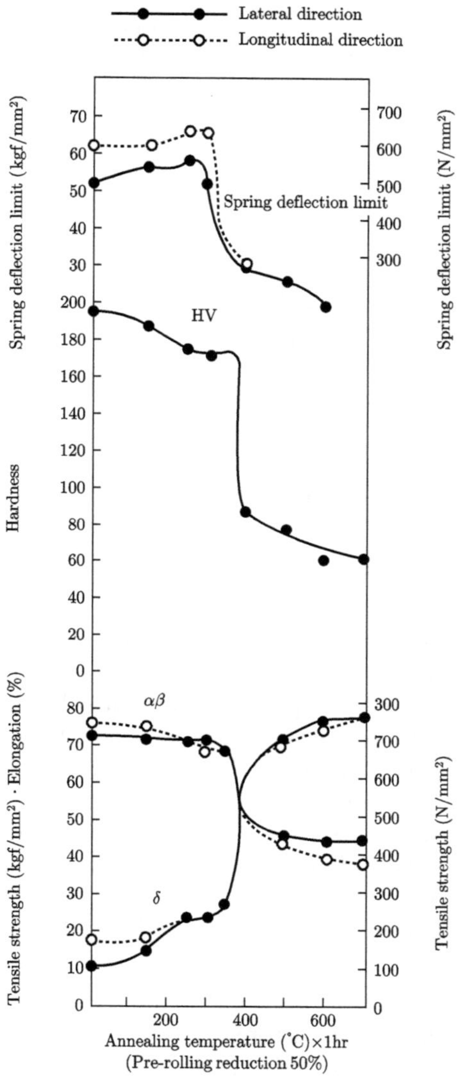

Fig. 2.226. Spring deflection limit of phosphor bronze (C5212) [10]

been standardized. However, Nishihata et al. have pointed out the difference of fatigue characteristics between testing methods and have also proposed a bending type testing method with pulley [12], which was standardized by JSMA. Two kinds of method among those can be shown in Fig. 2.227 [11].

(iv) Stress relaxation

The stress relaxation is such a phenomenon that the spring load decreases with time, even if the displacement is kept constant like a connector. Before carrying out evaluation tests of the product, design engineers can predict

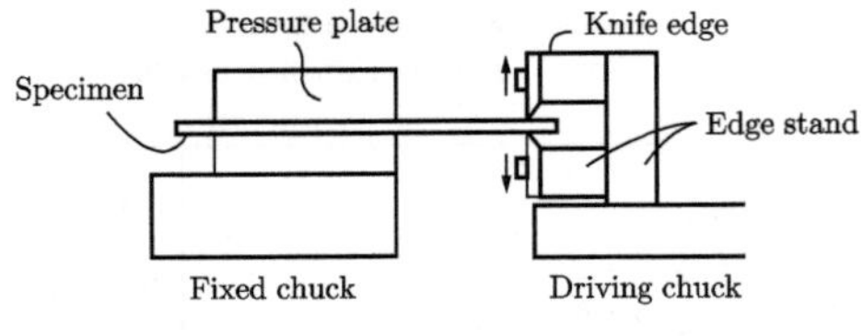

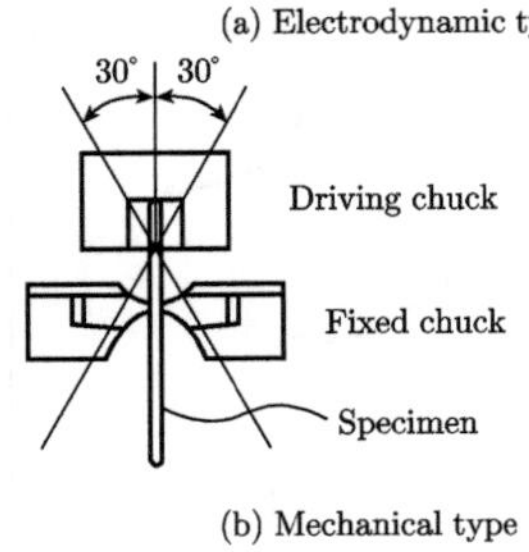

Fig. 2.227. Fatigue testing methods [11]

the results with this characteristic while sorting the material. The testing method is standardized in ASTM E328 and EMAS 3003. There are two kinds of testing method, one is measuring load directly, and the other is to measure permanent deflection to be converted to stress. The method of measuring permanent defection is more popular and practical because of its simplicity. In the latter method, there are still a few methods, having the surface stress constant (same curvature), or having the stress varied (cantilever). One of each example in ASTM and EMAS can be shown respectively in Fig. 2.228 [13,14].

There is a report which describes that even if the initial stress in the surface varies from the 30% to 80% of proof stress, the stress relaxation ratio shows the same value [15]. However, the 75% or 80% of proof stress can be normally used as the initial stress. This relation is shown in Fig. 2.229 [15]. Besides this, there are some research works concerning with the stress relaxation characteristic at the plastic deformation region and the bent portion [16].

(6) Remarks for applications

The length of electric conductive springs used in recent electronic equipments is sometimes less than 1mm. This kind of minute spring is easy to be affected by forming, to show different spring load from the calculated load. For example, after slightly bending the beam, then bending to the opposite direction, there is a case that the load is lower than the calculated value. This can be due to Bauschinger effect. This issue can be resolved with eliminating forming strains by heating or bending to extra amount, to obtain the specified dimension by bending back from the opposite side [17].

When bending stress is loaded in the sheared surface of blanking materials, and a different process between prototype and serial production is applied, there can be a case that the results become different. When choosing wire cutting or etching in prototype productions and stamping in serial productions,

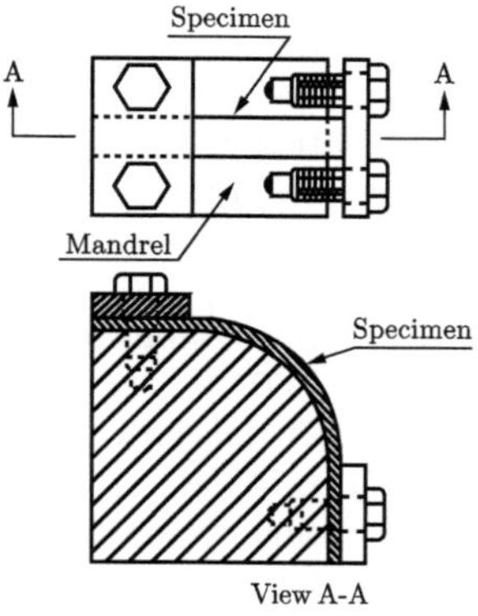

Fig. 2.228. (a) Stress relaxation testing method [13, 14]

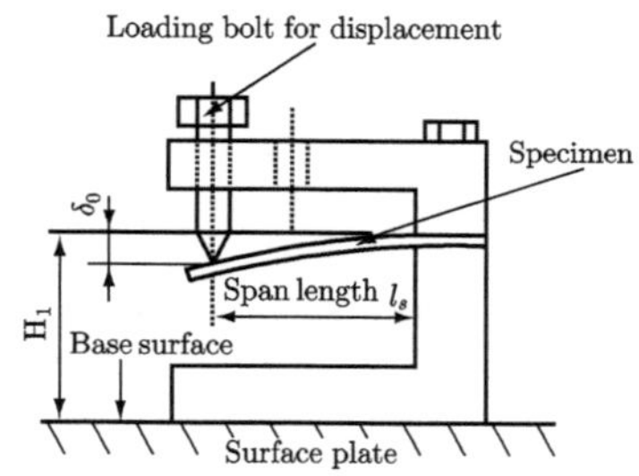

Fig. 2.228. (b) Cantilever method

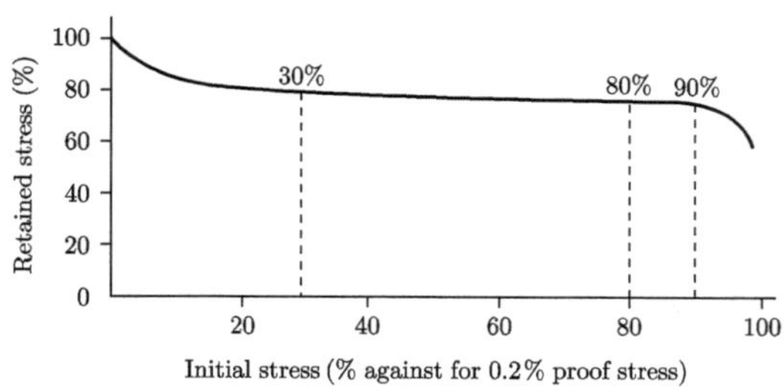

Fig. 2.229. Influence of initial stress for stress relaxation [15]

a special attention is required. In case of plating and soldering, it is necessary to make sure that no peeling from base material occurs due to the reaction with base material after long time passing.

References

1. Copper alloy data book compilation committee: Copper alloy data book Japan Copper and Brass Association (1997) p. 6

2. Copper alloy data book compilation committee: Copper alloy data book Japan Copper and Brass Association (1997) p. 27
3. Copper alloy data book compilation committee: Copper alloy data book Japan Copper and Brass Association (1997) p. 72
4. Copper alloy data book compilation committee: Copper alloy data book Japan Copper and Brass Association (1997) p. 7
5. Copper alloy data book compilation committee: Copper alloy data book Japan Copper and Brass Association (1997) p. 34
6. M. Sakamoto: Development of materials for communication and electronic parts, Homat Ad. corporation (1997) p. 74
7. Copper alloy data book compilation committee: Copper alloy data book Japan Copper and Brass Association (1997) p. 16
8. M.Sakamoto: Development of materials for communication and electronic parts, Homat Ad. corporation (1997) p. 75
9. JIS Hand Book Non-ferrous (1998)
10. Copper alloy data book compilation committee: Copper alloy data book Japan Copper and Brass Association (1997) p. 93
11. JSSR: Transaction of JSSR, 31 (1986) p. 84
12. JSMA: Study of evaluation methods of ultra-thin flat spring (1990) p. 20
13. ASTM E328, (1990) p. 431
14. Electronic material manufacturers association of Japan: The stress relief method by bent spring material (1991) p. 4
15. X.W. Horn et al.: Pro. of 15th IICIT Connector Meeting, Nov. (1982) p. 325
16. T. Ochiai et al.: Pro. of 18th IICIT Connector Meeting, Oct. (1990) p. 473
17. T. Ochiai et al.: Pro. of 31th IICIT Connector Meeting, Sep. (1998)
18. Y. Ishibashi et al.: Pro. of 29th IICIT Connector Meeting, Sep. (1990) p. 179

2.4.2 Brass

(1) Features

Brass is the alloy of Cu and Zn, and has unique yellow gloss. Being less expensive among copper alloys, it is widely used for precision instruments and parts for ships which cannot be coped with steel due to machinability and corrosion resistance to sea water. Brass has been used for radiators of automobiles, daily necessities such as buttons and name plates, and interior decoration goods, due to its good formability.

(2) Applications

Brass springs are used for lighting connectors and terminals of automobiles, lighting equipments and toys. In addition, although it is not a spring, brass wire with square or similar section is used for header pins of the base-plate of electronic parts.

(3) Chemical compositions

JIS standard of brass is designated in JIS H 3100 (Copper and copper alloy sheets, plates and strips), JIS H 3250 (Copper and copper alloy rods and bars), and JIS H 3260 (Copper and copper alloy wires). The chemical compositions are shown in Table 2.73 [1]. Among these materials, C2680 and

Table 2.73. Grades · chemieal compositions · features · applications of Brass [1]

Alloy No.	Shape	Chemical compositions %				Features	Applications
		Cu	Pb	Fe	Zn		
C2600	Sheet Strip Bar, Wire	68.5–71.5	0.05 max.	0.05 max.	Balance	malleability & ductility · deep drawability · plating ability	automobile radiator
C2680	Sheet, Strip	64.0–68.0	0.05 max.	0.05 max.	Balance		automobile radiator · wiring instrument
C2700	Bar Wire	63.0–67.0	0.05 max.	0.05 max.	Balance	cold forging workability	machine parts electric parts
C2720	Sheet Strip Wire	62.0–64.0	0.07 max.	0.07 max.	Balance	malleability & ductility · deep drawability	shallow drawing products
C2800	Bar Wire	59.0–63.0	0.1 max.	0.07 max.	Balance	hot workability	machine parts electric parts
C2801	Sheet, Strip	59.0–62.0	0.1 max.	0.07 max.	Balance	strength · malleability & ductility	wiring instrument · instrument board

C2700 (65/35 brass) and C2720 (63/37 brass) whose strength grade is 1/2 H or more, can be used for springs.

(4) Phase diagrams

The phase diagram of Cu–Zn alloy is shown in Fig. 2.230 [2]. The α brass containing 30–35% Zn shows good formability for deep drawing. When Zn exceeds 38%, β phase appears to show higher strength. Although the brass containing 40% Zn becomes $(\alpha + \beta)$ phase to show good workability at high temperature, it becomes hard and brittle when Zn exceeds 43%,.

(5) Manufacturing processes

Mechanical properties of brass can be determined by the content of Zn and cold working ratio of rolling and drawing. When low temperature stress relief is applied, the spring deflection limit can be increased as well as other alloys. Manufacturing processes of sheet and strip of brass are shown in Fig. 2.231 [3].

(6) Mechanical properties

(i) Tensile characteristics

Tensile strength increases with rolling ratio, as shown in Fig. 2.232 [4]. However, the elongations, δ, decrease. Since the JIS standard does not designate for spring applications, the 0.2% proof stress or spring deflection limit which can be a guide for designing springs, is not specified. However, the 0.2% proof stress can be approximately presumed from rolling ratio and tensile strength, as seen in Fig. 2.232 [4].

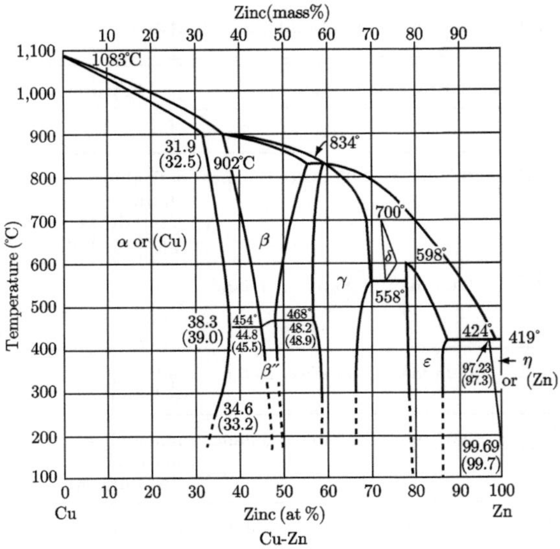

Fig. 2.230. Phase diagram of Brass [2]

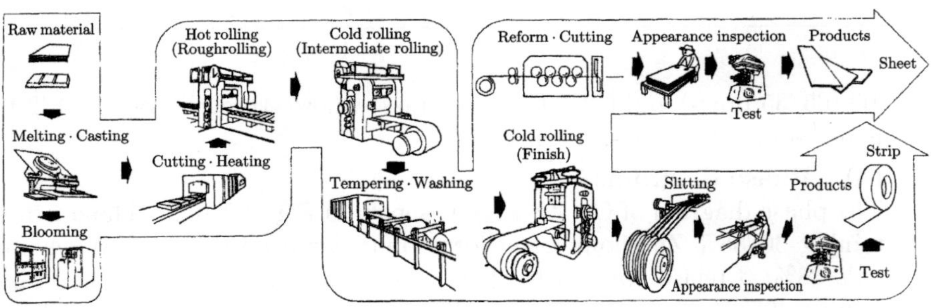

Fig. 2.231. Manufacturing process of Brass sheet and strip [3]

(ii) Spring deflection limit

Spring manufacturers can be normally supplied rolled material to produce springs. Since brass is also work hardening type alloy, spring deflection limit can be increased with stress relief (low temperature annealing), as shown in Fig. 2.233.

(7) Remarks in use

Since brass shows worse stress relaxation characteristic, lower fatigue life and higher stress corrosion crack sensitivity compared with other copper alloys, its applications to parts should be taken those features into consideration.

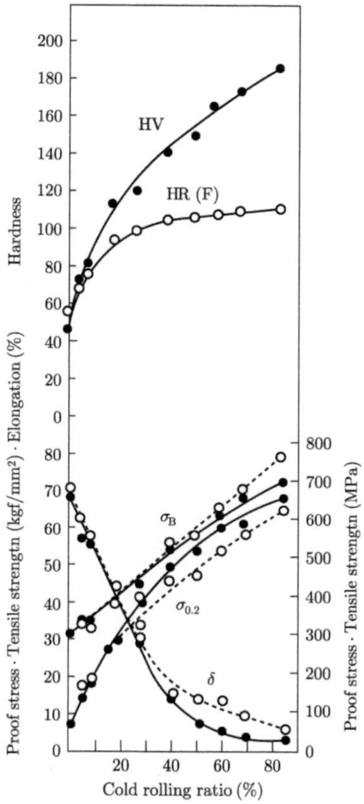

Fig. 2.232. Work hardeing characteristics of Brass [4]

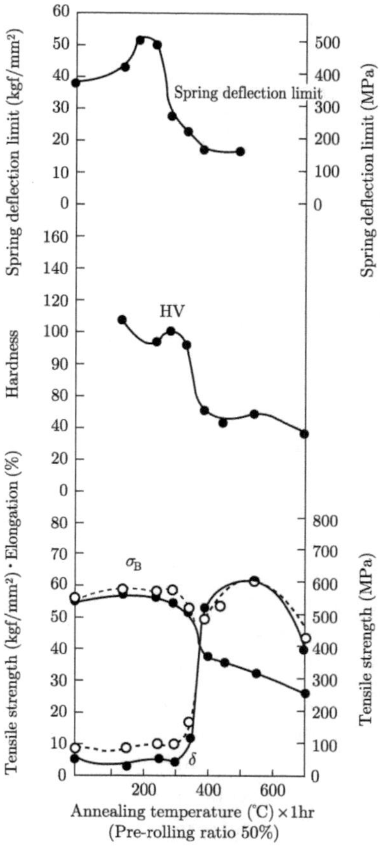

Fig. 2.233. Influence of low temperature annealing of Brass C2680 [5]

References

1. JIS Handbook Nonferrous p. 236, p. 238, p. 276, p. 277, p. 287, p. 288, (1998)
2. Hansen et al.: Constitution of Binary Alloys, 2nd ed. p. 650, Mcgraw Hill Book company (1958)
3. Japan Copper and Brass Association: Copper alloy data book, (1997) p. 21
4. Japan Copper and Brass Association: Copper alloy data book, (1997) p. 99
5. Japan Copper and Brass Association: Copper alloy data book, (1997) p. 91

2.4.3 Phosphor Bronze and Nickel Silver

Bronze, which has been used from the ancient times, and is relatively eco-friendly alloy, is alloy made naturally of Cu and Sn, due to their inherent affinity.

Phosphor bronze is the alloy added phosphorus to bronze which was invented and industrialized in the 19th Century. In near future by means of further theoretical and highly industrial studies, phosphor bronze can be said to have a possibility to be developed to next generation level. Phosphor bronze is the ternary alloy of which main compositions are Cu and Sn with adding P for de-oxidization. With the Sn content and making cold working ratio high, the strength of material becomes high.

Since phosphor bronze has such good characteristics as a good balance between electrical conductivity and strength, connecting characteristic, workability, plating performance and corrosion resistance under the highly dense condition in recent years for especially electronic parts, its application as advanced material has been extended to semiconductors, electronic equipments, electrical machineries, automobiles and information and communication equipments and so on. Especially because of computerization of automobiles (safety, reliability, comfort and convenience) and electronic civilizations, this material has been widely used as the worldwide fundamental alloy. Japan has become the worldwide supply base.

Nickel-silver called german silver in the old days, which is the silver white color alloy consisting of ternary alloy of Cu–Ni–Zn with the Cu as main composition and small amount of Mn for deoxidization, can be classified into several types by the percentage of Ni and Zn in alloy.

Having beautiful silver-white color, and superior corrosion resistance, its application is for ornaments, wind instruments, table-wares, medical equipments and glasses part etc., and being high in drawing ability, its application is extended to quartz oscillation case and transistor caps. In addition, it is also used for electronic parts with the same reason as phosphor bronze. Since phosphor bronze is much more commercially used than nickel silver, phosphor bronze is mainly described below.

(i) Features

Phosphor bronze has excellent characteristics as non-ferrous metal material such as 1) superior in spring characteristic, 2) high strength, 3) good bending and drawing ability, 4) high electrical conductivity, 5) high heat conductivity, 6) good abrasion resistance, 7) no season crack and no age hardening, 8) non-magnetic, 9) soldering easily, 10) plating easily, 11) good for die life, and 12) high resistance to chemical corrosion.

Main characteristics of phosphor bronze can be obtained by the following processes.

1) As-continuous cast dendrite structure is homogenized crystallographically by annealing.
2) Fine fibrous microstructure is obtained by repeating cold rolling and annealing.
3) By applying straightening and low-temperature annealing, internal stress can be reduced and stabilized.

(2) Applications

1) Electronics field
   Various connectors, Lead frames, Lead terminals, Transistor terminals
2) Electric field
   Various relay contact, Various switch contact, Fuse clips
3) Automotive field
   Bearing frames, Hydraulic caps, Packing seals, Roller bearings, Clutch plates, Electrical components
4) Others
   Various bellows, Diaphragms, Sliding contact, Blade material, Various springs, Anti-corrosion chemical machine devices

Other than the main spring use, phosphor bronze is also used for sliding contacts, roller bearings and blades due to its high mechanical strength and wear resistance. Furthermore, due to its high corrosion resistance, it is used for the parts of chemical apparatus that especially need corrosion resistance.

(3) Chemical compositions

Sheet, strip, bar and wire of phosphor bronze and nickel silver are standardized in JIS. Its type, chemical composition, features and applications can be shown in Table 2.74 [1] and Table 2.75 [2].

(4) Phase diagrams

The Cu-Sn binary phase diagram is shown in Fig. 2.234 [3]. The α solid solution, which can be basically the phase of phosphor bronze, is soft and has malleability and ductility. With the increase of Sn content in the alloy, the alloy color changes from copper reddish to bluish white and finally to copper gold color, and at the same time its mechanical strength is gradually increased with Sn content.

(5) Manufacturing processes

One example of manufacturing processes of phosphor bronze is shown in Fig. 2.235 [4]. Phosphor bronze is melted using tin crude metal and copper 15% phosphor alloy as raw material.

Among various kinds of melting furnaces, it is preferable to use an atmosphere adjustable furnace, because oxygen, hydrogen, and water (vapor) in the molten metal can be a cause of blow hole to give a serious impact on quality. The oxygen content in the molten bronze metal is inversely proportional to the square of hydrogen content. Namely, hydrogen can increase as oxygen decreases.

In case of bronze, containing no deoxidizer, hydrogen decreases as oxygen remains. As a result, only oxygen becomes a problem. In case of phosphor bronze, phosphor-copper alloy is added to the melt, by which oxygen content in the melt is decreased.

Hot forming ability of as cast phosphor bronze is not so good. Consequently, the transverse continuous casting method where the cast shape close

Table 2.74. Grades, Chemical compositions, Features, and Applications of Phosphor Bronze

Grade	Shape	Chemical Compositions, mass% Sn	P	Fe	Pb	Zn	Cu + Sn + P	Features	Applications
C5210 (for spring)	Sheet, Strip	7.0–9.0	0.03–0.35	0.1 max.	0.05 max.	0.20 max.	≧ 99.7	High forming-ability, and high corrosion resistance. High fatigue strength with low temperature annealing.	Switches, connectors, relays for electronic, communication, instrumental devices.
C5111	Sheet, Strip, Bar, Wire	3.5–4.5	0.03–0.35	—	—	—	≧ 99.5	High forming-ability, high fatigue strength and high corrosion resistance. For the applications of high spring limit, C5210 is recommendable.	Springs for electronic devices, switches, relays, lead frame, connector, diaphragm, bellows, fuse clips, bearings, bushes, wire net.
C5102		4.5–5.5	0.03–0.35	—	—	—	≧ 99.5		
C5191		5.5–7.0	0.03–0.35	—	—	—	≧ 99.5		
C5212		7.0–9.0	0.03–0.35	—	—	—	≧ 99.5		

Table 2.75. Grades, Chemieal compositions, Features, and Applications of Nickel-Silver

		Chemical Compositions, mass%							
Grade	Shape	Cu	Pb	Fe	Zn	Mn	Ni	Features	Applications
C7701 (for spring)	Sheet, Strip	54.0–58.0	0.10 max.	0.25 max.	Balance	0–0.5	16.5–19.5	Gloss, good forming-ability, high fatigue strength, high corrosion resistance.	Switches, connectors, relays for electronic, communication, instrumental devices.
C7351	Sheet, Strip	70.0–75.0	0.10 max.	0.25 max.	Balance	0–0.5	16.5–19.5	Gloss, good forming-ability, high fatigue strength, high corrision resistance.	Quartz oscillation case, volume movable contact, clock dial.
C7451		63.0–67.0	0.10 max.	0.25 max.	Balance	0–0.5	8.5–11.0		
C7521		62.0–66.0	0.10 max.	0.25 max.	Balance	0–0.5	16.5–19.5		
C7541		60.0–64.0	0.10 max.	0.25 max.	Balance	0–0.5	12.5–15.5		

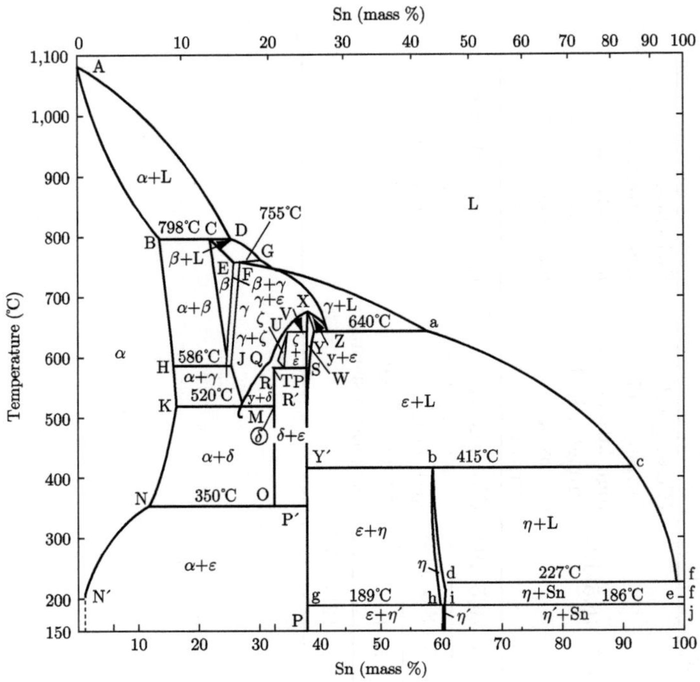

Fig. 2.234. Phase diagram of Cu–Sn base alloy

to the product can be produced, is normally adopted. Homogenizing annealing is carried out to change the solidification structure into α phase structure which shows high malleability and ductility in cold forming. In addition, since the inverse segregations of Sn and P to the ingot surface, occur during solidification in the continuous casting, the inverse segregation layer shall be ground off after rough and intermediate rolling. After that, cold rolling and annealing are processed repeatedly, and the specified size and strength can be obtained by finish rolling. For spring applications, roller leveler or tension leveler can be applied to improve flatness and straightness. Then, stress relief (low temperature annealing), which improves the accuracy of dimensions, spring characteristics, and fatigue life, can be performed.

(6) Physical characteristics

(i) Electric conductivity

The electrical conductivity of phosphor bronze depends upon the contents of Sn and P. However, the range of electric conductivity is from 11% to 23% IACS in general. Figure 2.236 [5] shows the relation between the electrical conductivity and Sn content.

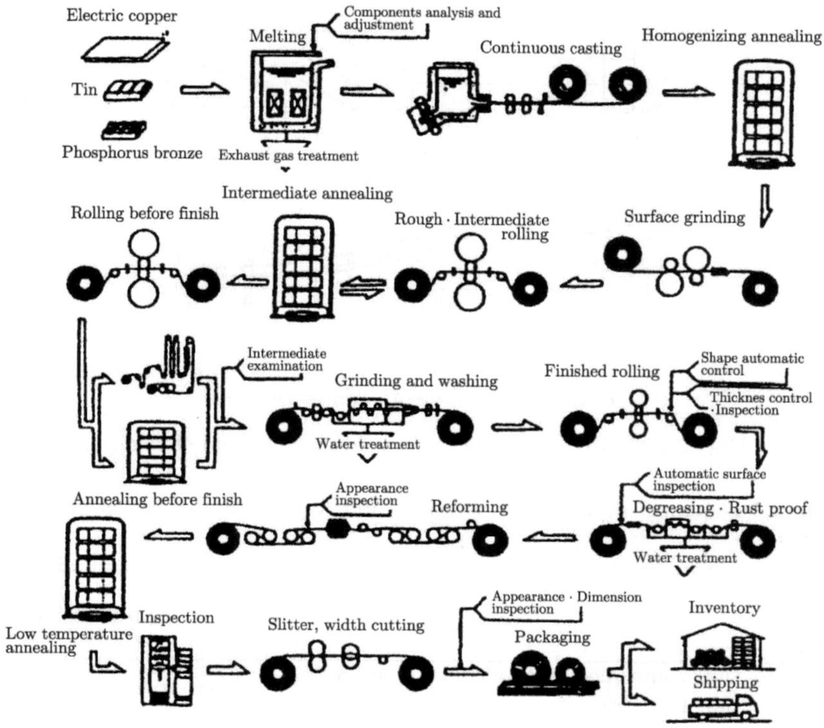

Fig. 2.235. One example of phosphorus bronze production process

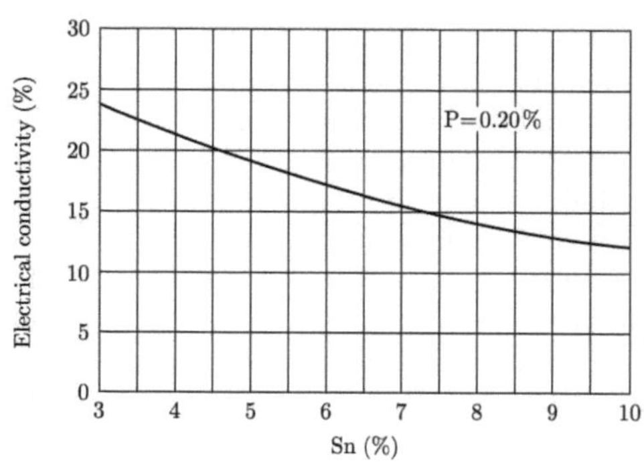

Fig. 2.236. Relation between electrical conductivty and Sn content

(7) Mechanical characteristics

(i) Relation between tensile strength and elongation

If the annealed and softened phosphor bronze is rolled at ambient temperature, it can be work-hardened in proportional to the working ratio subject to to increased tensile strength and decreased elongation. The tensile strength and elongation of the annealed O material is high as the Sn content increases.

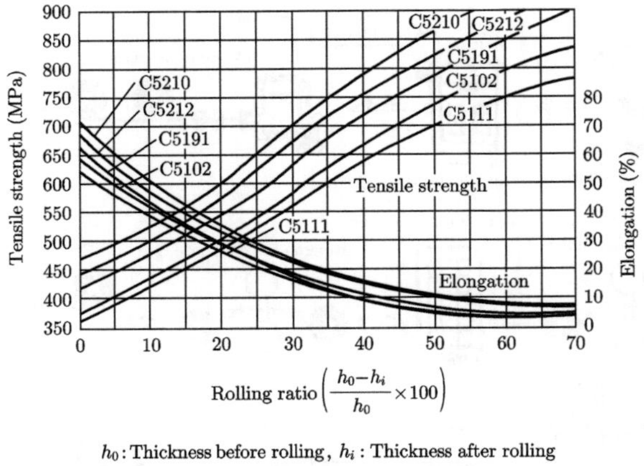

Fig. 2.237. Relation between tensile strength and elongation

Table 2.76. Temper designations, features, and Applications of Nickel-Silver sheet and strip

Temper designations	Features	Applications
O	Fully annealed and no cold rolling. Suitable for deep drawing or cold forging	Diaphragm, Bellows
1/4H	About 10% cold working after annealing. Suitable for drawing.	Fuse clips
1/2H	About 20% cold working after annealing. Good blanking-ability. Suitable for the products required strength after forming.	Connectors, Fuse clips, Lead terminals
H	About 30% cold working after annealing. Suitable for the products required bending-ability and spring characteristics.	Switches, connectors, Lead terminals, Fuse clips, Lead frame
EH	About 40% cold working after annealing. Suitable for the products required high spring characteristics.	Switches, Relays, Sliding contacts, Disc springs, Blade
SH	About 45% cold working after annealing. Suitable for the products required higher spring characteristics.	Sliding contact, Relays

This tendency can be remained after cold working. Figure 2.237 shows this relation [6]. The tensile strength and elongation of phosphor bronze are specified as the strength grade relative to the rolling ratio, in order to be made an appropriate material selection by users. Table 2.76 shows the temper designations of sheet and strip, features and applications.

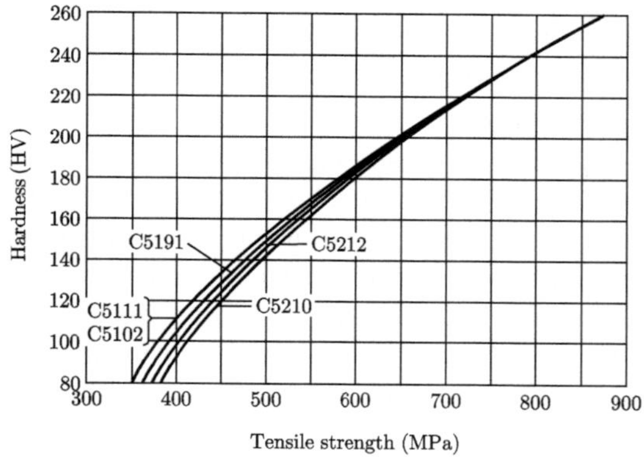

Fig. 2.238. Relation between tensile strengh and hardness

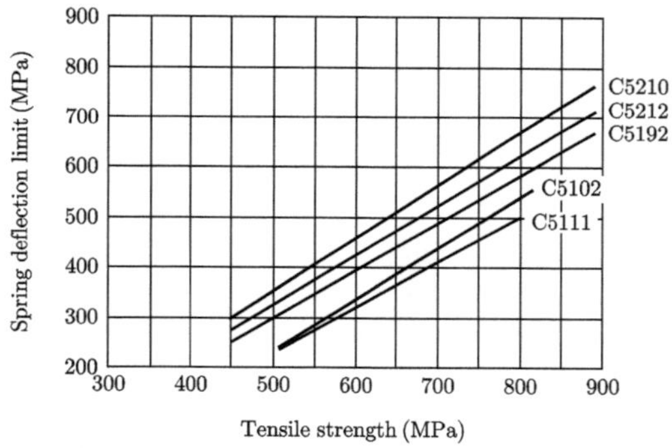

Fig. 2.239. Relation betweentensile strength and spring deflection limit

(ii) Relation between tensile strength and hardness

There is a proportional relation between tensile strength and hardness for phosphor bronze, as shown in Fig. 2.238 [7].

(iii) Relation between tensile strength and spring deflection limit

Phosphorus bronze has been widely used for spring material because of its excellent spring characteristics. The spring deflection limit can be used as the strength criterion to evaluate spring characteristics. The spring deflection limit increases proportionally as tensile strength increases, as shown in Fig. 2.239 [8].

(8) Remarks in use

(i) Design of flat spring

When a flat spring is subjected to repeated strain cycles ranging from zero to maximum deflection and the stress at the maximum deflection is within two-thirds of the spring deflection limit, several million cycles of fatigue life can be expected. However, when a spring has a a portion bent at around right angle or more than two complicated bent portions, it is recommendable to keep the stress within one-third of the spring deflection limit.

(ii) Change in color

Phosphor bronze does not easily change in color at normal ambient temperature and humidity. However, it should be cautious that touching with bare hands makes color change. In addition, when packed under insufficient drying after degreasing and washing with organic solvent, its color can be changed to bluish purple caused by the vaporised gas coming out of little solvent remained on the surface. Therefore, after degreasing and washing with organic solvent, it is necessary to pack the products after sufficient drying with ventilation.

(iii) Design of bending radius

Since there is a difference of workability between the bending in rolling direction (good way) and that in transverse direction (bad way), it is necessary to design a bending radius larger than the limit of respective bending ability.

References

1. JIS Hand Book, Nonferrous (1998)
2. JIS Hand Book, Nonferrous (1998)
3. Nishihata et al.: The Phosphorus bronze, basics and application (Japanese), The Nikkan Kogyo Shimbun Ltd. (1997) p. 3
4. Nishihata et al.: The Phosphorus bronze, basics and application (Japanese), The Nikkan Kogyo Shimbun Ltd. (1997) p. 20
5. Nishihata et al.: The Phosphorus bronze, basics and application (Japanese), The Nikkan Kogyo Shimbun Ltd. (1997) p. 11
6. Nishihata et al.: The Phosphorus bronze, basics and application (Japanese), The Nikkan Kogyo Shimbun Ltd. (1997) p. 8
7. Nishihata et al.: The Phosphorus bronze, basics and application (Japanese), The Nikkan Kogyo Shimbun Ltd. (1997) p. 8
8. Nishihata et al.: The Phosphorus bronze, basics and application (Japanese), The Nikkan Kogyo Shimbun Ltd. (1997) p. 9

2.4.4 Copper Beryllium

Copper beryllium is precipitation hardening type alloy adding 1.6% to 2.0% beryllium into Cu. While maintaining electric characteristic and corrosion resistance which Cu possesses, it shows mechanical properties, fatigue characteristics, and wear resistance comparable to steel.

Developments of copper beryllium started in the early stage of the 20th century. It was known that copper beryllium had been adopted for the German fighter plane during World War I and the full-scale practical use had been done in the U.S.. Many research works had been carried out by major computer manufacturers and connector manufacturers, in regard to the applications of connector sockets which is attached to the motherboard of super computers especially for military demands [1]. After that, copper beryllium has been used for the essential parts of communication equipments, industrial equipments, aircrafts, automobiles, personal computers and cellular phones. In Japan, the adoption of copper beryllium to the field of automotive electrical components and electronic parts started from the latter half of 1970's. In consumer applications, size of all parts was going to become smaller giving word of "Lighter and more compact". Since in many cases it has been difficult to design springs with phosphor bronze for springs because of insufficient strength, copper beryllium has been applied to switches and connectors for light-weight electronic equipments.

As for the conventional copper beryllium, although the price was relatively high due to many kinds of small quantity production, some cost down has been achieved by improving productivity with reducing kinds of material, and developing low copper beryllium based alloy, to become more familiar alloy.

(1) Features

Copper beryllium has the following features.

1) Tensile strength of 1500 MPa at the highest.
2) 20% to 60% electrical conductivity of Cu.
3) Good fatigue life
4) Good high temperature properties
5) Good forming ability for bending and deep drawing
6) Good corrosion resistance

These properties can be obtained with the solution heat treatment, cold working, and age hardening. The material without the cold working shows good forming performance and no anisotropy.

(2) Applications

(a) Electronic parts
Connectors, IC sockets, switches, relays, micro switches, mobile pieces of battery terminal
(b) Automobiles
Switches, relays, sensors, motor brushes, movable pieces of connector
(c) Others
Diaphragm, bellows, scale
(d) Not for spring use
Electrode, submarine cable repeater, bearing, metallic mold

(3) Chemical compositions

The chemical composition of typical copper beryllium is shown in Table 2.77 [2].

(4) Phase diagrams

The phase diagram of Cu–Be alloy is shown in Fig. 2.240.

Table 2.77. Chemical compositions · Features · Applications of copper beryllium [2]

Alloy No.	Shape	Chemical compositions %				Features	Applications
		Be	Ni+Co	Ni+Co+Fe	Cu+Be+Ni+Co+Fe		
C1700	Sheet Strip	1.60–1.79	0.20 min	0.6 max.	99.5 min	Corrosion resistance Ductility Durability Conductivity	High performance spring Connector, Socket Micro switch Relay
C 1720	Sheet, Strip Bar, Wire	1.80–2.00	0.20 min	0.6 max.	99.5 min		

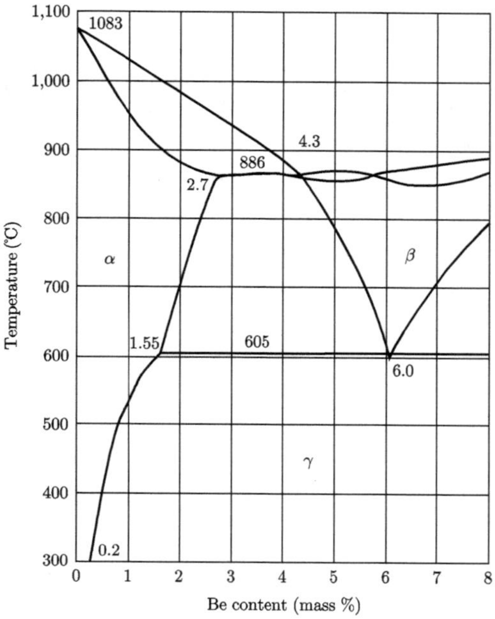

Fig. 2.240. Equilibrium phase diabram of copper beryllium

Copper beryllium in commercial use is 1) cooled rapidly to room temperature to make the supersaturated α solid solution (solution heat treatment). In this state, it becomes possible to form the alloy to bellows or diaphragms as being soft and rich in malleability and ductility. 2) Then, after precipitation hardening at around 315°C, the strength improves considerably with the precipitation of supersaturated beryllium. The 0.2%Co can be added to the copper beryllium in practical use, in order to refine crystal grain and give aging stability.

(5) Manufacturing processes

The base alloy containing approximately 4% beryllium is used as the raw material of copper beryllium. The base alloy is made by deoxidizing beryllium oxide with carbon in arc furnace around 2,000°C, and to melt the deoxidized beryllium into copper, as shown in the following equation.

$$2BeO + C + Cu \rightarrow (2Be + Cu) + CO_2 \uparrow$$

Manufacturing processes of copper beryllium sheet and strip are shown in Fig. 2.241. The melting of copper beryllium is performed to add the same amount of copper and the base alloy, and some elements together to make a billet through the vertical type semi-continuous casting process.

In the next process, the billet hot worked, after removing surface oxide layer generated at the hot-working process mechanically, is cald-worked repeatedly to pre-determined thickness coils. Annealing or solution heat treat-

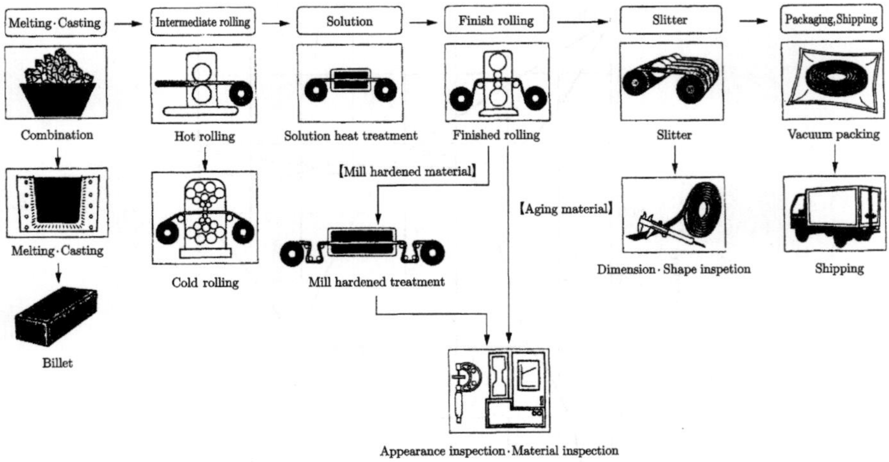

Fig. 2.241. Munufacturing process of copper beryllium sheet and strip

ment can be applied to the cold work-hardened material to get sufficient formability for making the next cold working process. This is the important process where annealing temperature and cooling conditions for controlling the grain size during the recrystallization, can give great influences on the characteristic of final products. After annealing or solution treatment, the surface oxide layer can be removed by sulfuric acid solution, etc. Then, the cold rolling process by multi-stage rolling mill is carried out to the specified thickness for users. The precipitation hardening treatment can be carried out by the spring manufacturer after forming springs. As for mill hardened material, since the appropriate precipitation hardening is processed by the manufacturer, the spring manufacturer carries out stress relief (low temperature annealing) after forming springs.

In case of wire, although cold drawing is conducted instead of rolling, the process is basically the same as that for sheet and strip.

(6) Physical properties

Physical properties of copper beryllium are shown in Table 2.78.

(7) Mechanical properties

(i) Tensile characteristics

Figure 2.242 [3] shows tensile strength, 0.2% proof stress, 0.01% proportional limit, and elongation after aging, for the strength grade of C1720. As the working ratio become high, the tensile strength increases and the elongation decreases. The anisotropy also becomes large.

(ii) Bendability

The bendability of typical copper beryllium is shown in Table 2.79 [4]. The GW (Good way) means the specimen with the rolling direction (0 degree direction) and BW (Bad way) does that at the right angle from the rolling direction.

Table 2.78. Physical characteristics of Beryllium copper

Items		C1700·C1720
Melting start temperature	°C	865
Solidification start temperature	°C	980
Density	g/cm^3 20°C	8.26
Specific heat	J/(kg/K)	419
Coefficent of thermal expansion	/°C 20–300°C	17.8×10^{-6}
Thermal conductivity	W/m 293 K	130
Electrical conductivity	%IACS	25
Modulas of elasticity	GPa	127
Modulas of rigidity	GPa	49
Poisson's ratio		0.3
Magnetic permeability		1.00004

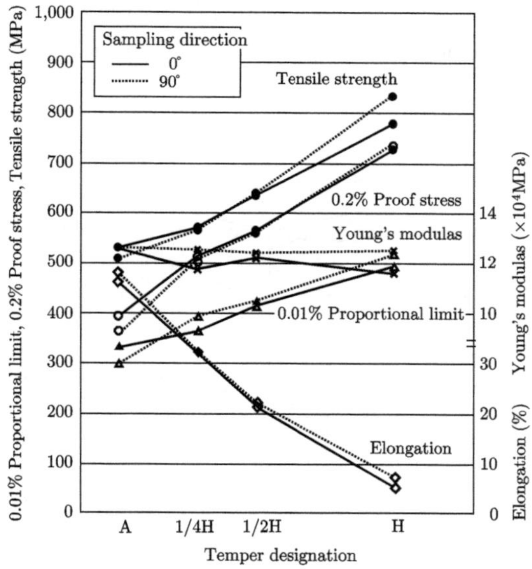

Fig. 2.242. Tensile characteristic of copper beryllium (C1720) [3]

(iii) Spring deflection limit

The spring deflection limit of copper beryllium can be improved greatly with the age hardening processing. The relation between the spring deflection limit and anisotropy, and cold working ratio is shown in Fig. 2.243 [5].

(iv) Fatigue characteristics

The fatigue characteristic of typical copper beryllium is shown in Fig. 2.244 [6].

In order to meet with the actual use, the fatigue tests were conducted for not only the completely reversed stresses but also the pulsating stresses, and also conducted beyond yield stress. The fatigue strength diagram where both

Table 2.79. Bendability of copper beryllium (R/t) [4]

Alloy No.	Strength grade	R/t of bending direction	
		Rolling direction	Perpendicular to rolling direction
C 1720	A	0	0
	1/4H	1.0	2.0
	1/2 H	2.0	3.0
	H	2.0	6.0
	AM	0.8	2.0
	1/4HM	1.0	3.0
	1/2HM	2.0	4.0
	HM	3.0	6.0
	XHM	4.0	—
	XHMS	6.0	—

R: Bending radius, t: Thickness, A: Equivalent with O (no hardening)
HM: mill harden, Hardness Level: HM < XHM < XHMS

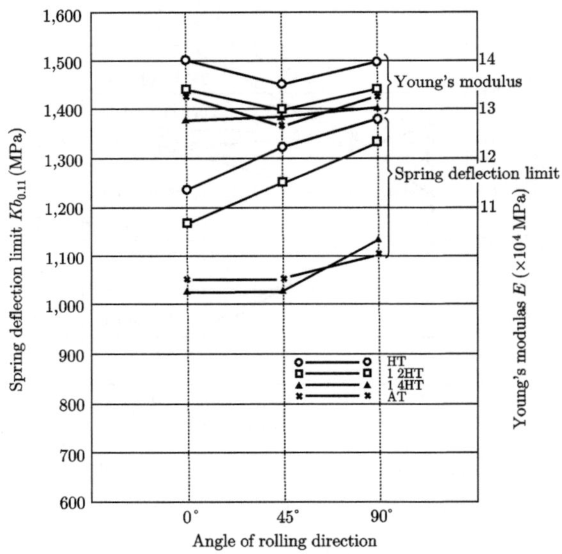

Fig. 2.243. Spring deflection limit and antisotropy of copper beryllium (C1720) [5]

mean stresses and stress amplitudes vary, is shown in Fig. 2.245 [7]. The tests were conducted with an electro-dynamic type fatigue testing machine.

(v) Stress relaxation

The stress relaxation characteristics of typical copper beryllium are shown in Fig. 2.246 [8]. The tests were carried out by the cantilever method standardized in EMAS3003.

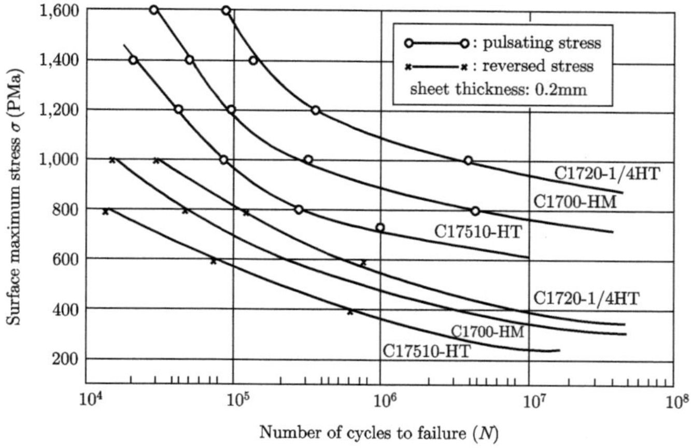

Fig. 2.244. Fatigue characteristics under pulsating and reversed stress test for Beryllium copper alloys [6]

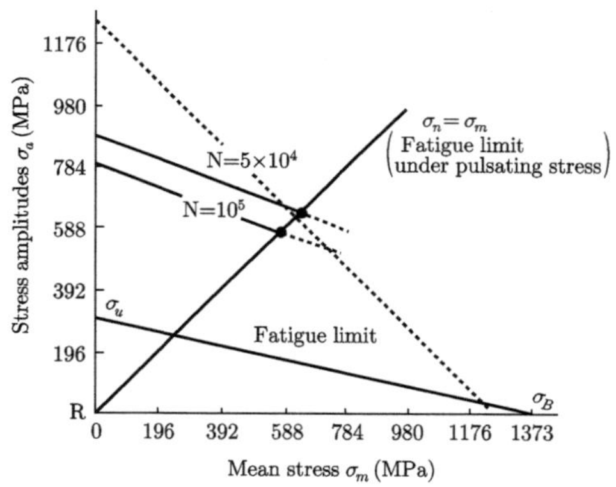

Fig. 2.245. Fatigue strength diagram of copper beryllium (C1720P–1/2HT) [7]

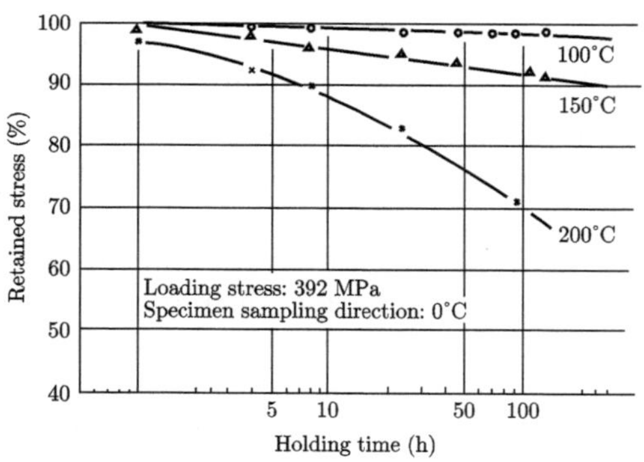

Fig. 2.246. Stress relaxation characteristics of copper beryllium (C1720P–1/4HT) [8]

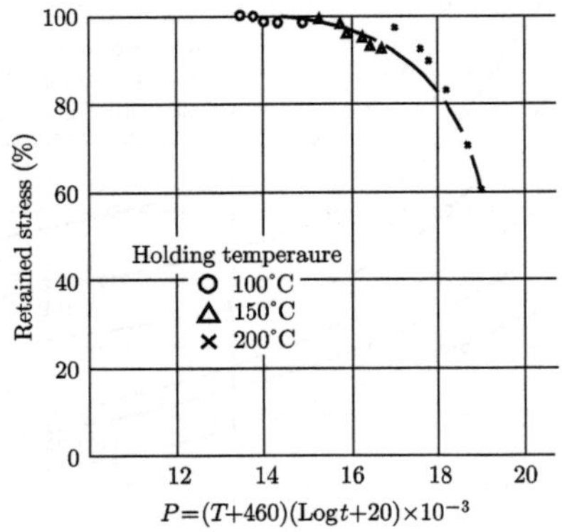

Fig. 2.247. Larson Miller parameter of copper beryllium (C1720P–1/4HT) [9]

It is possible to predict long-term reliability from short-term test results at high temperature. For examples, when predicting a spring load after 10 years usage under ambient temperature of 120°C, the Larson Miller's Parameter, P is calculated as 14.47 by substituting $T = 120°\text{C}$ and $t = 87600$ hours (10 years) to the Larson Miller's Formula, see Fig. 2.247 [9]. It can be read from the Fig. 2.247 that there would be no load loss under these conditions. When 200°C and 10 years, the P is calculated as 16.46. It can be also read from the Fig. 2.247 that there would be about 10% load loss under these conditions.

$$P = (T + 460)(\log t + 20)10^{-3}$$

P: Larson Miller's parameter

T: Temperature, °C

t: Time (Hr)

(8) Remarks in use

The potential ability of copper beryllium can be obtained with a suitable precipitation hardening processing. When applying this heat-treatment, it is essential to control the appropriate temperature and atmosphere.

(i) Temperature condition of heat treatment

The heat-treatment of copper beryllium is normally carried out at 315°C for 2 to 3 hours. It can be recommendable that the temperature should be controlled within ±5°C. Applying gaseous atmosphere such as nitrogen rather than vacuum, and inside forced agitation, the product with less hardness scatter can be obtained. The aging curves of this alloy in term of 0.2% proof stress and fatigue life are shown in Fig. 2.248 [10] and Fig. 2.249 [11], respectively. The temperatures in these figures indicate aging temperature.

(ii) Surface oxide layer

When carrying out age hardening of copper beryllium, it is necessary to avoid the occurrence of oxide film by adding a little hydrogen into nitrogen gas as required. Several hundred angstrom of oxide film is normally formed

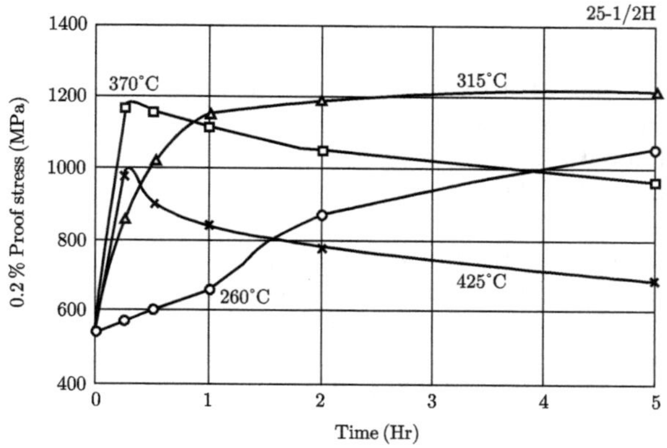

Fig. 2.248. Aging curve of 0.2% proof stress measured at room temperature (C1720–1/2H) [10]

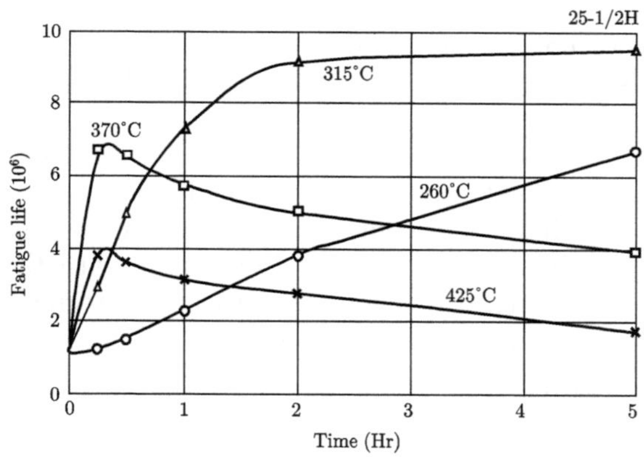

Fig. 2.249. Aging curve of fatigue life measured at room temperature (C1720–1/2H) [11]

on the heat-treated surface in atmosphere [12]. When plating, it is possible to remove the oxide layer only with approximately 20% warm sulfuric acid.

(iii) Safety and health concerns

As for beryllium, no cases of oral obstacle have been reported different from other harmful substances. However, when continuing to inhale the dust over a certain level, in such cases as handling intermediate products or melting a large quantity of alloy at the production mill, there would be some possibility of causing lungs functional disorder. It is required to install powerful dust collectors when re-melting, welding or dry grinding copper beryllium alloy. In other cases, it can be handled in the same manner as normal copper alloy.

References

1. Proceedings of 1st–31th IICIT Connector Meeting
2. JIS Handbook Nonferrous (1998)
3. NGK Insulators Ltd. http://www.ngk.co.jp
4. NGK Insulators Ltd. http://www.ngk.co.jp
5. NGK Insulators Ltd. http://www.ngk.co.jp
6. NGK Insulators Ltd. http://www.ngk.co.jp
7. Adachi et al.: NGK Metal review, 4th number (1981) p. 16
8. Adachi et al.: NGK metal review, 5th number (1984) p. 11
9. Adachi et al.: NGK metal review, 5th number (1984) p. 14
10. Ochiai: Trans. of JSSR 40 (1995) p. 33
11. Ochiai: Trans. of JSSR 40 (1995) p. 33
12. Ishikawa et al.: Japan Copper and Brass Association 26 (1987) p. 96

2.4.5 Other Copper Alloy

The copper alloys not yet standardized in JIS and not introduced in this book are discussed here. There are approximately 133 types of alloys in this classification [1].

The strength level of these alloys are at large between brass and phosphor bronze and electrical conductivity is mostly higher than brass and phosphor bronze. Copper titanium has become popular to cover the properties between phosphor copper and copper beryllium. Concerning with the strengthening methods, work hardening, dispersion strengthening, precipitation hardening and spinodal decomposition have been adopted.

(1) Applications

The products at a special field with a large amount of productions are lead frames and harness connectors for automobiles. Most alloys in this category have been applied to either of these two areas. As far as the strength level is permitted, these alloys can be applicable for other parts such as connectors, switches and relays.

(2) Chemical compositions

The chemical compositions of the developed copper based alloys registered by CDA are shown in Table 2.80 [2].

(3) Characteristics

As for this alloy category, the optimum alloy designs and evaluations have been carried out for the respective applications. Although some of those items are introduced, all the characteristics of each alloy can not be described. For more details, see the catalogs of manufacturers or refer to copper alloy data book [3].

(i) Lead frame material

In order to meet the bending test requirement of lead frames, it is not possible to make the material strength excessively high. As a result, its strength level is mostly lower than that of brass and phosphorus bronze. Normally, lead

Table 2.80. Chemical compositions of developed alloy [2]

Alloy base	CDA No	Alloy name	Principal compositions (%)
Copper-low Tin	C14410	EFTEC-3	Cu–Sn(0.15)
		SNDC	Cu–Sn(0.15)–P(0.010)
Copper-Zirconium	C15100	HCL-151	Cu–Zr(0.1)
		C151	
Copper–Beryllium	C17300	BeCu25K	Cu–Be(1.9)–Co(0.25)-Pb(0.8)
	C17510	11	Cu–Ni(1.8)–Be(0.4)
	C17530	7	Cu–Ni(2.2)–Al(0.6)–Be(0.3)
Copper-Chrome	C18030	EFTEC-6	Cu–Cr(0.15)–Sn (0.1)
	C18040	EFTEC-64/64 T	Cu–Sn(0.25)–Cr(0.3)–Zn(0.2)
	C18200	YCC	Cu–Cr(1)
	C18990	EFTEC-8	Cu–Sn(2)–Cr(0.15)
Copper-Nickel-Tin	C19020	NB-105	Cu–Ni(1)–Sn(0.9)–P(0.05)
Copper-Iron	C19210	EFTEC-7	Cu–Fe(0.1)–P(0.03)
		KFC	Cu–Fe(0.1)–P(0.034)
	C19220	DK-3	Cu–Fe(0.25)–Sn(0.16)–P(0.1)–NiB(0.2)
	C19280	EFTEC-5	Cu–Fe(1)–Sn(0.5)–Zn(0.5)
	C19400	HCL-194	Cu–Fe(2.35)–Zn(0.13)–P(0.8)
		KLF194	Cu–Fe(2.2)–Zn(0.13)–P(0.3)
		Olin194	Cu–Fe(2.4)–Zn(0.1)–P(0.1)
		TAMAC194	Cu–Fe(2.3)–Zn(0.12)–P(0.03)
	C19500	Olin195	Cu–Fe(1.5)–Co(0.8)–Sn(0.6)–P(0.1)
	C19700	Olin197	Cu–Fe(0.6)–P(0.2)–Mg(0.05)
	C19520	TACMAC5	Cu–Fe(0.85)–Sn(0.75)–P(0.03)
Copper-Titanium	C19900	TiCuRl	CuTi(3.1)
		YCuT–M	CuTi(3.2)
		YCuT–T	
Copper-Zinc	C42500	TB2	Cu–Zn(10)–Sn(2)
Copper-Tin-Nickel	C50510	LP004T	Cu–Sn(1.25)–Ni(0.2)–Zn(0.15)–P(0.04)
	C50500	S-1 R	Cu–Sn(1.1)–P(0.15 max.)
	C50710	NK 202 R	Cu–Sn(2)–Ni(0.2)–Zn(0.15)–P(0.04)
		MF 202 R	Cu–Sn(2)–Ni(0.25)–Zn(0.15)–P(0.15 max.)
		C5071W	
	C50715	KLF-5	Cu–Sn(2)–Fe(0.1)
Copper-Nickel-Silicon	C64710	KLF-1	Cu–Ni(3.2)–Si(0.7)–Zn(0.3)
	C64730	KLF-125	Cu–Ni(3.2)–Sn(1.25)–Si(0.7)–Zn(0.3)
	C70250	Olin 7025	Cu–Ni(3.2)–Si(0.7)–Mg(0.15)
Copper-Nickel-Tin	C72500	CAC92	Cu–Ni(9)–Sn(2)
		C725	Cu–Ni(9)–Sn(2.3)

frames do not require the properties which are essential to springs. However, while the lead distance of the semiconductor package becomes narrow and the material thickness also becomes thinner, there will be a problem of deformation in production process, unless the material does not have enough rigidity

Table 2.81. Material characteristics required for lead frame [4]

First characteristics	Second characteristics
Ⓐ Physical properties	Ⓒ Chemical characteristics
1. Themal conductivity	1. Plating
2. Electrical conductivity	2. Bonding
3. Magnetic susceptibility	3. Adhesiveness of oxide film
4. Coefficient of thermal expansion	4. Sealant
5. Density	5. Soldering
6. Elastic modulus	· Wettability
	· Strength, deterioration characteristic
Ⓑ Mechanical properties	
1. Tensile strength, proof stress, elongation	· Soldering fatigue
	· Silver brazing
2. Hardness	6. Corrosion resistance, others
3. Cyclic bending fatigue	· SCC (stress corrosion cracking)
4. Heat-resistant strength (softening resistance)	· Hydration oxidization
	· Gas corrosion
5. Stiffness	· Salt spray test
6. Spring back	· Hydrogen embrittlement
7. Stress relaxation	· Acid resistance
8. Fatigue limit	· Etching method
9. Bending ability	· Metal migration resistance
• Material properties needed for designing and forming lead frame	• Relate to packaging and reliability, and caused by surface physical science phenomenon

like spring material. Therefore, the mechanical characteristics equivalent to spring material can be required for the lead frame. Table 2.81 [4] shows the material characteristics to be required for the lead frame.

(ii) Harness connector material

Since the wire connecting method of a connector and wire is staking, high strength alloy cannot be applicable. Because of this reason, brass has been mainly used. However, in case of automobile, due to higher temperature at engine room and the adoption of electronic equipments, the material having higher quality has been required. The strength level of most alloys is still lower than phosphor bronze for springs. The requirements for harness connector material are shown in Table 2.82 [5].

(iii) High strength copper alloy

Various alloys have been developed to target the characteristics just second to copper beryllium. As the most common alloy in this strength level, copper titanium is presently in production with a certain amount. The chemical compositions, mechanical properties, and its applications are shown in Table 2.83 [6]

Copper titanium containing 3.2%Ti is the precipitation hardening type alloy, which can be classified between copper beryllium and phosphor bronze

Table 2.82. Subject of material for new spring and connector [5]

(a)	Proof stress-bendability and high electrical conductivity
(b)	Stress relaxation characteristic
(c)	Electro migration
(d)	Wettability to tin and solder
(e)	Adhesivness to tin and solder
(f)	Re-wettability to tin and solder
(g)	Growth suppression of acicular product to punch by tin and solder
(h)	Whisker growth suppression on shearing cross section of punch
(i)	Pre-treatment performance of plating
(j)	Punch life
(k)	Punching speed up

Table 2.83. Characteristics of copper titanium (3.2%Ti) [6]

CDA No.	Shape	Strength grade	Tensile strength	Elongation	Hardness	Electrical conductivity	Applications
			MPa	%	HV	%IACS	
C19900	Sheet	1/4HM	735–930	10 min.	250–300	15	Jack
	Strip	EHM	885–1080	5 min.	280–350	13	Connector
		1/4H	440–590	10 min.	140–190	3–4	Relay
		EH	685–885	1 min.	230–280	3–4	Terminal
		1/4HT	735–850	10 min.	250–300	14	
		EHT	930–1130	2 min.	310–360	10	

for springs in regard to both characteristics and prices. Since its precipitation hardening temperature is high compared with copper beryllium, the surface oxidization can be appeared during the heat-treatment. As this surface oxidization can deteriorate the plating quality, the mill hardened material has been in commercial uses. This is used for switches, motors, and jacks of home appliances and acoustic products. The production is now made only in Japan.

Copper–nickel–Sn based alloy having high strength characteristic is also classified between copper beryllium and phosphor bronze. The contents of Ni and Sn are 9 to 21% and 2 to 8% respectively, depending upon the alloy grade. Due to good stress relaxation characteristics, this material has been applied to connectors and sockets. However, because of its poor formability and lower fatigue strength, it seems to be used within limited applications.

(4) Remarks in use

(i) As for material selection

To find the optimum material that meets with the purpose, it should be necessary to make a close communication with material manufacturers. When using the material developed for lead frame and harness connectors to other application, design engineers need to have special knowledge for the material. Especially in case of newly developed alloy, sufficient studies and evaluations should be required.

(ii) Availability

It is often the case that lot size of rolling is large and thickness selection is limited for the material for particular applications. Therefore, careful attentions are required for using small quantity for a long term. If the quantity consumed is small, it is recommendable to study and understand a merit of using existing materials. Since there are many cases that each material manufacturer has their own alloy, the availability under the global base should be taken into consideration.

References

1. Japan Copper and Brass Association: Copper alloy data book (1997) p. 30
2. Japan Copper and Brass Association: Copper alloy data book (1997) p. 324
3. Japan Copper and Brass Association: Copper alloy data book (1997) p. 328
4. Mitsuo Sakamoto: Homat Ad. Corporation (1997) p. 185
5. Miyafuji et al.: Bulletin of Japan Copper and Brass Association, 36 (1997) p. 11
6. Japan Copper and Brass Association: Copper alloy data book (1997), p. 328

2.5 Titanium Alloys for Springs

The term "Titanium" is derived from the name of "Titan" who appears in Greek mythology. Titanium was discovered as early as the end of 18th Century. Titanium is a newcomer as metallic material. It was extracted firstly in 1910 [1]. Commercial titanium production started at 1948. Although titanium has short history as industrial material, it has gained the rapid progress of industrial applications because of its excellent strength, lightness and corrosion resistant property. Titanium alloy has been widely used for aerospace components and chemical processing equipments, however, its applications are spreading to consumer products like watches, eye-glasses, golf equipments, bicycles and automobile parts [2–4].

Titanium alloy also has the excellent properties of spring material [5], and is expected to be the next generation of spring material with the advantage of lightness and high-performance capability.

2.5.1 Resource and Refining Process of Titanium

Titanium is a chemical element in the periodic table that has the symbol Ti and atomic number 22. It is a light, strong, lustrous, corrosion-resistant (including resistance to sea water and chlorine) and transition metal with white-silvery-metallic color. Its concentration of about 0.6% makes it the fourth most abundant of structural metal (after Al, Fe and Mg) in the crust of the Earth.

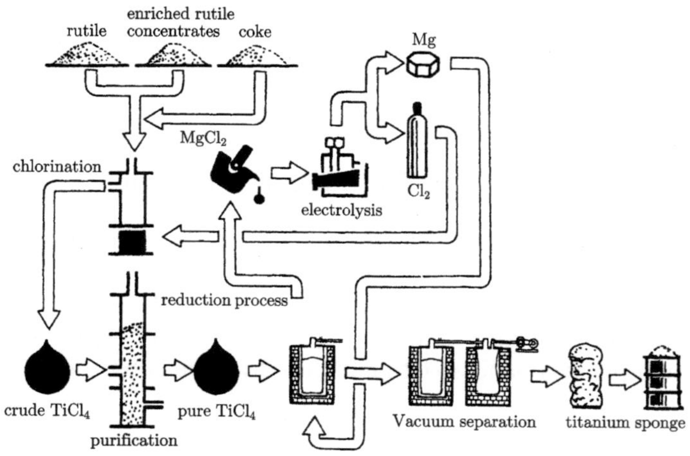

Fig. 2.250. Kroll's process [6]

Titanium is widely distributed and occurs primarily in the minerals anatase, brookite, ilmenite, perovskite, rutile, titanite (sphene), as well in many iron ores. Of these minerals, only ilmenite and rutile have significant economic importance, yet even they are difficult to find in high concentrations. Because it reacts easily with oxygen and carbon at high temperatures, it is difficult to prepare pure titanium metal, crystals, or powder. Therefore, the extraction of titanium metal from titanium mineral requires a large amount of energy. The enormous energy-consumption is one reason for higher material cost of titanium.

The most well-known method of titanium production is the Kroll process (Mg reduction), which involves the reduction of $TiCl_4$ with molten Mg in an argon atmosphere. Figure 2.250 [6] shows the schematic diagram of the Kroll process. Large-scale titanium production adopts this process. Extracted titanium (over 99% purity) is named as titanium sponge because of its porous appearance.

2.5.2 Melting Process of Titanium

The production of titanium ingot from titanium sponge is accomplished by a consumable-electrode arc furnace where molten titanium is poured into a water-jacketed metallic crucible. Titanium is extremely reactive with all potential crucible materials, so conventional melting processes is useless.

Melting shall be executed in vacuum or under a low partial pressure of inert gas such as argon or helium. The use of vacuum melting has the advantage of eliminating hydrogen and magnesium impurities in the titanium ingot preventing the transfer of those element existing in the titanium sponge.

Consumable-electrode arc melting method, non-consumable arc melting method with rotating water-jacketed copper electrode, electron-beam melting method and plasma melting method are used in the industrial production of titanium ingots.

2.5.3 Forging, Rolling and Finishing of Titanium

Manufacturing process of titanium wrought products is shown in Fig. 2.251 [7]. The manufacturing process of titanium is similar to that of steel or other metals, so the facilities for iron products can be used for titanium products. Heating can be performed in an atmospheric furnace, however a slightly oxidizing atmosphere is preferable to minimize hydrogen absorption. Heavy scale

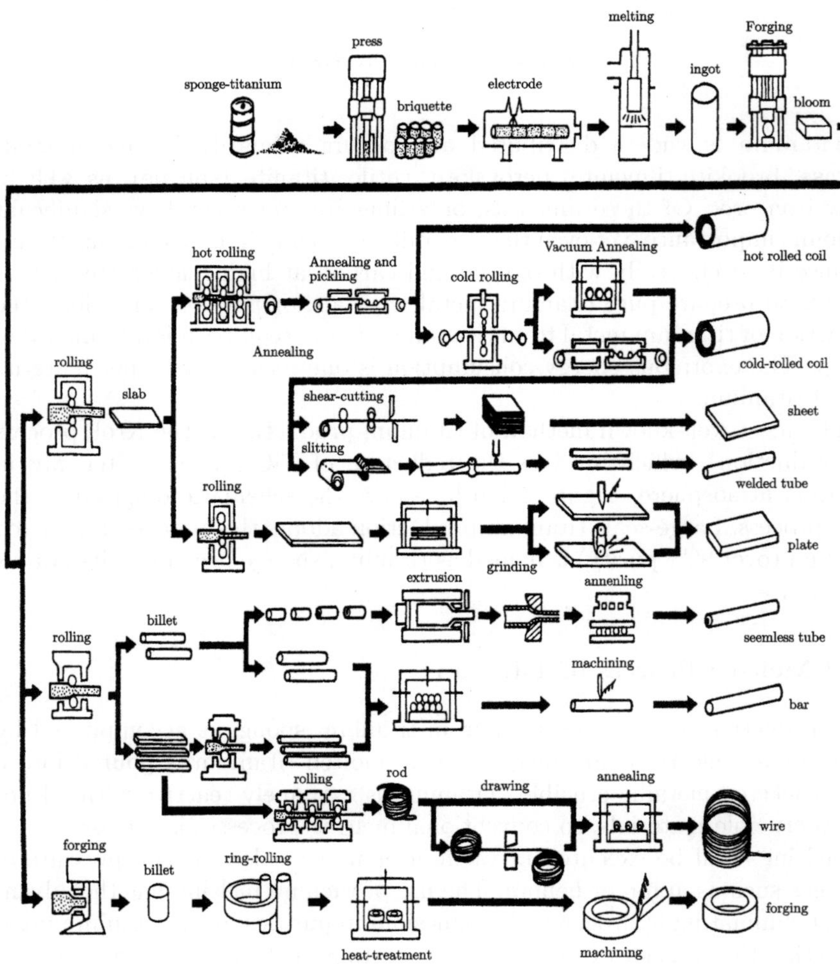

Fig. 2.251. Manufacturing process of titanium products [7]

formed on the surface and the oxygen-contaminated metal beneath the scale should be removed by sand blasting, acid pickling or surface grinding.

In the process of cold-drawing or coiling of titanium wire, severe metallic contacts should be avoided. Deep scoring named "tool mark" caused by seizing might take place on the surface of bare titanium wire if proper care were not provided. Solid lubrications of nickel-coating or thin-oxide film [8] are one of the solutions. Roller die is popular to produce titanium wire effectively avoiding those surface defects.

Surface defect of the titanium wire affects seriously the quality of titanium spring.

2.5.4 Physical Properties of Titanium

The physical properties of pure-titanium, Ti-6Al-4V alloy and Ti–15V–3Cr–3Sn–3Al alloy are listed in Table 2.84 together with other alloys. The features of titanium are as follows;

① Melting temperature of titanium is 1668°C, higher than that of iron.
② Atomic number of titanium is 22, and it is a transition element belonging to group IV. The allotropic transformation of titanium takes place at 885°C.
③ Titanium exists in two crystallographic forms. At room temperature, titanium has a hexagonal close-packed crystal structure which is called an α phase. At 883°C, the crystal structure transforms to a body-centered cubic structure which is called a β phase.

Table 2.84. Comparison of Physical properties between Titanium alloy and various materials

	pure-Ti	Ti–6Al–4V	Ti–15V–3Cr–3Sn–3Al	Iron	18Cr–8Ni Stainless steel	Al alloy
Melting temperature °C	1668	1540–1650	1524–1620	1530	1400–1427	476–638
Structure	HCP < 885°C < BCC	HCP < 990°C < BCC	HCP < 760°C < BCC	BCC < 830°C < FCC	FCC	FCC
Specific gravity	4.51	4.42	4.71	7.86	8.03	2.70
Young's modulus $\times 10^4$ (MPa)	10.2	11.3	7.9 (ST) 9.8 (STA)	19.2	20.0	7.1
Poisson's ratio	0.34	0.30–0.33	0.3	0.31	0.29	0.33
Electrical resistivity ($\mu\Omega \cdot$ cm)	47–55	171	147	9.7	72	5.8
Thermal conductivity (cal/cm^2/s/°C/cm)	0.041	0.018	0.017	0.145	0.039	0.294
Coefficient of thermal expansion $\times 10^{-6}$ (cm/cm/°C)	8.4	8.8	8.4	12.0	16.5	23.1
Specific heat (cal/g/°C)	0.13	0.13	0.12	0.11	0.12	0.23

ST; Solution-treament STA; Solution treatment and aging

④ β transformation temperature has a great influence over the mechanical properties of the titanium alloys, so it is regarded as a key index figure.
⑤ The relative density of titanium alloys are 4.4–4.8, 60% that of steel. Since tensile strength of titanium alloys are similar to that of steel, titanium alloys have a high strength-to-density ratio.
⑥ Young's modulus of titanium alloy is about half of steel's value.
⑦ Electrical conductivity and thermal conductivity of titanium are very low, similar to that of austenitic-stainless steel.
⑧ Thermal expansion coefficient of titanium is about 2/3 of steel, and about half of austenitic-stainless steel and copper.
⑨ Specific heats of titanium alloys are comparable with that of austenitic stainless steel. Because of small specific gravity, titanium alloy products are easily heated up and cooled down.
⑩ Magnetic permeability of titanium is 1.0, meaning that titanium is non-magnetic material.

The cold-drawing and machining of titanium wire are relatively difficult compared with other metals. Galling (metallic sticking) and wearing between wire tend to take place easily [9,10]. Low thermal conductivity and low specific heat of titanium could be the major cause of those troubles. Since heat created by friction does not dissipate rapidly, high temperature at the contact area would accelerate the chemical reaction between active titanium and tool metals. In case of machining, it is therefore necessary to keep low-temperature at the contact area using coolants or slow-rate cutting method. The solid lubrication film is particularly important in wire-drawing or other forming processes.

2.5.5 Properties of Titanium and Titanium Alloys

Titanium material is classified into four categories. Pure-titanium, α alloys, α-β alloys and β alloys. Typical properties and chemical compositions are listed in Table 2.85 [11].

Pure titanium means commercially pure titanium which is unalloyed grades. α alloys has stable α crystal-structure at room temperature. α-β alloys contain the mixture of α-phase and β-phase. β alloys has single β-phase matrix which is obtained by quenching the alloy from above the β-transformation temperature.

In titanium alloys, the principal effect of alloying elements is their effect on β-transformation temperature. Some elements stabilize α crystal structure by raising β-transformation temperature, while the other elements stabilize β crystal structure by lowering β-transformation temperature. Al, O, N, and C elements are called α stabilizer, and V, Cr, and Fe elements are called β stabilizer. Tin and Zirconium has considerable solid solubility in both α and β phases, and they are sometimes called neutral element.

Table 2.85. Various properties of titanium and titanium alloys [11]

TYPE	Composition (mass%)	Heat treatment	Tensile properties: Tensile strength (MPa)	Proof stress (MPa)	Elongation (%)
unalloyed Ti	JIS 1 Grade 1	A	270–410	165	≧ 27
	JIS 2 Grade 2	A	340–510	215	≧ 23
	JIS 3 Grade 3	A	480–620	345	≧ 18
	JIS 4 Grade 4	A	685–830	485	≧ 15
α alloy	Ti–5Al–2.5Sn	A	862	804	16
α-β alloy	Ti–8Al–1Mo–1V	A	1000	951	15
	Ti–6Al–2Sn–4Zr–2Mo	A	980	892	15
	Ti–3Al–2.5V	A	686	588	20
	Ti–6Al–4V	A	980	921	14
		STA	1170	1100	10
	Ti–6Al–2Sn–4Zr–6Mo	STA	1270	1180	10
	Ti–10V–2Fe–3Al	STA	1270	1200	10
β alloy	Ti–13V–11Cr–3Al	STA	1220	1170	8
	Ti–3Al–8V–6Cr–4Zr–4Mo	STA	1440	1370	7
	Ti–15Mo–5Zr–3Al	STA	1470	1450	13
	Ti–15V–3Cr–3Sn–3Al	STA	1230	1110	10

α phase (top) ↔ β phase (bottom):
large ←— specific gravity —→ small
good ←— heat-treatability —→ bad
decrease ←— creep strength —→ increase
increase ←— strain-rate sensitivity —→ decrease
decrease ←— Young's modulus —→ increase
decrease ←— oxidization resistance —→ increase

A: Annealing STA: Solution treatment and aging

(1) Commercially-pure (CP) titanium

Pure titanium means unalloyed titanium, of which strength mainly depends upon the quantity of O and Fe element. Because of high purity in titanium sponge, oxygen and iron are added to control the mechanical property of CP titanium product. CP titanium is different from "pure" titanium which is used in academic field. As shown in Table 2.86, it is classified into four types in JIS. They are used in the application where optimum corrosion resistance and good formability are desired compromising the mechanical strength. JIS grade 2 is the most popular in this category. Mechanical properties of JIS 3 grade CP titanium which has higher strength than JIS 2 grade are shown in Fig. 2.252 [12]. It has relatively good spring quality among CP titanium grades.

(2) α alloy

Since α alloy has a single phase, it cannot be strengthened by a heat treatment process. α alloys are slightly less corrosion resistant but higher in strength than the other type of titanium alloy. α alloy is quite ductile and tough with good creep strength at elevated temperature. α alloy such as Ti–5Al–2.5Sn retains sufficient ductility and toughness at cryogenic temperature such as liquid helium.

(3) α-β alloys

α-β alloy which contains α phase and β phase can be strengthened by heat-treatment. Its mechanical properties vary widely by the ratio of α-phase and β-phase. The mechanical property of α-rich alloy is similar to that of α alloy showing high creep strength at elevated temperature. On the other hand,

Table 2.86. Standards and chemical compositions of titanium and titanium alloys

Category	Common name	Standards	Chemical compositions										mass %
			Al	Sn	Zr	V	Cr	Mo	Fe*	O*	N*	H*	Ti
Pure Ti	Commercially pure Ti	JIS Grade 1	—	—	—	—	—	—	0.20	0.15	0.05	0.015	Bal
		JIS Grade 2	—	—	—	—	—	—	0.25	0.20	0.05	0.015	Bal
		JIS Grade 3	—	—	—	—	—	—	0.30	0.30	0.07	0.015	Bal
		JIS Grade 4	—	—	—	—	—	—	0.50	0.40	0.07	0.015	Bal
α alloy	Ti–5Al–2.5Sn	AMS 4926	4.0–6.0	2.0–3.0	—	—	—	—	0.50	0.20	0.05	0.015	Bal
α-β alloy	Ti–8Al–1Mo–1V	AMS 4972	7.35–8.35	—	—	0.75–1.25	—	0.75–1.25	0.30	0.12	0.05	0.020	Bal
	Ti–6Al–4V ELI	JIS 60E	5.50–6.50	—	—	3.5–4.5	—	—	0.25	0.13	0.05	0.0125	Bal
	Ti–6Al–4V	JIS 60E	5.50–6.75	—	—	3.5–4.5	—	—	0.30	0.20	0.05	0.015	Bal
	Ti–6Al–2Sn–4Zr–6Mo	AMS 4981	5.50–6.50	1.75–2.25	3.50–4.50	—	—	5.50–6.50	0.15	0.15	0.04	0.0125	Bal
	Ti–10V–2Fe–3Al	AMS 4983	2.60–3.40	—	—	9.0–11.0	—	—	1.60–2.20	0.13	0.05	0.0150	Bal
β alloy	Ti–13V–11Cr–3Al	AMS 4959	2.50–3.50	—	—	12.5–14.5	10.0–12.0	—	0.35	0.17	0.05	0.0250	Bal
	Ti–3Al–8V–6Cr–4Zr–4Mo (Beta C)	AMS 4957	3.00–4.00	—	3.50–4.50	7.50–8.50	5.50–6.50	3.50–4.50	0.30	0.12	0.05	0.0300	Bal
	Ti–15V–3Cr–3Sn–3Al	AMS 4914	2.50–3.50	2.50–3.50	—	14.0–16.0	2.50–3.50	—	0.25	0.13	0.05	0.0150	Bal

ELI: Extra Low Interstitial

*: upper limit value

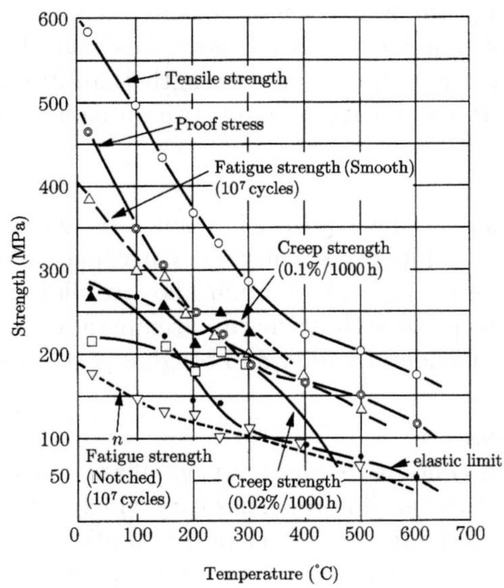

Fig. 2.252. Strengths of CPT (JIS Grade3) at various temperatures [12]

β-rich alloy can be hardened by heat-treatment and shows high strength and high toughness that are characteristics of β alloy.

Ti–6Al–4V alloy is a typical α-β alloy. Since it shows best combination of strength and toughness by applying suitable heat treatment, it has been widly applied in commercial productions.

α-β alloy has also good weldability and good cutting performance.

(4) β alloy

Since β alloy is sufficiently rich in β stabilizers, it retains β phase at room temperature by applying appropriate heating and cooling process. The features of β alloy are good formability at room temperature, and the highest strength among titanium alloys. Having relatively low Young's modulus and high fatigue strength, β alloy shows excellent properties for spring materials.

Ti–15V–3Cr–3Sn–3Al, Ti–13V–11Cr–3Al, Ti–3Al–8V–6Cr–4Zr–4Mo (beta-C) are typical β titanium alloys. Some β alloys containing Mo element, such as Ti–15Mo–5Zr–3Al alloy, have good corrosion resistance under reducing atmosphere.

2.5.6 Spring Characteristic of Titanium

As spring materials, the properties of titanium alloys are as follows: The titanium alloys retain good spring characteristics at elevated temperature as high as 400°C.

① Low Young's modulus (Fig. 2.253 [13])
② The specific weight of titanium is approximately 60% of steel.
③ Tensile strength is relatively high. (Fig. 2.254 [14])
④ High strength-to-density ratio and low modulus-to-density ratio
⑤ High fatigue strength (Fig. 2.255 [15])
⑥ High fatigue strength-to-density ratio (Fig. 2.256 [16])
⑦ High fatigue strength-to-tensile strength ratio (Fig. 2.257 {17})
⑧ Small permanent set (Fig. 2.258 [18])

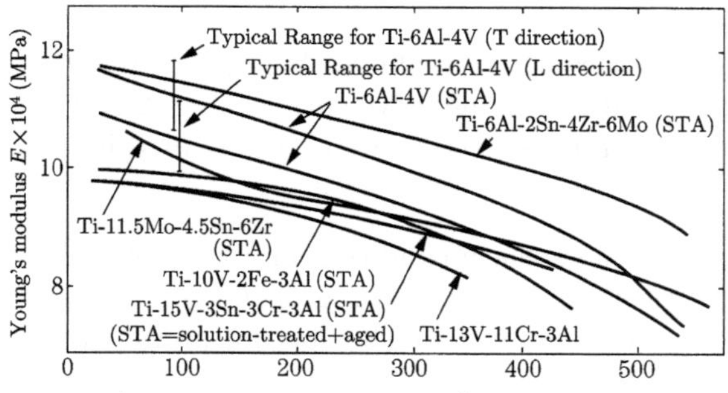

Fig. 2.253. Relation between Young's modulus and temperatures in Ti alloys [13]

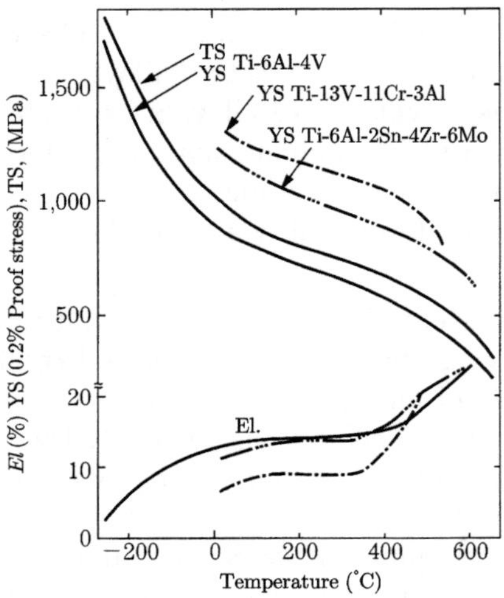

Fig. 2.254. Temperature dependency of mechanical properties in titanium alloys [14]

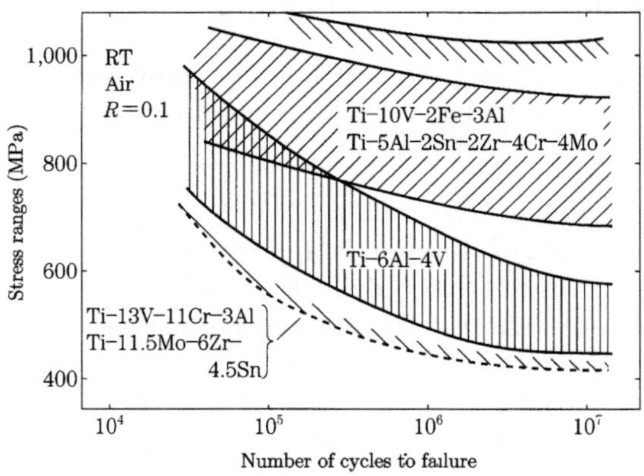

Fig. 2.255. High-cycle fatigue strength in titanium alloys [15]

⑨ Relatively small surging stress due to high characteristic frequency of vibration when used for engine valve springs (Fig. 2.259 [10])
⑩ Excellent corrosion resistance and low stress corrosion cracking susceptibility

The influence of the surface condition and the residual stress on the fatigue strength of titanium is shown in Fig. 2.260 [16]. It shows that the compressive residual stress attained by shot-peenig plays a key role to improve fatigue strength. Fatigue strength decreases remarkably when oxidized scale or oxygen-contaminated metal beneath the scale (alpha-case) are formed on the surface of spring [2, 4, 9]. It is necessary to remove those scale and con-

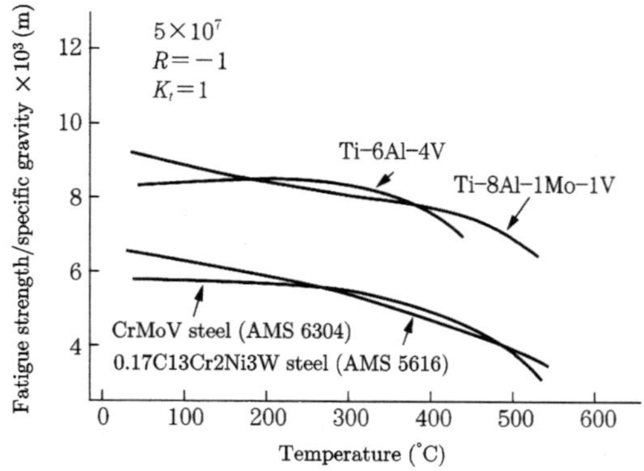

Fig. 2.256. Temperature dependency of high-cycle fatigue strength in titanium alloys [16]

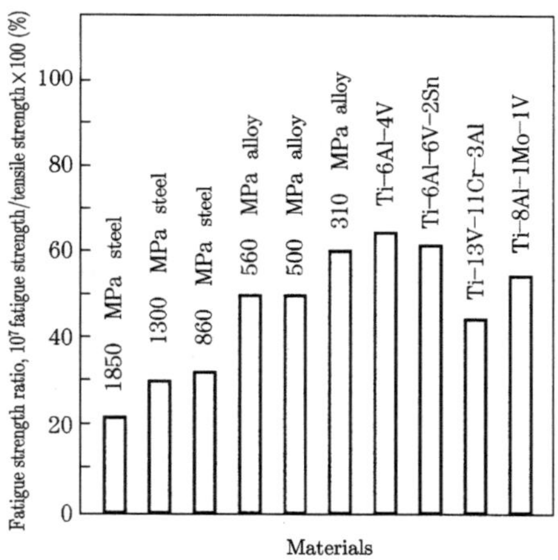

Fig. 2.257. Endurance ratio in titanium alloys and other metallic materials [17] ($K_t = 1.0$, $R = 0.1$, RT, 10^7 cycles)

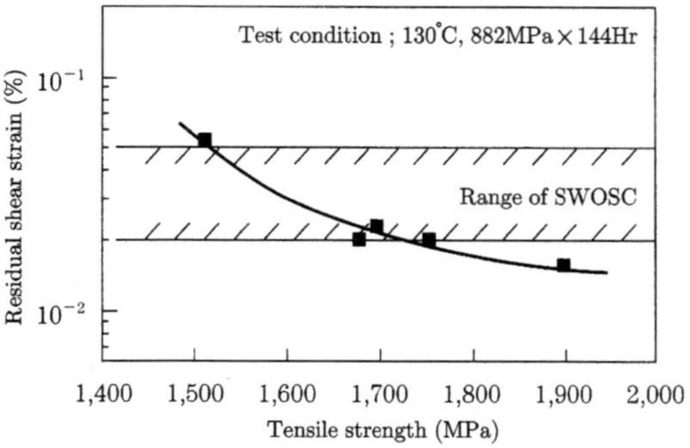

Fig. 2.258. Relation between tensile strength and permanent set of valve spring in Ti alloy [18]

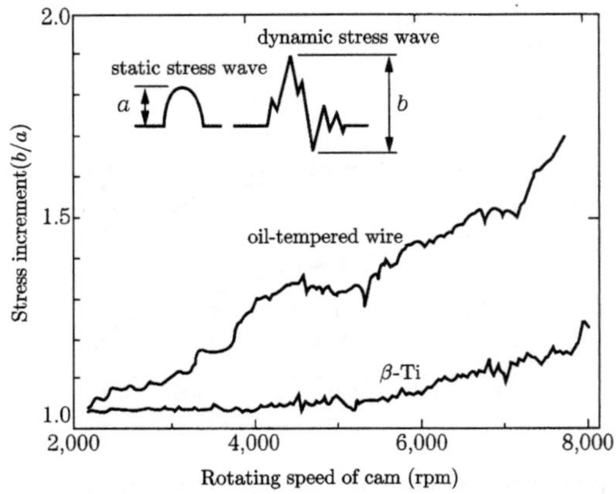

Fig. 2.259. Dynamical properties of titanium valve springs and steel valve springs

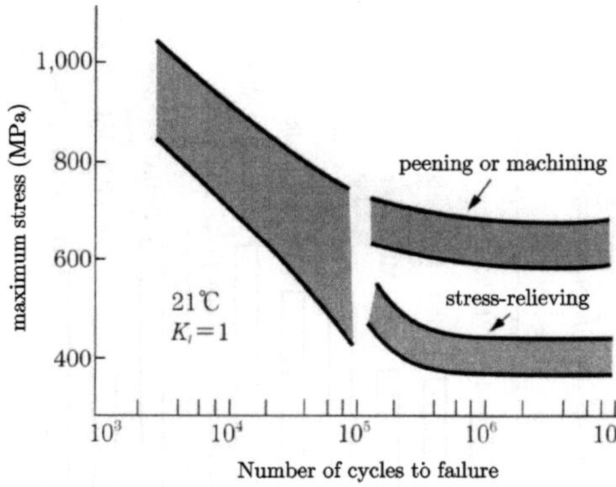

Fig. 2.260. Effect of residual stress and surface conditionings on fatigue strength of Ti–6Al–4V [19]

taminated layer sufficiently by using sand-blasting or acid pickling in order to retain desirable fatigue strength. Nitriding is effective for the improvement of sliding wear resistance, however it may bring the decrease of fatigue strength of titanium spring [20].

Table 2.87 [18] shows examples of valve spring design comparison between current SWOSC-V steel and titanium alloy. By adopting titanium alloy, following advantages can be obtained; 13% reduction in free height and 57% reduction in weight. This light-weight valve spring can contribute to the high performance of automobile engine. The largest problem of titanium as spring material is abrasion. Titanium is very reactive metal, so it is easy to stick to the other metal at the contact point. In the case of spring, the contact between coils could be a fatal problem. Various coating process [10,18,20] has been investigated for spring applications.

Table 2.87. An example of Ti valve spring and its effect [18]

material	Dia of wire (mm)	Spring rate (N/mm)	Outside dia of coil (mm)	Maxmum deflection (mm)	Free hight (mm)	Weight (g)
steel (SWOSC-V)	4.2	7.6	43	66	103	118
Ti–13V–11Cr–3Al	4.3	7.6	43	66	90	51
Effects	—	—	—	—	13% down	57% down

References

1. M.A. Hunter: J. Amer. Chem. Soc., 32 (1910), p. 330
2. Y. Hagiwara, K. Takahashi: Titanium and zirconium (in Japanese), 35 (1987), p. 69
3. M. Mushiake, K. Asano, N. Miyamura and S. Nagano: SAE paper 910428
4. A. Kimura, S. Nakamura, S. Isogawa, T. Matsubara, K. Kimura and Y. Satou: Denki-Seiko in Japanese, 63 (1992), p. 33
5. A.M. Sherman: SAE paper 800481
6. T. Nishimura: Forming technology of titanium (in Japanese), Nikkan Kogyou Shinbun, (1992), p. 3
7. Kobe Steel Data Book: Titanium
8. W. Takahashi, H. Maeda, H. Nagai and T. Fukuda: Sumitomo Kinzoku (in Japanese), vol. 42 (1989), p. 182
9. N. Maruyama, H. Kakuda and K. Nakazawa: Testu-to-Hagane in Japanese, 76 (1990), p. 262
10. T. Igarasi, T. Ozone: WIRE 90 SPRING TECHNOLOGY, ESF INTERNATIONL CONFERENCE (1990)
11. S.R. Seagle: Principles Alloying Titanium, 1968, ASM
12. Japan Titanium Society ed.: Titanium and its application (in Japanese), Nikkan Kogyo Shinbun, (1992), p. 44
13. MIL-HDBK-697 A, (1974), p. 36
14. MIL-HDBK-697 A, (1974), p. 35
15. MIL-HDBK-697 A, (1974), p. 45
16. M.J. Donachie Jr. (H. Suzuki, and K. Harada K. translated): Titanium, Technical Guide, Utida Roukakuho (1993), p. 181
17. H. Urusihara: Titanium and zirconium (in Japanese), 35 (1972), p. 189
18. T. Yabusita, N. Takamura and A. Tange: Transactions of JSSR, 41 (1996), p. 1
19. M.J. Donachie Jr. (H. Suzuki and K. Harada K. translated): Titanium, Technical Guide, Utida Roukaku (1993), p. 182
20. K. Tokaji, T. Ogawa, H. Shibata and M. Kamiya: Transactions of JSME, vol. 57 (1991), p. 2293
21. T. Abumiya, K. Yasui, K. Suzuki, H. Ooyama and H. Nishimoto: Kobe Steel Technical Report (in Japanese), 42 (1992), p. 64

2.6 Ni–Ti Shape Memory Alloys and Super-Elasticity

Shape memory alloys are characterized by their unique properties; "shape memory effect" and "superelasticity". The main usage of the alloys is superelasticity, though there are a few application of shape memory effect.

The shape memory effect can be schematically explained with the help of the stress-strain curve in Fig. 2.261 comparing with that of conventional metallic materials. In the case of conventional metallic materials schematically shown in the Fig. 2.261(a), the stress-strain relation is linear in the elastic region and the strain within this region returns to zero or original point after unloading. When sufficient stresses are applied to reach the tensile yield strength, permanent deformation called plastic strain occurs.

When straining shape memory alloy under its condition of martensite structure at low temperature, elastic deformation occurs at first, then yielding and plastic deformation follows. Unloading at this point leaves tentative plastic strain. This plastic strain recovers to zero after being heated up to above the reverse transformation (i.e., martensite to parent phase or austenite) temperature, as shown in the Fig. 2.261(b).

In the case of superelastic alloys, yielding and plastic deformation after elastic strain region trace a peculiar curve as shown in the Fig. 2.261(c). When unloaded, it recovers its original geometry like rubber materials. The maximum recoverable strain could reach as far as 5 to 7%. This phenomenon is called superelasticity or pseudoelasticity.

Since the first finding of shape memory effect in 1950 [1], more than ten alloys have been reported as exhibiting the shape memory effect. Among them, nickel-titanium alloys have been the most popular materials, and have been used for various applications since they show the best performance from the point of shape memory characteristics (especially, excellent stabilities for repeated actions), workability, corrosion resistance and so on.

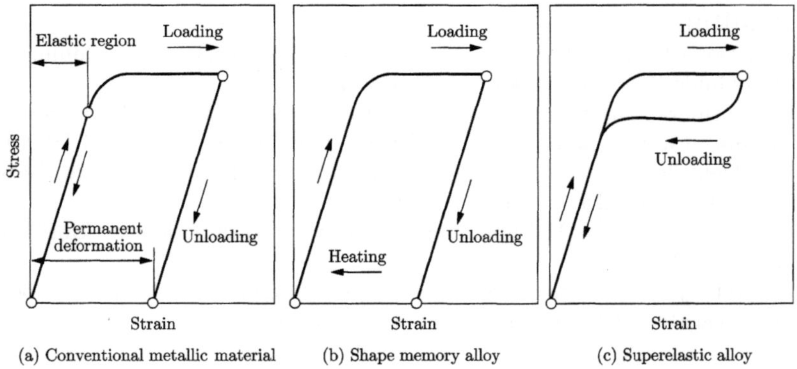

Fig. 2.261. Stress-strain curves for alloys

The favored application of nickel-titanium alloys is switches and actuators of industrial apparatus, and medical equipments to which reliability is strongly demanded. For this reason, the following section covers nickel-titanium alloys.

2.6.1 General Principles

(1) Martensite transformation

Martensite transformation (parent phase to martensite) and reverse martensite transformation (matensite to parent phase) are the cause of shape memory effect and the superelasticity. Martensite transformation is a kind of solid-state diffusionless phase transformation where a quasi-shear deformation with a quite small volume change takes place maintaining lattice correspondence.

There is a type of martensite called the thermo-elastic martensite where the martensite grows gradually from the parent phase through cooling process and subsides gradually to the parent phase (austenite) through heating process. The thermo-elastic martensite plays an essential role in the shape memory effect and the superelasticity.

The temperature range for the martensite-to-austenite transformation that takes place upon heating is somewhat higher than that for the reverse transformation upon cooling. The difference between the transformation temperatures upon heating and cooling is called temperature hysteresis.

The transformation temperatures are defined as follows:

M_s: The temperature at which parent phase starts to transform to martensite upon cooling.

M_f: The temperature at which parent phase completely transforms to martensite upon cooling.

A_s: The temperature at which martensite starts to transform to parent phase upon heating.

A_f: The temperature at which martensite completely transforms to parent phase upon heating.

These transformation temperatures are defined under the condition where no stress is applied. When stress is applied, the transformation to martensite can take place above A_s or A_f.

Shape memory effect:

The stress-strain curves for Ni–Ti alloys below their transformation temperatures are given in Fig. 2.262 [2]. The dotted lines in the Fig. 2.262 indicate recovered strains by heating. The pre-strain of around 6% may be completely recoverable.

The schematic illustration of shape memory effect in Ni–Ti alloy is given in Fig. 2.263 using a two-dimensional crystal model [3]. The deformation of conventional metal takes place along with a slip plane, and shear strain occurs as is shown in the Fig. 2.263(a). Since the atomic displacements during the

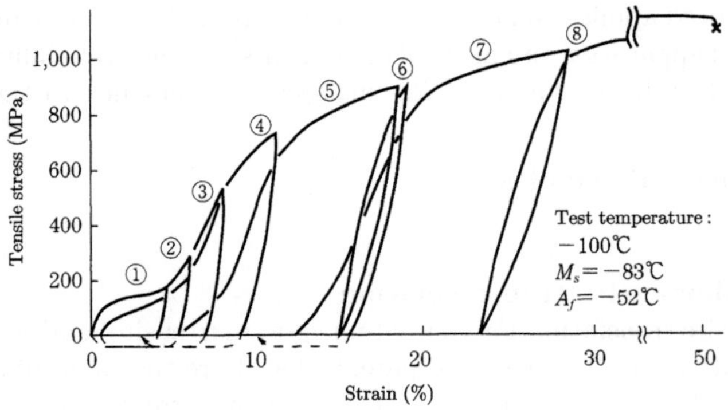

Fig. 2.262. Stress-strain curves for Ni–Ti alloy showing shape memory effect [2]

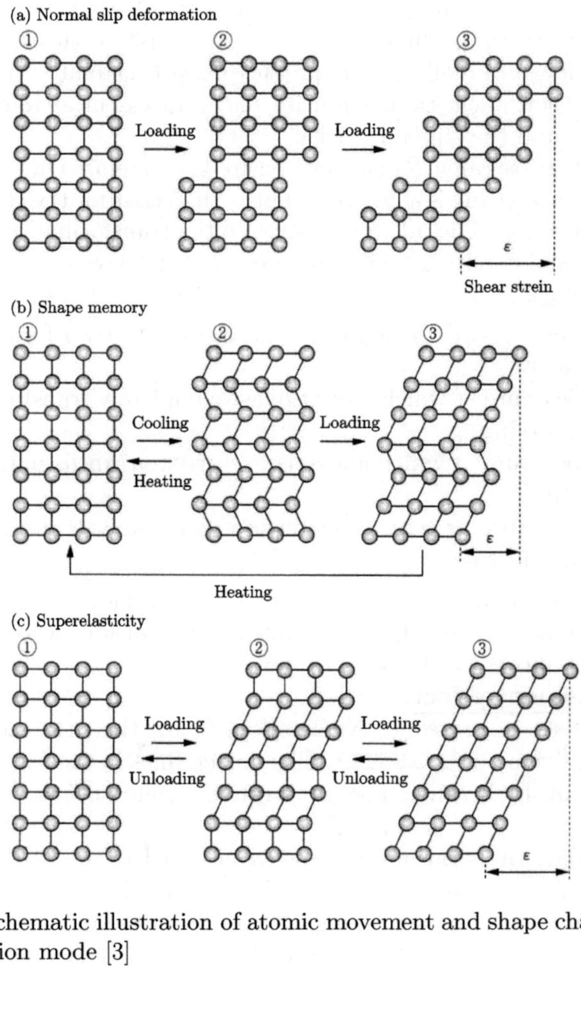

Fig. 2.263. Schematic illustration of atomic movement and shape change according with deformation mode [3]

deformation exceed an atom to atom distance, permanent strain of ε remains after the deformation.

In contrast, for the shape memory alloys, a parent phase schematically shown in the Fig. 2.263(b) ① transforms to martensite upon cooling as is shown in (b) ②. Martensite consists of crystallites called variants which are mirror symmetry with neighboring variants (twined crystal relationship between neighboring variants). The boundary between such variants is called a twin plane with the thickness of single crystal lattice (actually, a variant may have the thickness of more than one lattice spacing). The crystal structure of martensite is composed of variants having different orientations. When martensite is unidirectionally loaded as is shown in the Fig. 2.263(b) ③, the shear deformation of ε progresses by twinning deformation mechanism where a group of variants having orientation that relax the load propagate at the expense of another group of variants with different orientations. Upon heating to a temperature above A_f, the martensite phase transforms to the parent phase (i.e., reverse transformation), hence the shear deformation of ε disappears as is shown in the Fig. 2.263(b) ①. These are brief explanations of mechanism for the shape memory effect. Since the twin planes can move easily under a comparatively low stress, the martensite phase is generally more deformable than the parent phase (face centered cubic structure).

Superelasticity:

When a shape memory alloy of parent phase is deformed at a temperature higher than A_f, a super-elastic behavior similar to rubber is observed. Figure 2.264 shows stress-strain curves for Ni–Ti alloys strained above A_f temperature. Upon loading up to the yield point, an elastic deformation of the parent phase proceeds. Beyond the yield point, a flat stage appears where martensitic transformation proceeds with strain. After the stress induced martensite transformation is completed, an elastic deformation of the martensite takes over. During unloading from point ④ in the figure, the reverse transformation

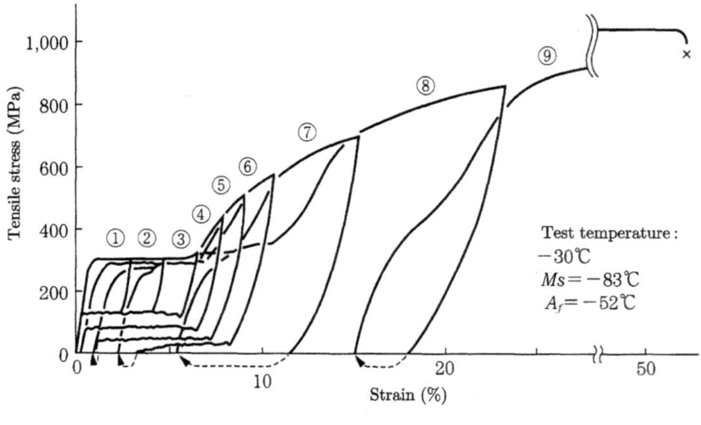

Fig. 2.264. Stress-strain curves of Ni–Ti superelastic alloy [2]

from the martensite phase to the parent phase takes place tracing back the flat stage of stress, follwed by elastic recovery of parent phase to the original shape. A crystal model of superelasticity is shown in Fig. 2.263(c). The macroscopic deformation is accommodated by the formation of martensite even at temperature higher than A_f. When the stress is released, the martensite transforms back into austenite and the specimen returns back to its original shape. Super elastic Ni–Ti can be strained several times more than ordinary metals without being plastically deformed.

(2) Transformation temperatures and their control

Ni–Ti alloys exhibiting shape memory effect have the composition of around fifty-fifty atomic% Ni and Ti, which is equivalent to Ni 55 mass% and Ti 45 mass%. The transformation temperature is affected by chemical composition, heat-treatment, and working history where the specimen has been subjected to. The effect of chemical composition is the most influential factor among them. As shown in Fig. 2.265, the A_f point of the Ni 50.0 atomic%-Ti alloy is at around 70°C, and that of the Ni 51.0 atomic%-Ti alloy is at around 0°C [4]. This indicates that the 0.1 atomic% difference brings about the 70°C drop of the transformation temperature.

The transformation temperature can also be shifted by adding a third element to the Ni–Ti binary alloys. For example, as shown in Fig. 2.266, Fe, Al, Co, Mn and Cr can decrease the M_s transformation temperatures drastically with quite a small amount of their additions [5].

There are some cases where an intermediate phase appears in some Ni–Ti alloys with a specific metal element composition when subjected to a specific heat-treatment. The intermediate phase is called R-Phase (Rhombohedral crystal). Upon cooling, the specimen transforms in two stages from parent phase to R-Phase and then to martensite phase. The first stage transformation yields around one-tenth of the 2nd stage martensite transformation strain. It exhibits quite a stable repeatability and a very small temperature hysteresis

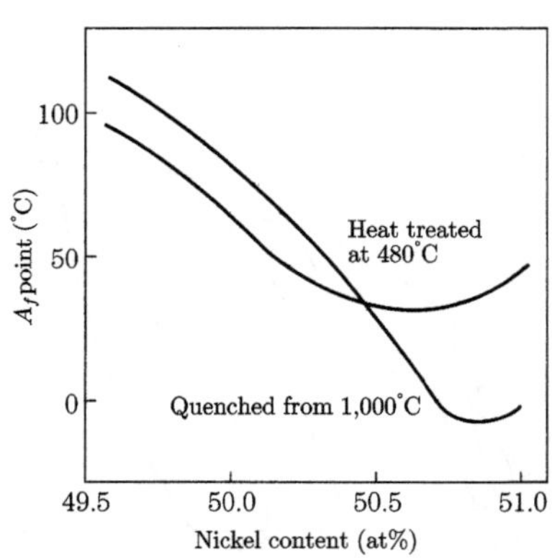

Fig. 2.265. Relation between Ni–Ti alloy compositions and transformation temperature [4]

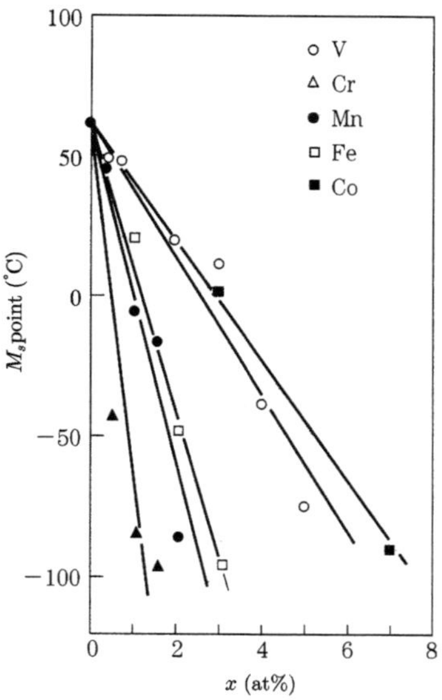

Fig. 2.266. Effect of alloy element addition on transformation temperature of Ni–Ti alloy [5]

of 1–2°C. From these points, Ni–Ti alloys with R-Phase have some advantage for practical applications. The R-Phase appears when a heavily cold-worked specimen is heat-treated at around 400°C [6].

The transformation temperature is also affected by the microstructure variation such as precipitates generated by aging in Ni rich alloy and the introduction of dislocations generated by working. In Fig. 2.265, transformation temperature curves for Ni–Ti alloys having been subjected to different heat-treatments are presented. Aging at 480°C makes A_f temperature significantly higher than as-quenched condition at a rather high Ni content region. Figure 2.267 shows the effect of aging temperature on the load-temperature curves for pre-strained Ni–Ti compression coil springs [7]. The transition points in the figure are corresponding to the transformation temperatures. Upon increasing the aging temperature (heat treatment temperature), the transformation temperatures tend to be decreased.

(3) Phase transformations and their characteristics

(i) Shape memory alloys

The applications of Ni–Ti alloys are categorized by the extent of transformation temperature hysteresis that is characteristic to each alloy [8].

In Table 2.88, the classified applications of Ni–Ti alloys based on transformation temperature hysteresis and the categories of shape memory effect are summarized together with superelasticity. The extent of temperature hysteresis ranges from 2°C to 100°C. The type of crystal structure that could form during transformation is as follows:

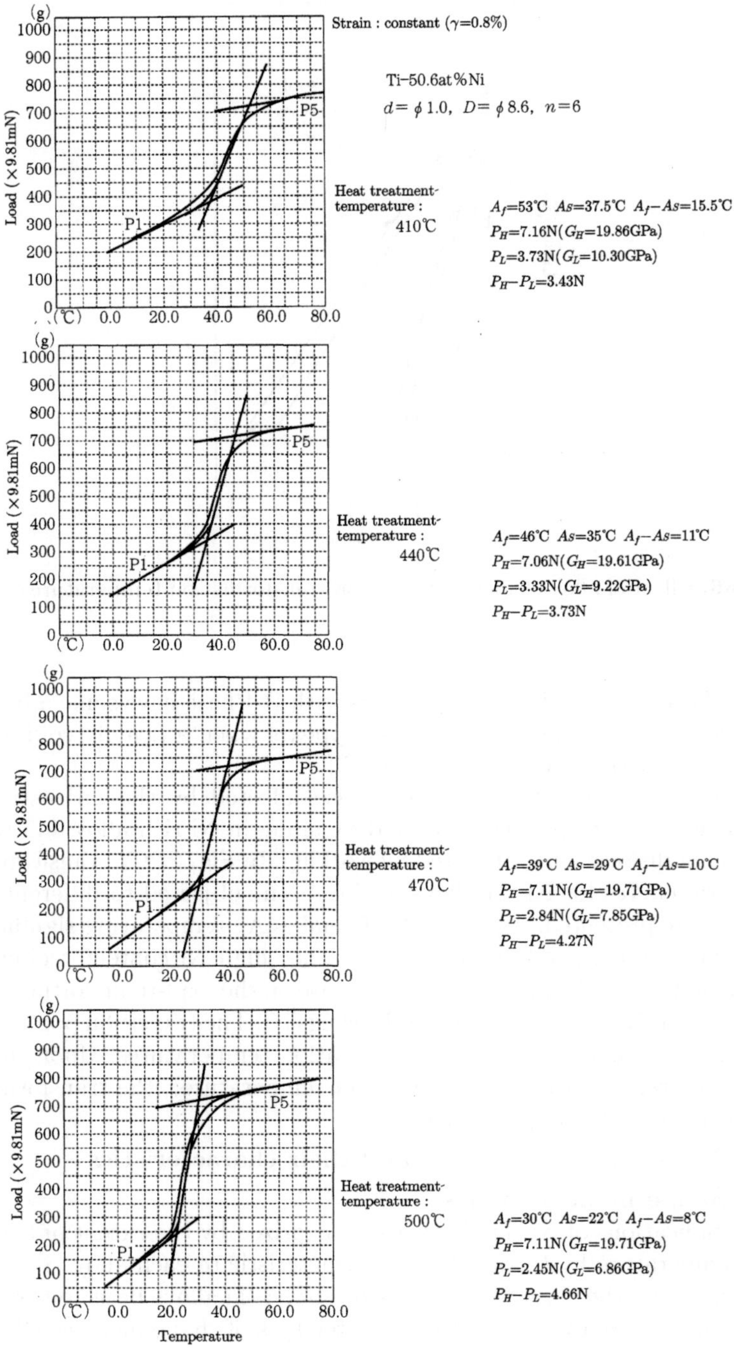

Fig. 2.267. Effect of heat treatment temperature on force-temperature curve of coil spring made of Ni–Ti alloy (parent phase to/from R-phase transformation)

Table 2.88. Characteristics and applications of shape memory alloys

Effect	Temperature hysteresis °C	Type of phase transformaton	Alloy system	Shape recovery strain %	Limit cycle of hysteresis	Applications
Two way shape recovery	2–3	P ⇔ R	Ni–Ti, Ni–Ti–Fe	1	>1,000,000	Sensors, actuators
	10–15	P ⇔ O	Ni–Ti–Cu	5–7	10,000–50,000	Sensors, actuators
One way shape recovery	20–40 60–100	P ⇔ M	Ni–Ti, Ni–Ti–Fe, Ni–Ti–Nb	6–8	<100	Connector, pipe coupler

Effect	Stress hysteresis MPa	Type of phase transformation	Alloy system	Superelastic strain %	Applications
Super-elasticity	200–400	P ⇔ M	Ni–Ti, Ni–Ti–Cr/Fe	6–8	Various types of springs
	100-300	P ⇔ O	Ti–Ni–Cu–Cr/Fe	5–7	

Remarks: P: Parent phase, R: Rhombohedral phase, O: Orthorhombic phase, and M: Monoclinic phase.

R-Phase
Orthorhombic martensite (Ni–Ti–Cu alloy)
Monoclinic martensite (Ni–Ti–Cu alloy)
Monoclinic martensite (Ni–Ti–Nb alloy)

The shape memory effect is a one-way action. Upon heating to a temperature higher than their transformation temperature the specimen recovers its original shape, however, it does not change their shape upon cooling. In order to achieve a two-way action, a bias spring is utilized to exert required force for sensor or actuator applications. Most of the sensors and actuators based on R-Phase or orthorhombic martensite transformation are two-way type. The characteristics of these transformations are as follows;

① R-Phase transformation

When R-Phase transformation takes place, the lowest level of temperature hysteresis is obtained in Ni–Ti alloys. Since the shape recovery strain is as small as 1% at the largest, this type of Ni–Ti alloys are usually shaped into coil springs in order to get maximum deflection. The feature of this types of springs are small temperature hysteresis (within 2°C) and enhanced durability. As for the durability against temperature cycles, the spring can deform quite a same amount as initial state after the repetition of five hundred thousand

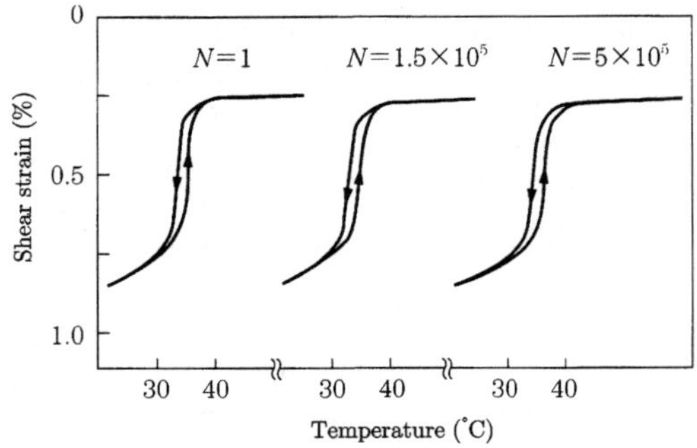

Fig. 2.268. Transition of shear strain-temperature curve for coil springs utilizing R-phase transformation with temperature cycles N [9]

temperature cycles [9] as shown in Fig. 2.268. Therefore, this transformation is suitable for the application of sensors and actuators.

② Orthorhombic martensitic transformation

Ni–Ti–Cu alloy in which some part of Ni atoms are substituted by copper atoms presents unique property. More than 30% of copper is in solid solution, and the transformation temperature curve shows a sharp vertical line. It also shows sufficient shape memory effect [10]. The temperature-strain chart of Ni–Ti–Cu alloy with orthorhombic structure in martensite state is given in Fig. 2.269 (b). A chart for Ni–Ti alloy is also shown in (a) and (c) for the specific alloy composition in the same Figure [8]. Compared with (a) and (c), the Ni–Ti–Cu alloy (b) reveals quite a unique hysteresis chart. The hysteresis of around 10°C for the Ni–Ti–Cu alloy shows the hysteresis level between R-Phase transformation and orthorhombic martensitic transformation. The orthorhombic and monoclinic martensite phases give larger transformation strains than R-Phase, so the shaped compressive coil spring could be deformed until solid contact stage of coils at lower temperature. Therefore the Ni–Ti–Cu alloy coil springs can be used effectively for applications where large shape recovery strains are demanded. In actual services, it is possible to design devices smaller than R-Phase transformation devices. As for the durability against temperature cycles, the Ni–Ti–Cu alloys can be used for devices up to ten thousand to fifty thousand cycles requirement.

③ Monoclinic martensitic transformation

In the case of monoclinic martensitic transformation shown in the Fig. 2.269(c), the transformation temperature hysteresis reaches over 20 K and the temperature cycle characteristics are inferior to other alloys. For these reasons, this type of alloy cannot be used for temperature sensors. For the application pipe-coupler fittings, however, the relatively large shape recovery strain could be advantageous, since they do not require extended number of strain cycles. The Ni–Ti–Nb alloys have larger temperature hysteresis than

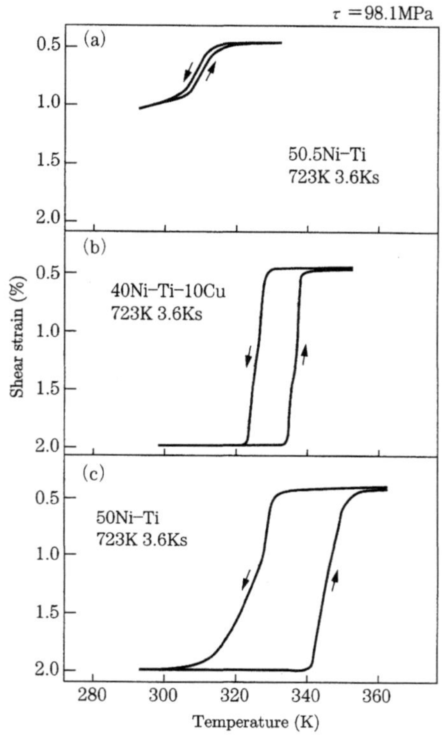

Fig. 2.269. Shear strain-temperature curves of Ni–Ti alloys coil springs under a constant load (shear stress: 98.1 MPa) [8]

Ni–Ti alloys [11]. In the Ni–Ti–Nb alloys, niobium rich precipitation causes a large temperature hysteresis [12].

(ii) Superelastic alloys

When a Ni–Ti alloy is used as a superelastic material, it becomes an absolute condition that the service temperature shall be higher than its transformation temperature. As shown in the Figures 2.265 and 2.266, Ni–Ti alloys have Ni content close to 50 atomic percent, and in some cases, approximately 1 atomic% of third element additive are also contained. Figure 2.270(c) is a typical stress-strain characteristic for monoclinic phase. Figure 2.270(b) shows smaller hysteresis of Ni–Ti–Cu alloys (of which martensite phase is orthorhombic). The latter is suitable for applications where a restoring force is a priority matter upon unloading, such as arch wires in the field of orthodontics. In order to decrease transformation temperatures, superelastic Ni–Ti–Cu alloys are modified as Ni–Ti–Cu–X alloys, where X means Fe or Cr addition [13]. Figure 2.270(a) shows R-Phase transformation, and this alloy is difficult to be put into practice as a superelastic material.

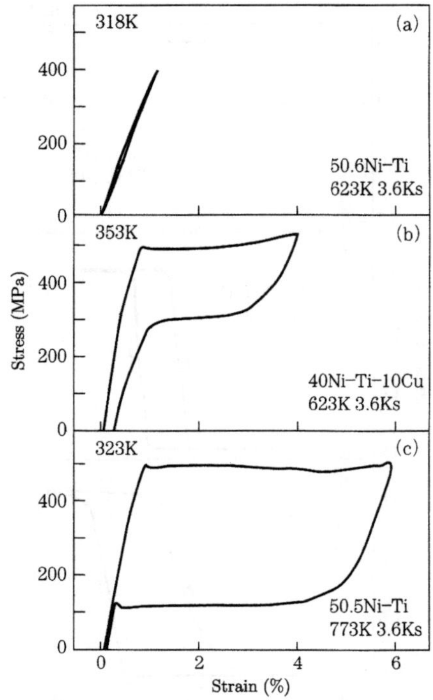

Fig. 2.270. Stress-strain curves of superelastic alloys [8]

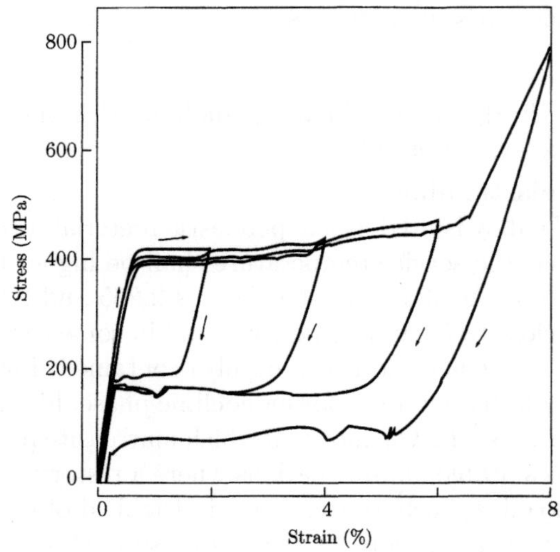

Fig. 2.271. Stress-strain curves at 23°C (296 K), for superelastic alloy wire practically used

Superelastic alloys are usually used in room temperature environment. Figure 2.271 shows stress-strain curves for Ni–Ti alloy superelastic wires plotted upon loading and unloading up to the strain of 2%, 4%, 6% and 8% respectively. It is conceivable that the strain up to 6% is completely recoverable, and the maximum strain of 8% leaves only 0.2% of permanent strain upon unloading [14].

(4) Durability

Two types of durability must be taken into consideration when the properties of Ni–Ti alloys are discussed. The first type is the durability against thermal cycles. The degradation of restoring force caused by thermal cycles is a priority matter for applications like temperature sensors. The second type is the durability against failures caused by mechanical fatigue. This fatigue durability must be considered for superelastic alloy components.

① Durability against thermal cycles

Even for R-Phase alloys having the best durability among Ni–Ti alloys, the condition of use must be strictly selected. Figure 2.272 shows the relations between temperature and shear stress under different constrained shear strains revealing the degradation curve of constraining force after the temperature cycles of ten thousand times [15]. When the constrained strain is increased, high temperature side forces tend to decrease after a certain number of temperature cycles. In addition, with the constrained strain increment, hysteresis starts to appear at high temperature side. Consequently, springs shall be designed to avoid such a hysteresis keeping the constrained shear strain under 1%.

When R-Phase transformation is applied for use, it should be taken into consideration that the R-Phase alloy may transform to a monoclinic martensite phase due to lowered surrounding temperature below M_s temperature. In Fig. 2.273, symbol "○" indicates normalized force against temperature cycles for parent phase and R-Phase transformation. The curve of symbol

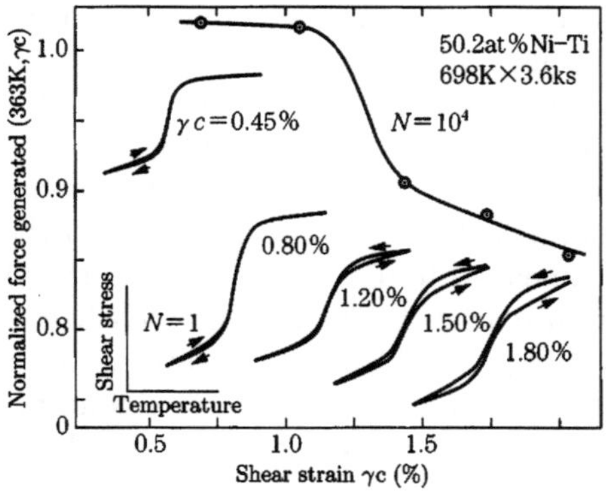

Fig. 2.272. Effect of constraint strain on stress-temperature curve and fatigue life dependent characteristics

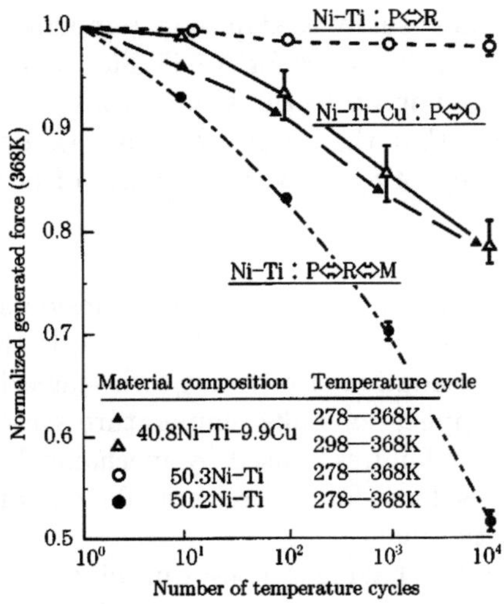

Fig. 2.273. Fatigue life of Ni–Ti alloys coil springs due to repeated temperature cycles under a constraint strain (shear strain: 1%) (P: Parent phase) [8]

"●" is for the tri-phase transformation of parent phase, R-Phase, and monoclinic martensite. In the latter case, force level is decreased to the half of intial value after the repetition of ten thousand thermal cycles. Therefore, the service temperature shall be higher than the martensite transformation temperature [8].

The curves of symbol "▲ & △"are for Ni–Ti–Cu alloys with parent phase and orthorhombic martensite transformation. The performance characteristic of this material goes between preceding two materials. The deterioration curve of normalized force against thermal cycles is shown in Fig. 2.274 for a biased spring device made of Ni–Ti–Cu alloy. In this case, 88% of initial force is kept for use after the thermal cycles of ten thousands [16]. Small sized devices of this kind are popularly used for electric home appliances requiring durability cycles over tens of thousand [17].

② Durability against mechanical stress cycles

Figures 2.275 shows S-N curves for 50.8 atomic% Ni–Ti alloys. The characteristic of the curves are similar to conventional metals. The flexure point of S-N line is around 343 MPa which corresponds to the yield point of this alloy [18]. It is recommended to use the alloy below the yield point if long life is required. The strain reaches to the level of 1% when the superelastic alloy is elongated to the yield point. This value is several times larger than conventional metal yield strain. As long as the service condition is set to keep the stress level lower than the yield point, the long durability of the superelastic alloy device can be guaranteed.

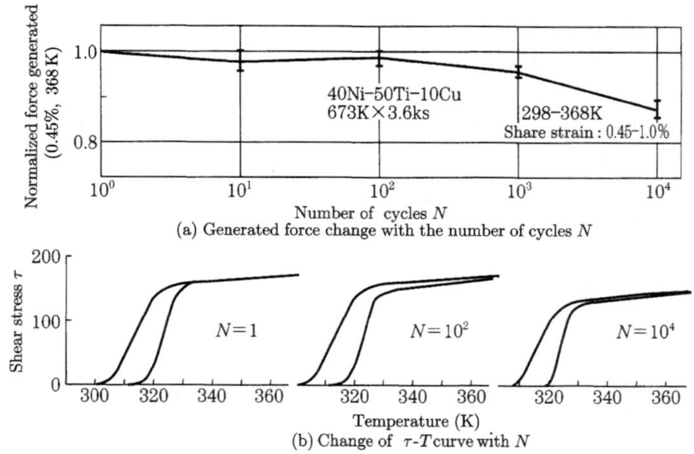

Fig. 2.274. Life property of Ni–Ti–Cu coil springs as two-way device (constant shear strain: 0.45%) [16]

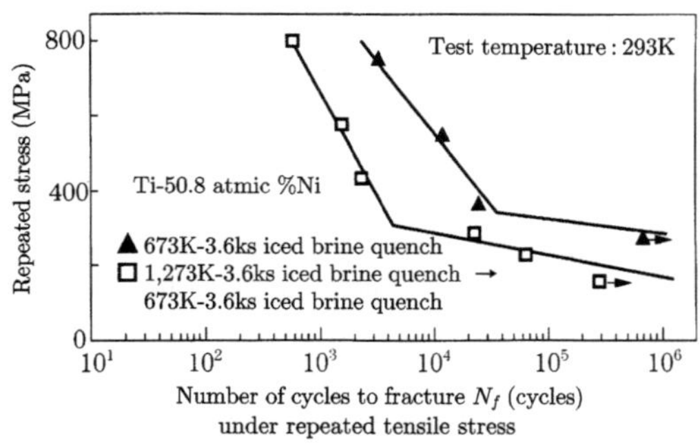

Fig. 2.275. S-N curves for superelastic Ni–Ti alloys [18]

2.6.2 Manufacturing and Evaluation Method

The manufacturing process of Ni–Ti alloy is similar to that of titanium alloys represented by Ti–6Al–4V following the step of raw material melting, casting into ingot, and hot-rolling to bars. Then the process of cold drawing and annealing shall be repeated to obtain wire with a specified diameter. In manufacturing sheet materials, ingot shall be hot-rolled to slabs, then cold-rolling and annealing shall be repeated to obtain sheet with specified thickness. Ni–Ti alloy has inter-metallic compound matrix, and its cold workability is extremely inferior to stainless steel. Vacuum melting and vacuum casting is the best process sequence for the stability of mechanical properties [19].

(1) Melting and casting

Large sized Ni–Ti alloy ingot shall be manufactured in a high performance melting furnace like high frequency vacuum induction furnace, arc furnace, or plasma arc furnace. What is the most important factor in Ni–Ti alloy melting is the reaction of alloying elements. Since the molten titanium is extremely reactive to oxygen, the melting of Ni–Ti alloys shall be conducted in vacuum or inert gas environment. In addition, the chemical reaction of alloy elements and crucible material must be minimized when a crucible is used for melting.

The high frequency induction furnace has an advantage that the molten metal is agitated by high frequency induction effect, resulting in homogeneous chemical composition in the ingot. Using graphite crucible has a merit of low carbon percentage (less than 0.05%) in the ingot, since Ni–Ti alloy with the composition of shape memory metal has a limited solubility with carbon [19].

(2) Heat treatment and forming

Figure 2.276 shows an example of working process for Ni–Ti wire products. The ingot shall be hot-rolled and hot-swaged to bars, and then cold-drawing and annealing shall be repeated to obtain specified diameters. The manufacture of shape memory device requires a piece of wire to be formed into desired shape. The characteristic heat treatment is then done after setting the material to its final shape. Adequate heat treatment parameters (temperature and time) are needed to set the shape and properties of the item. Normally, the temperature is 400–500°C and the time is a few minutes to one hour.

In the case of coil spring production for small quantity, coiling shall be done around a cylindrical fixing jig, and then the jig having coil specimen around it shall be heat treated to put shape memory effect. When the quantity is increased, the use of automatic spring former is inevitable. Precaution against failure shall be kept in mind as follows; Since cold-worked Ni–Ti alloy has a significant work-hardening property, spring back strain could reach to the level as high as three times of conventional metals. Since tools used for spring coiling tend to suffer a heavy wear and seizure, careful tool life management is required. Formed coil springs are then constrained to jigs to be put into a furnace for the required heat treatment [19].

(3) Evaluation methods

① Measurement of transformation temperatures

The most popular transformation temperature measuring equipment is Differential Scanning Calorimeter (DSC). The merits of using this equipment

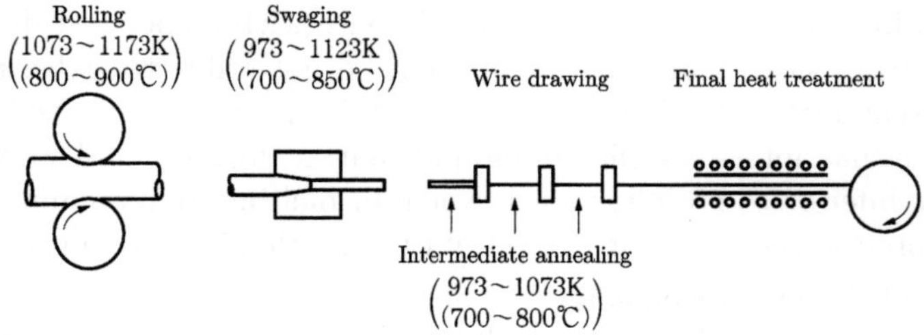

Fig. 2.276. Manufacturing process of Ni–Ti alloy wire

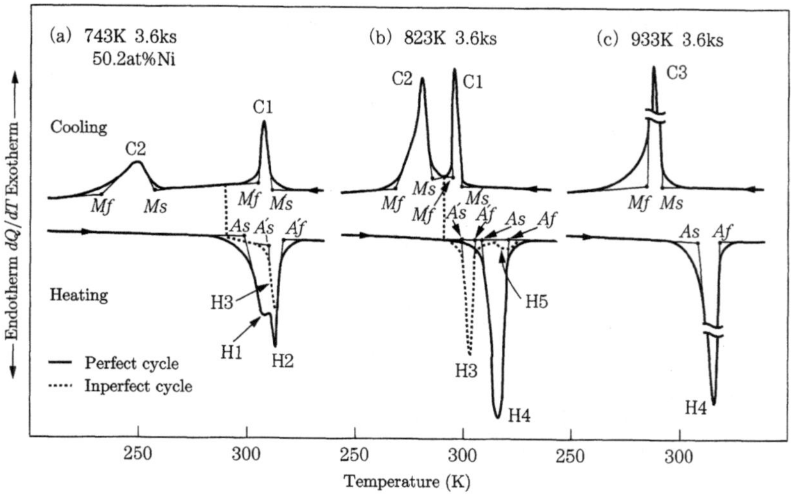

Fig. 2.277. DSC curves for Ni–Ti alloy [20]

are, a small sample size, an easy sample preparation, low equipment cost, and an excellent reliability. Figure 2.277 gives some result of transformation temperature measurements for Ni–Ti alloy [20]. The characteristic transformation temperature is indicated as the crossing point of extrapolated base straight line and maximum inclined straight line. As for the method of measuring transformation temperatures, JIS has designated it in 1989 [21].

② Constant strain method for the evaluation of coil spring

For the evaluation of Ni–Ti alloy springs, it is necessary to measure the temperature dependence of spring force under constant strain. For this purpose, JIS has designated it in 1993 [22]. An example of the testing apparatus is schematically illustrated in Fig. 2.278. A coil spring compressed to a con-

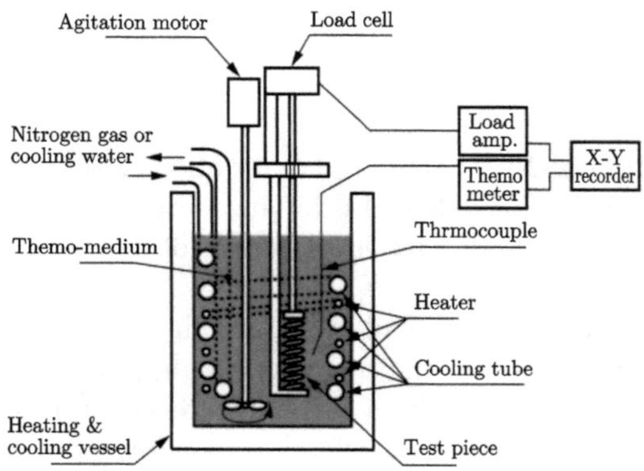

Fig. 2.278. Constant strain test apparatus

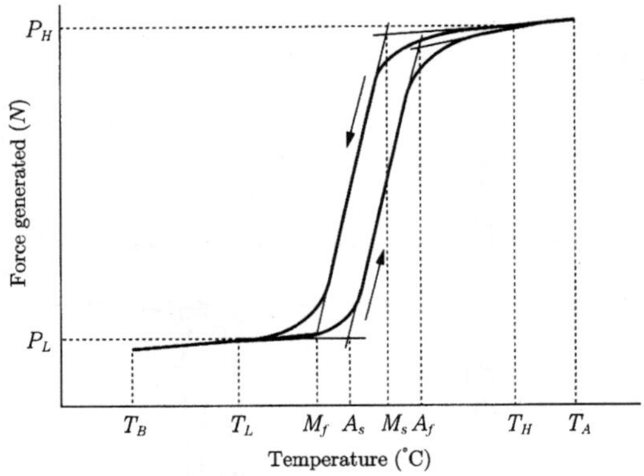

T_A: Maximum temperature in the temperature range designated by supplier and accepter (°C)
T_B: Minimum temperature in the temperature range designated by supplier and accepter
T_H: Temperature to determine generated force (P_H) in helical spring during heating, that is designated by supplier and accepter (°C)
T_L: Temperattre to determine generated force (P_L) in helical spring during cooling, that is designated by supplier and accepter (°C)
P_H: Force generated by helicai spring at temperature T_H during heating (N)
P_L: Force generated by helicai spring at temperature T_L during cooling (N)

Fig. 2.279. Force generated-temperature curve and characteristic value

stant deflection is dipped in temperature controlled liquid being connected to a load-cell. The temperature is increased and decreased at the slow rate of 2°C/minute, and spring force reading shall be plotted against temperature by X-Y recorder. An example curve is shown in Fig. 2.279.

2.6.3 Applications

The first application of Ni–Ti shape memory alloy (1970) was a pipe coupler in the US as shown in Fig. 2.280. In Japan, various applications have been developed since around 1980. The development of new applications has been still active in recent years.

For a shape memory effect, a coil spring for air conditioner (using R-Phase transformation effect) may be the most famous application (see Fig. 2.281). There are other applications like a temperature control device installed in a faucet (R-Phase transformation effect) and a pressure device in electric rice cookers (orthorhombic transformation effect) as successful applications of shape memory effect.

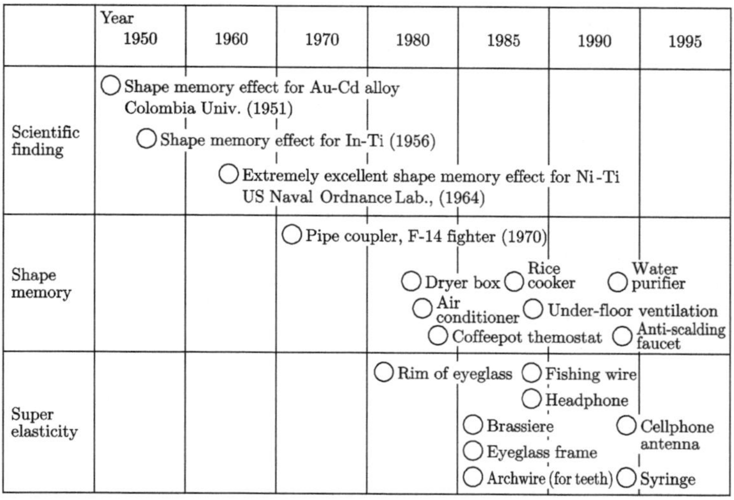

Fig. 2.280. History of shape memory alloys [23]

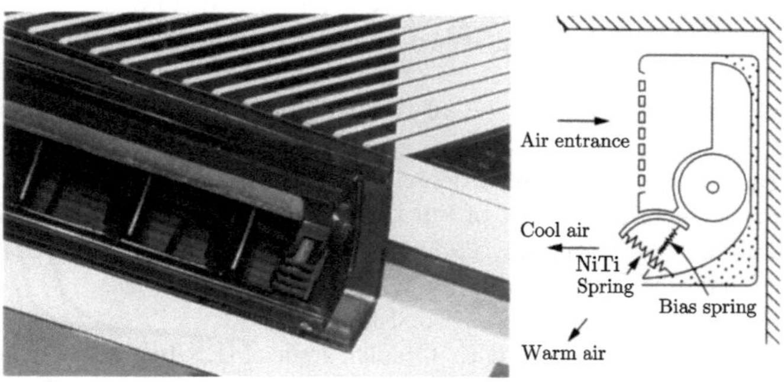

Fig. 2.281. Air conditioner equipped with Ni–Ti alloy coil springs

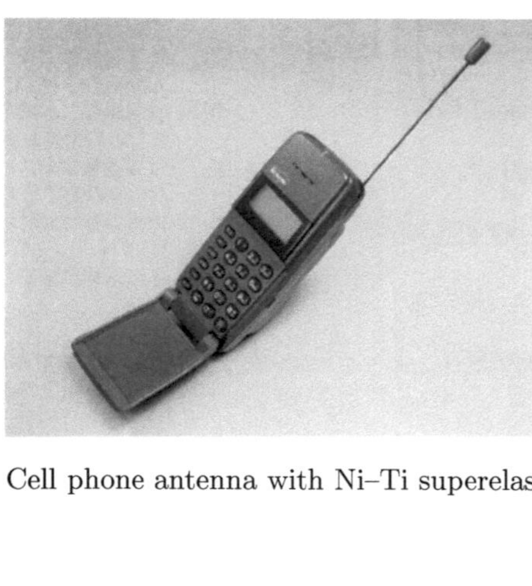

Fig. 2.282. Cell phone antenna with Ni–Ti superelastic alloy wire

For a superelastic property, there are applications like brassiere wire, eyeglass frames, and the antenna of cell phone (see Fig. 2.282).

References

1. L.C. Chang and T.A. Read: Trans Am. Inst. Met. Eng. 189 (1951) p. 47
2. S. Miyazaki, K. Otsuka and Y. Suzuki: Scripta Met., 15 (1981) p. 287
3. S. Miyazaki, K. Otsuka: Kinzoku (Japanese) 59 8 (1989) p. 2
4. Y. Suzuki: Jitsuyou keijo kioku goukin (Japanese) Kogyo Chosakai Publishing (1987) p. 25
5. T. Honma et al.: Proc. 4th int. conf. On Ti, Kyoto May (1980) p. 1455
6. Y. Suzuki: Jitsuyo keijo kioku goukin (Japanese) Kogyo Chosakai Publishing (1987) p. 43
7. T. Ishii: Seminar Text Shape Memory Alloy Asso. (1998) p. 28
8. Y. Suzuki and H. Horikawa: Mat. Res. Soc. Symp. Proc., 246 (1992) p. 389
9. T. Todoroki, K. Fukuda, T. Soun, and Y. Suzuki: Proc. of Japan Institute of Metals Oct (1983) p. 130, April (1984) p. 311
10. N. Melton and O. Mercier: Met. Trans. 9A (1978) p. 1487
11. H. Horikawa et al.: Technical report of Furukawa Electric Co. Ltd. (1991) p. 23
12. L.C. Zhao et al.: Scripta Met., 24 (1990) p. 221
13. H. Horikawa and T. Ueki: Proc. of IUMRS int. Conf. on Ad. Metals18B (1993) p. 1113
14. Technical report of Furukawa Electric Co. Ltd.
15. Y. Tamura: Technical report of Furukawa Electric Co. Ltd. 81 (1987) p. 53
16. H. Tsuzuki and H. Horikawa: Technical report of Furukawa Electric Co. Ltd. 86 (1990) p. 54
17. Nikkei Mechanical: (1988 Sep. 5) p. 130
18. S. Mizukami, Y. Sugaya, and K. Otsuka: Proc. MRS Int. Mtg on ad. Metals 9 (1989) p. 251
19. Y. Suzuki: Proc. of Japan Institute of Metals 9 (1989) p. 251
20. T. Todoroki and Y. Tamura: Proc. of Japan Institute of Metals 50, 1 (1986) p. 1
21. JIS H 7101
22. JIS H 7105
23. S. Komatsu: Physical Education 45 2 (1997) p. 92

3

Polymeric Materials

3.1 Plastics

3.1.1 Introduction

The polymeric material is a common name of plastics having relatively large molecular weight which is either natural product or synthetic chemicals. They are also called polymers.

Plastics develop plasticity when heat and/or force are applied. Organic polymeric materials are usually classified into this category. There are two different kinds of plastics like thermoplastics and thermosetting plastics depending on polymer structure and/or thermal behavior. Among other organic polymeric materials, fiber, rubber, paint, and adhesive are excluded from this section.

The feature of plastics is desirable as the spring materials since it is light, easy to shape, and rustless. On the other hand, much care shall be taken when we consider plastic springs. The low modulus of elasticity, and tendency to creep even at room temperature are limiting factor not to make springs from plastics. Therefore, it should be noted that plastics are not suitable to the springs which preserve energy for a long time, since the load decreases gradually by the stress relaxation.

Needless to add, by taking those factors into consideration, we can put a proper design of plastic springs into place for proper applications. Recently, the recyclable products have been getting more important in the event of global environmental concerns. The application of plastic springs should be taken the recyclability into consideration.

3.1.2 Comparison of Typical Material Properties

The plastic materials are classified into thermoplastics and thermosetting plastics. The thermoplastics are categorized into general-purpose plastics and engineering plastics as shown in Fig. 3.1. There are two sort of engineering

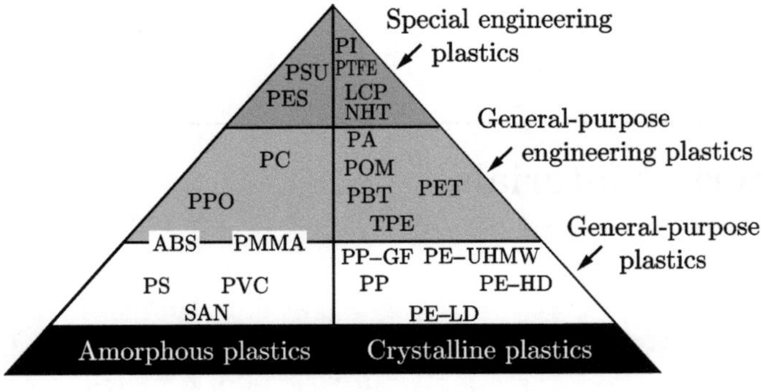

Fig. 3.1. Classification of thermoplastics

plastics like general-purpose engineering plastics and special engineering plastics.

The fundamental characteristic of plastic materials depends on its polymer structure (cross-linked or not), and whether or not the plastic is crystalline. Thermal behavior, flow behavior, and mechanical property are affected by those factors.

The examples of crystalline and amorphous plastics are as follows;

(1) Crystalline plastics

Special engineering plastics: PI (polyimide), PTFE (polytetrafluoroethylene), LCP (liquid crystalline polymer), NHT (semi-aromatic high temperature nylon)

General-purpose engineering plastics: PA (polyamide), POM (polyacetal), PBT (polybutyleneterephtalate), PET (polyethyleneterephtarate), TPE (thermoplasitic-elastmar)

General-purpose Plastics: PP (polypropylene), PE (polyethylene), UHMW-PE (ultra-high molecular weight polyethylene), HDPE (high density polyethylene), LDPE (low density polyethylene), PPGF (glass fiber reinforced polypropylene)

(2) Amorphous plastics:

Special engineering plastics: PSU (polysulfone), PES (polyethersulfone), PPS (polyphenylene sulfide)

Table 3.1. Qualitative comparison of plastic materials [3]

	PA	POM	PC	PPO/PPE	PBT/PET	TPE
Waterproof	D	B	A	A	A	A
Mold shrinkage factor	B	B	A	A	A	A
Fatigue strength	A	A	C	C	A	A
Toughness	A	B	A	B	B	A
Creep resisitance	C	A	A	B	B	B
Solvent resistance	A	A	D	C	A	A
Weather resistance	C	C	B	B	A	B
Electrical properties	C	B	A	A	A	B
Abrasion resistance	A	A	C	C	B	B
Flame resistance	B	D	A	B	B	B

A: Excellent B: Good C: Unsatisfactory D: Inferior

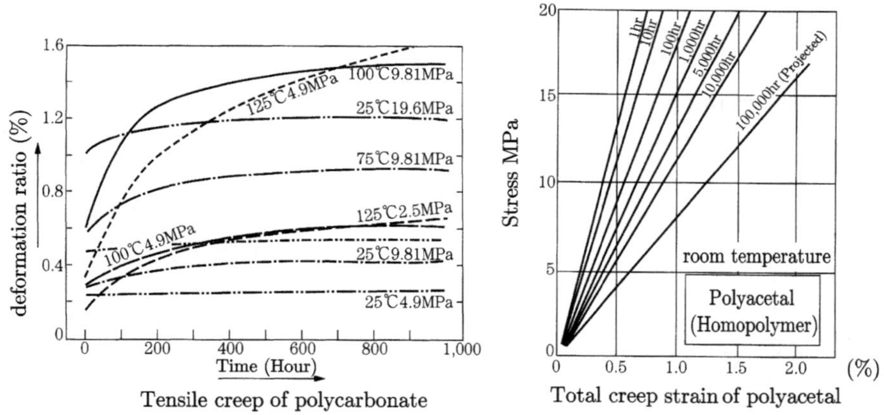

Fig. 3.2. Tensile creep of polycarbonate and polyacetal [1]

General-purpose engineering plastics: PC (polycarbonate), PPO (polyphenylene oxide)

General-purpose plastics: ABS (acrylonitrile butadiene styrene), PMMA (polymethyle methacrylate), PS (polystyrene), PVC (polyvinyl-chloride), SAN (styrene-acrylonitrile)

Among these plastics, the general-purpose engineering plastics are the most popular for the springs of industrial use. Qualitative assessment is summarized in Table 3.1. It is necessary to note that there could be an exception by the difference of grades. The creep resistance and the fatigue strength are especially important to secure the long life of springs.

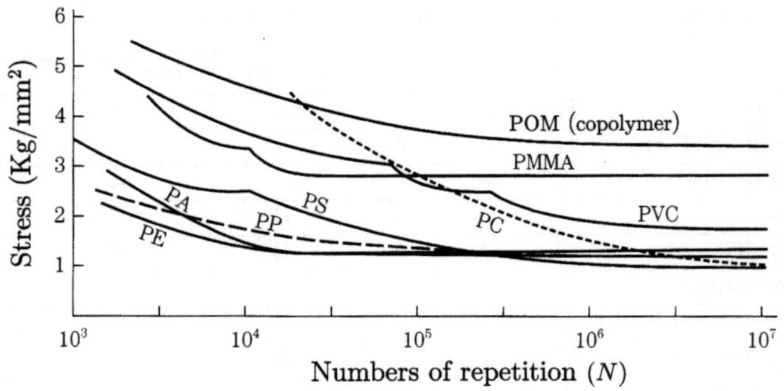

Fig. 3.3. Fatigue S-N curve of engineering plastics [2]

Table 3.2. Property of typical plastic materials for springs

Plastic type	Specific gravity	Yield stress (Elongation at yield) MPa (%)	Tensile modulus MPa	Fatigue strength MPa $\times 10^6$ cycles	Creep modulus*3 MPa 1 Hr/1,000 Hr
Nylon 66	1.14	85 (4.4)	3,300	–	–
Nylon 66 (50%RH)*1	1.14	58(24)	1,600	24	1,450/930
POM	1.42	72(15)	3,200	35	2,900/1,700
PBT resin	1.31	58 (3.6)	2,700	–	2,600/1,800
PBT elastomer	1.22	19 (31)	280	–	248/182
Nylon 66 (30%GFRP)	1.37	208 break (3.3)	10,000	–	–
Nylon 66 (30%GFRP) (50%RH)*1	–	135 break (5.1)	7,500	45	7,250/6,000
PET (30% glass fiber reinforced)	1.56	122 break (2.0)	11,000	41	10,800/8,800
LCP*2	1.62	158 break (1.5)	16,800	–	–

Note *1: At equilibrium in relative humidity 50% *2: Liquid crystal polymer *3: Ratio of stress to creep strain

Fatigue strength: Durability up to 10^6 listed in this table could be used for the selection of spring material, however, it is recommendable to refer to specific S-N curve under the actual service condition.

Creep properties: Creep moduli listed in this table are at the stress level of 1/3 tensile strength at room temperature. It can be observed that reinforced plastics are better than normal plastics.

The creep data of polyacetal and polycarbonate which are commonly used for springs are shown in Fig. 3.2. Fatigue S-N curves are also shown in Fig. 3.3. The typical properties of plastic materials for springs are shown in Table 3.2.

3.1.3 Consideration of Shape

The most effective method to enhance the strength of the structural member is to increase the thickness improving the cross-sectional property (section modulus and geometrical moment of inertia). However, in the case of springs, this will cause the increase of stress in order to obtain the given deformation. For instance, in the spring with simple rectangular section, the stress will increase in proportion to the thickness since the deflection is in inverse proportion to the cube of thickness and the stress is in inverse proportion to the square of thickness. It could be a better idea to broaden the width of the spring instead of thickness. By doing so, larger spring rate will be achieved avoiding stress increase for the given deflection. Resin mold parts with integrated hinge are the best use of this idea. Moreover, in some cases, it can be done to improve the material strength with optimum molecular orientation by pressing the thin potion further to increase the fatigue strength. This effect might appear by merely bending the hinge potion of parts. Polypropylene is suitable to do this, however similar effect is expectable for nylon and polyacetal resin.

A flat metallic spring normally has a uniform thickness, however plastic springs can achieve any 3-D shape to optimize strength distribution.

The most efficient thickness shape in the cantilever beam configuration is a parabolic curve to its longitudinal direction. Since section modulus increases by the second power of plate thickness, the stress level will be constant along with the longitudinal direction.

Since a design freedom is enabled by integral molding manufacturing process, plastic material has a great advantage to allow spring designers selecting any 3-D shape like laterally curved surface structure and/or framed construction in addition to the longitudinal thickness distribution.

3.1.4 Consideration of Fabrication

The spring rate is governed by the shape of the molded parts and the modulus of elasticity of plastic materials. The modulus of elasticity is listed in the product catalogue. However, it is necessary to note that the value is provided under a certain condition, and that the filling and the shape (wall thickness) of the molded parts are other governing factors. Especially, at the thick wall area, it is difficult to obtain homogeneous molding through the process in which the molten resin solidifies after filling in the mold. The modulus of elasticity at the thick wall area is different from the one in the catalogue which is measured in the specific condition.

The modulus of elasticity of plastic material is defined in ASTM standard. The definition is that the value shall be calculated from load-deflection

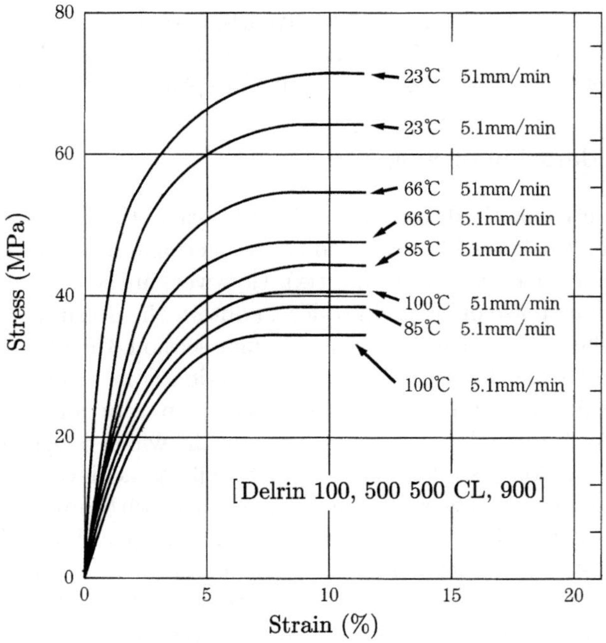

Fig. 3.4. Stress-Strain curves of polyacetal

curve under the condition where the specimen of 3.2 mm thickness is supported in 50 mm span, and load rate to the center of the specimen shall be 1.3 mm/min. (ISO: 2 mm/min.) at room temperature (23°C). If the testing condition changes, the modulus of elasticity will vary. The variation of the modulus of elasticity under the different measurement condition is shown in Fig. 3.4 (tensile test condition).

Naturally the modulus of elasticity will decrease when the temperature rises. The influence of other factor like wall thickness, support span, and loading rate is a bit difficult to understand. When the ratio of wall thickness and support span become greater than 1/16 defined by the ASTM standard, the influence of shearing stress becomes unnegligible, and the modulus of elasticity will decrease. In the thick wall mold parts, due to the problem of generating voids, resulting shear strain might lead to the smaller reading of the modulus of elasticity.

The modulus of elasticity is affected by the influence of the orientation of fiber reinforcement. While fiber may orient to the direction of the resin flow with directional dependency (anisotropy) in the surface layer, the reinforcement fiber may disperse irregularly resulting in less reinforced effect in the inner cross section area. In the thin wall molded parts, while the ratio of surface layer is large, the influence of inner area become greater in the thick wall mold parts resulting in the smaller modulus of elasticity.

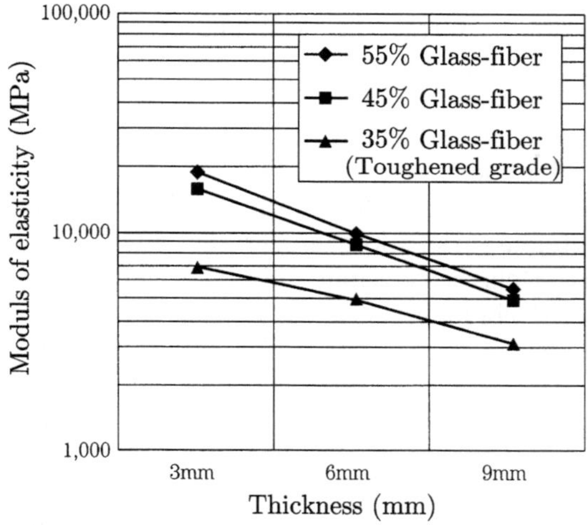

Fig. 3.5. The modulus of elasticity by wall thickness for glass fiber reinforced PET

The modulus of elasticity by wall thickness for glass fiber reinforced PET is shown in Fig. 3.5.

3.1.5 Application Examples

A formed coil made of thermoplastic polyester elastomer is used for the reinforcement of hose and telephone wire. It is also used for the keypad spring of a remote control switch that needs appropriate click feeling caused by buckling phenomenon.

In the other applications, a set of stacked disc springs made of tough grade polycarbonate enters between the horseshoes and the hoof making horse's footwork light.

As for the application of polyacetal, there are clothes pegs used at light load, fasteners for vehicle and consumer electronics, seat belt buckles with snap action.

The disc springs have been used as a strong spring for machine elements from old times. Plastic disc springs are integrated to fasteners on many occasions making the best use of the flexibility to shape. The secure effect of the fixation of the fasteners will be increased by repulsion force of the disc spring. In the case of the plastics, the wall thickness and the shape can be achieved more arbitrarily than metals making it possible to design bellows structure in stead of simple cone.

3.1.6 Instruction for Plastic Use

Since there is no exclusive plastics for spring use, it is advisable to follow the instructions for the common usages. In general, the following attention is necessary from the characteristics of plastics.

(1) The degradation by ultraviolet rays: The polymeric material has a tendency to suffer a large decrease of elongation. The weatherproof grade is especially desirable for outdoor use. In case of the glass fiber reinforced plastics, the problem seems less substantial compared with the non-reinforced grade.
(2) Chemical resistance: Since the amorphous polymers such as polycarbonate with high-glass-transition temperature (The temperature at which the modulus of elasticity remarkably changes. The modulus of elasticity is comparatively high below this temperature.) are excellent in creep-resistance, they are popular for spring application. However, since their resistance to the organic solvents is rather poor, it is unsuitable for the application machine elements in oil contaminated area.
(3) Heat resistance: The modulus of elasticity of the plastic materials varies remarkably depending on the temperature. Heat distortion temperature (HDT) of each material is not necessarily suitable as the standard of material selection. It is necessary to confirm the data of each material.
(4) Stress crack and solvent crack: Those are a kind of delayed fracture, and crack caused under the loading stress and/or the residual stress in the particular environment. It is necessary to confirm the characteristics of each material like polyethylene and amorphous polycarbonate under the environmental requirements. The internal stress or residual stress at the time of molding, or the existence of early cracks by the presence of insert and/or the screw clamping could affect the durability of plastic parts.

References (Japanese)

1. Enpra dokuhon, Japan High Polymer Center, p. 6
2. T. Yoshi: Machine design, 1998, Aug, Nikkan Kogyo Shinbun, p. 47
3. H. Maki, T. Kobayashi: Engineering plastics, Sangyo Tosho Publishing, p. 28

Bibliography (Japanese)

4. Koubunshi Zairyou Kyodogaku, Ohmsha
5. Koubunshikotai no Oyokaiseki to Hakai, Baihukan
6. Purasutikkuzairyo no Jyumyou-Taikyusei to Hakai, Nikkan Kogyou Shinbun
7. Enjiniaringupurasutikku Katuyou Gaido, Nikkan Kogyou Shinbun
8. Plastic Age Encyclopedia, Advanced Edition, Plastic age Publishing
9. Koubunshisozai Gaido, Japan High Polymer Center
10. Koubunshi-Sozaibinran, Society of Polymer Science Japan, Maruzen
11. Purasutikku-Seikeihin-Sekkei, Kogyo Chosakai Pubulishing.

3.2 Rubber

3.2.1 A Choice of Rubber Materials for Springs

Rubber materials are classified broadly into the natural rubber category picked from plants and synthetic rubber category. Particular selections are being done by required characteristics, intended use, and the difficulty level of workability.

The comparative properties of key rubber materials are shown in Table 3.3 [1]. Rubber materials for vibration isolator are selected by following guideline.

(1) Rubber for general-purpose
NR (Natural Rubber)
SBR (Styrene-Butadiene Rubber)
BR (Butadiene Rubber)
IR (Isoprene Rubber)
(2) Oil-resistant rubber
NBR (Acrylonitrile-Butadiene Rubber)
(3) Weatherproof rubber with slight oil-resistance
CR (Chloroprene Rubber)
(4) Rubber with large vibration damping capability
IIR (Butyl Rubber)
(5) Rubber with heat-resisting property
EPDM (Ethylene-Propylene-Diene-Monomer elastomer)

When EPDM are selected in consideration of elevated temperature property, it should be advised that the strength at normal temperature sometimes become insufficient. In this case, it might be advantageous to use category (1) with improved heat-resistance by vulcanization. Additionally, various combinations of raw materials and compounding ingredient have been conceived to control dynamic properties, creep properties, low-temperature properties, adhesiveness with metals, crashworthiness, fatigue strength, and so on.

3.2.2 Compounding Ingredients

In general, raw rubber materials are combined with reinforcing ingredients and vulcanizer. A vulcanization process is required for practical use. A variety of compounding ingredients are shown in Table 3.4. Principal mechanical properties like static and dynamic modulus of elasticity, mechanical strength, and fatigue properties are influenced by the combination of compounding ingredients. Therefore, it is necessary to select the optimum raw rubber materials and compounding ingredients in order to obtain the product demand. Natural rubber is often used for general rubber spring application in consideration of required mechanical properties. Examples of composition for natural rubber springs are shown in Table 3.5.

Table 3.3. Comparative properties of key rubber materials [1]

A: Excellent B: Good C: Unsatisfactory D: Inferior

	Name (ASTM code)	Natural rubber (NR)	Styrene-butadiene rubber (SBR)	Butadiene rubber (BR)	Chloroprene rubber (CR)	Acrylonitril Butadiene rubber (NBR)	Butyl rubber (IIR)	Ethylene-propylene rubber (EPDM)	Fluoroelastomer (FKM)	Silicone rubber (Q)
	Chemical structure	Polyisoprene	Butadiene-stylene copolymer	Polybutadiene	Polychloroprene	Butadiene-acrlonitrile copolymer	Isobutylene-isoprene copolymer	Ethylene-propylene-diene terpolymer	Hexfluoropropylene-fluo-rovinylidnene copolymer	Polysiloxane
	Principal features	The most popular rubber with good mechanical properties.	Better abrasive resistance and aging resistance than NR.	Better abrasive resistance than NR.	Weatherproof, ozone-proof best resistance, and chemical resistance.	Oil resistance, wear resistance, aging resistance	Weatherproof, ozone-proof, and good airtightness	Aging resistance, ozone-proof, durability in polar solvent	Highest heat resistance and chemical resistance	Better beat resistance, and cold resistance
	Specific gravity	0.92	0.93–0.94	0.91–0.92	1.15–1.25	1.00–1.20	0.91–0.93	0.86–0.87	1.80–1.82	0.95–0.98
Physical properties	Hardness [JIS A]	30–90	40–90	40–90	40–90	40–90	35–90	40–90	60–90	30–80
	Tensile strength (MPa)	7–28	5–23	5–23	6–25	5–25	5–15	5–20	7–15	3–9
	Elongation (%)	100–600	100–500	100–500	100–500	100–500	100–600	100–500	100–300	100–500
	Rebound resilience	A	B	A	A	B	C	B	C	A
	Temperature range Upper (°C)	60	90	90	100	100	120	140	220	200
	Temperature range Lower (°C)	−40	−35	−45	−30	−25	−40	−40	−25	−75
	Abrasion resistance	B	A	A	B	A	C	C	B	C
	Flex cracking resistance	A	B	C	B	B	A	B	B	C
	Ozone resistance	D	D	D	B	D	A	A	A	A
	Permanent compression set	B	B	B	B	B	C	B	B	B
	Airtightness	C	C	C	B	B	A	C	B	D
	Fire resistance	D	D	D	B	D	D	D	A	C
	Electric insulation performance volume resistivity (Ω cm)	10^{10}–10^{15}	10^{10}–10^{15}	10^{14}–10^{15}	10^{10}–10^{12}	10^{2}–10^{10}	10^{16}–10^{18}	10^{12}–10^{15}	10^{15}–10^{18}	10^{11}–10^{15}
Oil and chemical resistance	Lubricant	D	D	D	B	A	D	D	A	B
	Gasoline	D	D	D	C	B	D	D	A	D
	Aliphatic hydrocarbon	D	D	D	B	A	D	D	A	D
	Aromatic hydrocarbon	D	D	D	D	D	D	D	B	D
	Chlorinated solvent	D	D	D	D	D	D	D	A	D
	Alcohol	A	A	A	A	A	A	A	B	C
	Ketone	B	B	B	C	D	B	A	D	C
	Water	A	A	A	A	A	A	A	A	A
	Dilute acid	B	B	B	A	B	A	A	A	B
	Concentrated acid	D	D	D	C	D	B	B	A	D
	Alkaline solution	B	B	B	A	B	A	A	C	B
	Applications	General and industrial proproducts such as Automotive tires (especially heavy-duty truck and industrial tire), Foot wears, Belts, Hoses, Air springs.	General and industrial products such as Automotive tires and parts, Foot wears, Rubber coated fabrics, Sporting goods, Belts.	Automotive tires, Landing gear tires, Foot wears, Springs, Roller conveyor, Hose.	Wire coating, Conveyer belt, Springs, Window frame rubber, Rubber coated fabric, Adhesive.	Oil seal, Gasket, Oil resistant hose, Conveyer belt, Print roll.	Rubber inner tube for automotive tire, Varuum bag, Roofing, wire coating, Window frame rubber, Steam hose, Conveyer belt.	Wire coating, Automotive weather strips, Steam hose, Conveyer belt.	High performance gasket for roket motor, chemical machine gasket, Diaphragm.	Sealant application in cold on high temperature environment.

Table 3.4. Compounding ingredients in rubber materials

Type of ingredients	Examples	Purpose
Raw rubber material (Polymer)	NR, SBR, BR, IR, EPDM, NBR, IIR.	Principal performance
Reinforcing material	Carbon black	Improvement of physical properties Control of modulus of elasticity
Filler	Calcium carbonate	Improvement of workability, cost cutting Improvement of dimensional stability
Softening material	Petroleum oil	Improvement of Plasticity
Antioxidant	Amine reactants	Improvement of aging resistance and ozone-proof property
Vulcanization accelerator	Sulfenic amide reactants	Acceleration of vulcanization reaction
Vulcanizing agent	Sulfur	To make the rubber harder, durable, and more resistant to chemical attack.

Table 3.5. Examples of composition for natural rubber spring

Composition	A	B	C
Shear modulus (MPa)	0.59	0.98	1.37
Raw rubber	100 phr	100 phr	100 phr
Stearic acid	1	1	1
Antioxidants	2	2	2
Zinc oxide	5	5	5
Carbon black	15	30	50
Calcium carbonate	10	8	4
Sulfur	2.5	3	3
Vulcanization accelerator	1	1	1

phr: Weight of each compounding ingredient when raw rubber materials is assumed to be 100

By controlling particle size, surface-activity and mixture amount of carbon black and $CaCO_3$, the static shear modulus of rubber materials can be optionally adjusted from 0.5 to 2.0 MPa. However, it should be noted that dynamic properties might be different even if static properties are identical depending on the amount and particle size of carbon black.

3.2.3 Manufacturing Process of Anti-Vibration Component

Manufacturing process of an anti-vibration component is illustrated in Fig. 3.6. The raw rubber material shall be masticated in the first place. The purpose of this process is the control of raw rubber's plasticity by cutting coagulated molecules and molecule chain. (This process is mainly done to the natural rubber materials.)

The kneading and mixing is the very important process in which compounding ingredients are uniformly distributed in the masticated rubber material by mixing machines (Banbury mixer, Kneader, etc.) or roll mills. It is very important to control the plasticity in order to achieve the desired spring constant of the product.

Moreover, it is necessary to note that the dispersion condition of the filler materials and the compounding ingredients is the key influential factor on the durability. A simple method to confirm the dispersion condition is a visual inspection, though various test methods are provided like the one where a thin film of vulcanized rubber is expanded and checked for example. To improve the dispersion of vulcanizing agents, many kinds of pre-kneaded polymer with vulcanizing agents has been in the market.

An anti-vibration component is a combination of metal fitting and rubber; thus adhesiveness between metal and rubber is an important factor to guarantee product life. As a surface treatment of metal, chemical treatment by alkali or acid, and physical treatment by shot blasting or wet abrasive blasting are the popular way of practice. Moreover for the product under severe environment such as an outdoor use, chemical film by phosphoric acid etching process might provide a good corrosion resistance. In the bonding process, after the adhesive is applied 1 or 2 ply (undercoat and overcoat) on the metal fitting which went through surface treatment beforehand and dried, then the metal fittings is inserted into the mold, and the vulcanization bonding is executed through the molding process. It is also necessary to use proper adhesives matching to the rubber materials and the expected environment.

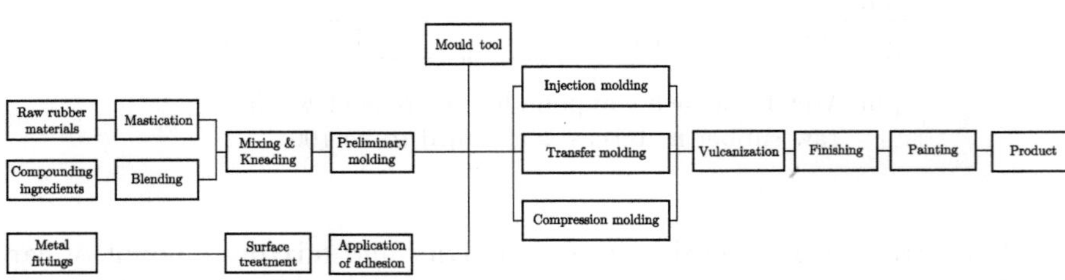

Fig. 3.6. Manufacturing process of anti-vibration piece

3.2.4 Examples of Molding

As for general vulcanization molding process, there are three types of process like compression molding, transfer molding, and injection molding. The compression molding is used for simple shape, small, thin-wall products. The transfer molding is used for large size products with strict dimensional tolerance, and is suitable for small lot production. The injection molding is the most frequently adopted vulcanization process, and there are many alternative types.

It is the feature of this process that raw rubber ribbon with vulcanization ingredient is extruded by screw, and injected into the mold through a narrow orifice causing heat build-up in the mold. During this process, the material self-heated properly can be vulcanized in a short time. So the injection molding process is suitable for mass-production.

3.2.5 Material Properties for Vibration Isolator

Test items for rubber isolators are defined in JIS K 6386, and the excerpt from this standard is shown in Table 3.6.

(1) Static shear modulus

The stress is measured in the range of the tensile strain of 100% or less, and the static shear modulus is calculated from the following equation.

$$\sigma = G_s \left[\left(1 + \frac{\varepsilon}{100}\right) - \left(1 + \frac{\varepsilon}{100}\right)^{-2} \right]$$

G_s: Static shear modulus

σ: Stress

ε: Strain

Usually G_s is determined at the strain condition of $\varepsilon = 25$. Substituting 25 for ε, in the above equation, the following equation is used to calculate static shear modulus for practical purpose, by using the measured value of $\sigma_{0.25}$.

$$G_s = 1.639\sigma_{25}$$

There would be a difference between the actual modulus of elasticity in the product and material testing result, since the strain dependency of the static shear modulus in the rubber material exists.

(2) Elongation

The test shall be conducted under the condition that the specimen is stretched at constant rate, and the elongation at breakage shall be recorded. Moreover, tensile strength at breakage and stress at 100% strain are sometimes recorded.

(3) Heat aging test

The rubber material has the nature of being deteriorated by atmospheric oxygen. Therefore, the heat resistance of rubber material is being evaluated

Table 3.6. Test items and ranks of rubber material (excerpt of JIS K 6386)

Test term	General test					Specialized testing			
Items	Static shear modulus (Mpa)	Elongation (minimum) (%)	Heat aging test (70°C, 70 Hrs)		Permanent compression set (%) (Maximum) (70°C, 22 Hrs)	Permanent compression set (%) (Maximum) (70°C, 22 Hrs)	Rebound resilience (minimum) (%)	Percent change of hardness after aging test (70°C, 70 Hrs) (JIS A)	Percent change of elongation after aging test (70°C, 70 Hrs)
Rank			Percent change of stress at 25% elongation	Percent change of elongation (minimum)					
A 06	0.59 ± 0.10	500					75		
A 10	0.98 ± 0.10	400	−10 ∼ 30	−50	50	25	65	0 ∼ 7	−25
A 14	1.37 ± 0.14	300					55		

by comparing the property of fresh material to the aged material under the prescribed condition.

(4) Compression set test

In order to evaluate the creep property in a short time, this test condition has been set. The short cylindrical specimen (ϕ29 mm × 12.7 mm) shall be compressed to 25% thickness in a thermo chamber for prescribed time. The compression set is calculated using the thickness measured at 30 min. after the removal of compression force.

$$C_s = \frac{l_0 - l_2}{l_0 - l_1} \times 100$$

Where, C_s: Compression set (%)
l_0: Thickness (original)
l_1: Thickness (right after the removal of force)
l_2: Thickness (30 min. after the removal of force)

(5) Rebound resilience test

This test indicates the damping performance of rubber. An iron rod drops to bump on rubber specimen. The ratio of the rebound height and initial height indicates the rebound resilience factor.

(6) Ozone deterioration test

An ozone attack crack might take place in polymer materials with double bonding polymer structures (NR, SBR, NBR, etc.). This deterioration can be prevented by anti-ozone ingredient. The test shall be conducted under the condition that the specimen is subjected to 20% tensile strain, and exposed in the ozone tester at 40°C and 50 pphm (part per hundred million) ozone concentration. The static ozone test is to observe the presence of cracks on the surface by visual inspection after a predetermined time has passed. Moreover, the dynamic ozone test where the specimen undergoes repeated extension could be used for the evaluation of anti-ozone grade.

(7) Immersion test

This test shall be conducted to the rubber materials that need the resistance to liquid. The specimen shall be immersed in the liquid at predetermined time and temperature. The percent change of weight and volume, etc. shall be measured thereafter. Rubber materials such as CR and NBR are tested by this way to confirm the oil-resistance in many cases.

3.2.6 Testing Methods for Rubber Vibration Isolators

Testing methods for rubber vibration isolators are defined in JIS K 6385. The test items are as follows;

(1) Static spring rate;

In order to obtain the load-deformation curve, the load shall be applied twice from zero to the test load preliminary, then the third loading curve shall be recorded. The static spring rate shall be determined within a limited load range as indicated in the following equation.

$$K_S = \frac{P_2 - P_1}{\delta_2 - \delta_1}$$

K_S: static spring rate

P_1: lower load

P_2: upper load

δ_1: deflection by P_1

δ_2: deflection by P_2

(2) Dynamic properties test

The dynamic properties of rubber shall be determined by load-deflection curve under the sinusoidal waveform load. The dynamic property like dynamic spring rate and loss factor are to be determined by this test. Those dynamic properties vary with different test conditions. So, it is necessary to set the test condition matching to the actual service conditions such as temperature, frequency or strain rate, etc.

(3) Hardness test

To obtain the spring rate in a simple manner, the hardness of rubber is measured by spring type hardness tester. This test is effective when the relationship between the static spring rate and the hardness is properly correlative under the same testing condition of temperature and measurement point.

(4) Tensile strength test

The test shall be conducted under tensile strain condition, and fracture load shall be measured. Since a rubber spring has a bonded construction with a metal fitting, adhesive strength shall be evaluated at the same time. In the bonding strength test, the conditions, of failure are classified as follows;

R: Fracture of rubber

RC: Fracture of adhesive on rubber side

CP: Fracture of adhesive itself

M: Fracture of adhesive on metal side

An optimal combination of rubber material and adhesive shall be selected to cause the fracture of rubber body. As bonding strength might be a key factor of products, it is necessary to evaluate the reliability by running the bonding test under the condition that the product should withstand.

(5) Endurance test

The test shall be conducted under cyclic load in order to evaluate the crack and fracture of rubber body. The test conditions shall be determined in consideration of the service environment of the product. For the accelerated test, the endurance test might be run in faster repetition cycles than actual condition. It is necessary to note that the result might be different in some cases because of the heat build-up in rubber materials.

(6) Salt water spray test and salt water immersion test

The metal fitting and/or bonded area between metal and rubber might be damaged by salt. Therefore, salt water spray or salt water immersion test shall be conducted to evaluate the soundness of the product.

In addition to the tests from (1) to (6) as have been mentioned here, heat aging test and ozone deterioration test described in section 3.2.5 are recommended to the rubber vibration isolator according to the situation.

References

1. Gomugijyutsu no ABC, Society of Rubber Industry, Tokai Branch.
2. New Edit. Rubber Isolators, Japan Association of Rolling Stock Industries

3.3 Composite Materials

3.3.1 Various FRP Springs

FRP (Fiber reinforced plastic) is the most popular type of composite material for springs. There are a variety of examples like GFRP (glass fiber reinforced plastic) spring, CFRP (carbon fiber reinforced plastic) spring, hybrid spring (a combination of GFRP and CFRP), and hybrid form of metal and plastic, etc.

Speaking of spring shape, there are simple leaf springs, coil springs, wave springs, and specialized springs.

The GFRP spring has been used for leaf springs for automobiles since the 1980's. After the 1990's, GFRP springs became much popular since significant weight reduction, low spring rate, space allocation freedom, and comfortable riding quality were recognized to be superior to metals. E-glass continuous fiber (13 to 23 μm) is widely used for reinforcement.

Epoxy resin, excellent in strength and weather resistance, is mainly used as the matrix resin for FRP. Unsaturated polyester could be used when strength is not a demanding factor.

As for CFRP spring, leaf springs composed of high-strength, high-modulus continuous polyacrylonitrile-based carbon fiber (5 to 10 μm) and epoxy resin is the typical product. By virtue of high-specific strength, high-specific modulus, and excellent fatigue resistance, the CFRP spring has a remarkable feature of lightweight, compactness, and long service life. This merit could also be advantageous for the other high standard application like a spring which needs dimensional accuracy, chemical resistance, and complex shape.

We can count feeder transportation machine springs, vibrator springs, and engine valve springs as typical FRP spring applications other than automotive suspension springs.

3.3.2 Material Composition of FRP Spring

FRP springs are composed of fiber and matrix resin. The properties of the springs are mostly governed by the mechanical property of fiber material.

On the other hand, the matrix resin has the secondary properties such as environmental resistance and durability of springs. Moreover, the bonding quality of fiber and resin has a key role of spring characteristics.

High tensile strength and high modulus of elasticity are regarded as the most popular merit of FRP springs. Therefore, glass fiber or carbon fiber has been selected as a suitable fiber material to the spring application.

High-modulus Aramid fiber such as Kevlar was also studied for spring application, however, the problem of creep and hygroscopic property remains to be solved. There is a promising trial though where the woven cloth material of Kevlar fiber and glass fiber has excellent characteristics in respect of high-strength and the high energy absorption power [1].

As FRP materials, glass fiber has an advantage over carbon fiber from cost and productivity standpoint. Carbon fiber is to be used in the field of high performance applications such as aerospace industries.

Matrix resin, though it has only a secondary role for the mechanical property of the spring, governs the property of toughness, heat-resistance, moisture resistance, oil resistance, and fatigue resistance, etc. Generally, epoxy resin is widely used. However, it is also possible to use other resins such as unsaturated polyester resin, phenol resin or thermoplastic resin for a proper application.

3.3.3 Manufacturing Process of FRP Spring

Filament winding method (FW method), pre-preg hand lay-up method (HL method), and pultrusion method (PL method) are put in place for FRP spring manufacturing process.

In the FW method, long glass fiber is impregnated to liquid resin with hardener, and then wound on a mandrel. Thickness of the leaf spring shall be adjusted by changing the number of windings. The mandrel is made of aluminum or steel. Thereafter, the shaped FRP spring shall be cured in a heat chamber while staying on the mandrel.

In the HL method, pre-pregs are piled up to necessary thickness, and then cured under high pressure and specified temperature. Pre-preg is a term for "pre-impregnated" composite fibers. These usually take the form of a weave or are uni-directional. They already contain an amount of the matrix material used to bond them together and to other components during manufacture.

In the PL method, a bundle of long glass fiber impregnated to liquid resin and hardening agent is drawn through the heated die to obtain cured and shaped products.

The manufacturing process of CFRP spring is basically identical with the GFRP. They prefer HL method to other methods for CFRP springs since small-lot production is the normal case for the high performance springs.

3.3.4 Material Properties for Design

The following formulas are given in mechanics of material to calculate spring rate for a beam (both ends supported).

$$\delta = \frac{wl^3}{6EI}$$
$$I = \frac{bh^3}{12}$$
$$Z = \frac{bh^2}{6}$$
$$k = \frac{w}{\delta} = \frac{6EI}{l^3}$$
$$M = \frac{wl}{4}$$
$$\sigma = \frac{M}{Z} = \frac{3\delta EH}{4l^2}$$

Where,

b: beam width, h: beam thickness, l: support span, w: load,
E: modulus of elasticity, I: moment of inertia of section
z: section modulus, k: spring rate, M: bending moment,
δ: deflection at center, σ: stress

A comparison table of spring performances based on those equations is shown in Table 3.7 [2] for CFRP, GFRP, and steel spring.

Fiber orientation of FRP material has a great effect on modulus of elasticity and shear modulus. The laminated structure of [0/90] or [0/+45/−45/0] are usually chosen. By selecting optimum placement of fiber orientation, we can design a high performance spring with lighter weight (60% weight reduction compared with metal spring) and excellent fatigue strength against large amplitude vibration having less space occupation. However, in the case of design by longitudinal orientation-rich ratio of composition, it is necessary to take care not to cause an unexpected breakage at bolt-hole caused by stress

Table 3.7. Comparison of spring performances for CFRP, GFRP, and steel spring

	Material	CFRP (Pirofil AS/210)	GFRP	Spring steel
Items	Unit			
Size	mm	$200L \times 50B \times 4H$	$200L \times 50B \times 4.55H$	$200L \times 50B \times 2.6H$
Modulus of elasticity E	GPa	57.5	42.7	205.9
Spring rate K	N/mm	11.5	12.6	11.3
Amplitude δ	mm	17	17	17
Stress σ	MPa	146	124	341
Fatigue strength at 10^7 σ_f	MPa	400	300	180–320
Specific gravity		1.56	1.92	7.8
Weight ratio		0.31	0.43	1

Table 3.8. Material properties of FRP and metals

Properties / Materials	Specific gravity g/cc	Tensile strength MPa	Specific tensile strength 10^6 cm	Modulus of elasticity GPa	Specific modulus 10^6 cm	Coefficient of linear thermal expansion $10^{-6}/°C$
CFRP*1	1.5	1,700	11.3	135	900	0.2
GFRP*1	2.0	1,100	5.5	42	210	20
Hight tensile strength steel	7.8	1,400	1.8	210	270	14
Super duralumin	2.8	550	2.0	70	250	23
Aluminum	2.7	170	0.6	70	260	24

*1: Standard grade; 60% volume content of fiber; fiber direction measurement

concentration. Design of mounting is a key factor for this problem. Material properties of FRP and metals are shown in Table 3.8.

3.3.5 Evaluation of FRP Spring Products

The product characteristic of the FRP spring is almost determined at the stage of the selection of materials and the configuration design. Basic characteristics shown in Table 3.8 and specific characteristics for each product should be evaluated by lot inspection or random inspection. Since spring rate of product varies with dimensional accuracy, dimensional inspection becomes an important process. Random or total inspection of thickness, width and bolt-hole distance of the spring shall be checked according to the given standard. Since internal strain might take place during resin molding, it might be a good idea to put load beforehand for a reliable dimensional inspection.

3.3.6 Tips on FRP Materials

Resin is sensitive to the environmental condition. As for temperature condition, the properties of FRP spring changes along with the property change of resin when used at high temperature. Moreover, resin deteriorates under wet condition, strong acid or strong alkaline environment. Avoiding these influences is strongly recommended.

The breakage of leaf spring often originates at the vicinity of bolt-hole. It is necessary to pay attention to this fact. The fiber orientation of [0/90] combination is recommended to avoid bolt-hole crack. It might be a good idea to put flexible spacer at the bolt-hole location to avoid fretting corrosion on the surface and the delamination of inner structure.

An example of GFRP strength retention curve against increased temperature is shown in Fig 3.7 [3]. Generally speaking, modulus of elasticity and bending strength starts deteriorating at around the temperature of deflection

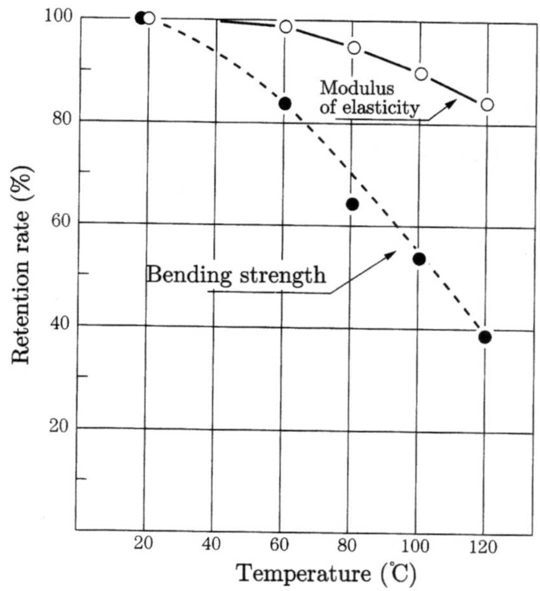

Fig. 3.7. Retention of strength properties (GFRP)

under load, T_s, minus 30°C. It is necessary to pay attention to the cumulative heat storage in service.

A proper heat release design is desirable.

References (Japanese)

1. H. Aoki: Study of design No. 100, 1993, pp. 43–50
2. I. Chikanishi: Plastics Age (9), 1980, p. 70
3. Spring Technology vol. 2, 1998, p. 57, JSSR

4. Inorganic Materials

4.1 Ceramics

4.1.1 Ceramics for Spring

A spring is defined as a machine element which store or absorb energy caused by elastic deformation. The stored energy is released when the applied force is removed. The spring is also defined as a machine element which has the capability of easy deformation by using materials with enough strength for the stress [1]. Therefore any elastic materials can be used as a spring with proper shape design. However, on practical uses, high elasticity limit is desirable for better fatigue strength and smaller permanent set as spring material.

When it comes to ceramics, the material has been thought to be unsuitable for spring use with regard to this respect. However, recent study has made a breakthrough to manufacture ceramic springs through a refinement of raw powder material, and sintering process technology improvement.

Ceramics that can be used to spring application are PSZ (Partially Stabilized Zirconia) in oxide system and silicon nitride (Si_3N_4) in non-oxide system. The physical properties of PSZ and silicon nitride are shown in Table 4.1 [1,2]. The feature of those material are as follows;

(1) PSZ (Partially Stabilized Zirconia)

Zirconia has three kinds of crystal structure and transforms from high temperature to low temperature by taking the structure of cubic, tetragonal, and monoclinic crystal. At cooling stage after sintering, the crystal structure of zirconia transforms from cubic to tetragonal, and then from tetragonal to monoclinic crystal. The last stage is called martensitic transformation. The volume expansion of 4% takes place during this martensitic transformation, and a lot of cracks propagate in a sintered body causing self destruction. To avoid this phenomenon, additives of stabilizer such as CaO, MgO, and Y_2O_3 have been adopted to expand temperature range where cubic crystal structure keeps stabilized condition down to room temperature. This material is called stabilized zirconia and being used for fire resistant materials and oxygen sensor

Table 4.1. Properties of PSZ and Silicon nitride

	PSZ (ZrO_2)	Silicon nitride (Si_3N_4)
Specific gravity	6.05	3.2
Bending strength (MPa)	750 ~ 1,300	900 ~ 1,300
Fracture toughness (MPa$\sqrt{\text{m}}$)	7 ~ 14	6 ~ 10
Hardness (GPa)	15	16 ~ 20
Young's modulus (GPa)	150 ~ 200	300
Coefficient of linear expansion (10^{-6}/°C)	8.7 ~ 11.4	3.2
Thermal conductivity (cal/cm °C)	0.0045 ~ 0.008	0.03

etc. In addition to stabilized zirconia, partially stabilized zirconia (PSZ) has become another candidate of ceramic material of spring use.

Partially stabilized zirconia (PSZ) of the CaO system with reduced amount of CaO has precipitated fine tetragonal zirconia grain in cubic zirconia matrix. The tetragonal crystal exists in meta-stable state thermodynamically at room temperature. This material structure will be strengthened by applying external force through stress-induced martensitic transformation [4]. On the other hand, PSZ of Y_2O_3 system is manufactured by using fine raw powder material at low sintering temperature, and has only tetragonal crystal structure without applying heat treatment [5]. The effect of this stress-induced martensitic transformation is greater than CaO or MgO system, so it has stronger mechanical properties. When the amount of Y_2O_3 additive is 3 mol%, the maximum bending strength of sintering body can be obtained. The bending strength of 1,500 MPa [6] by hot press sintering process, and 1,000 MPa [7] by standard sintering process are reported.

The bending strength of PSZ compared with other materials is shown in Fig. 4.1. Mechanical and thermal characteristics of PSZ look like SUS304 where strength decreases as temperature rise, though the strength at room temperature is fairly high.

Since tetragonal crystal structure stabilizes at high temperature, martensitic transformation does not take place. From such a reason, the temperature range in which PSZ can be used is in the neighborhood of room temperature.

(2) Silicon nitride (Si_3N_4)

The covalent bonding of silicon nitride is so strong that sintering process is not so easy to perform. Therefore, it is general to use sintering accelerators. Y_2O_3-Al_2O_3 family additives developed by Komeya receive wide recognition for this use [8, 9].

When α-Si_3N_4 powder is sintered with Y_2O_3, Al_2O_3 additives, spherical α-Si_3N_4 transforms into tall column β-Si_3N_4 by partially liquid phase sintering process [10] (see Fig. 4.2). The silicon nitride shows the highest strength among many non-oxide ceramics by the shape effect of this crystal grain.

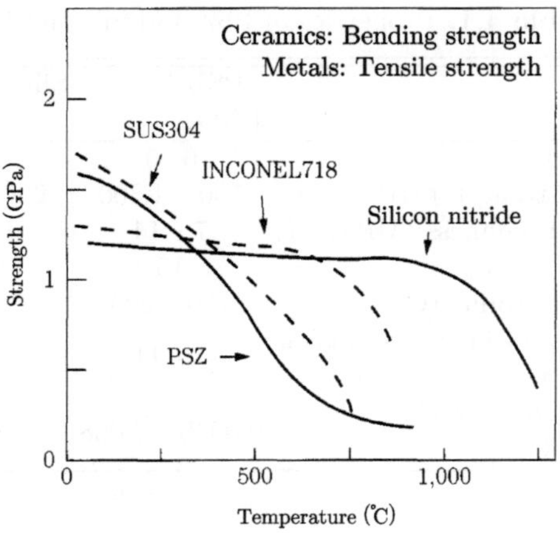

Fig. 4.1. High temperature strength of ceramics and metals

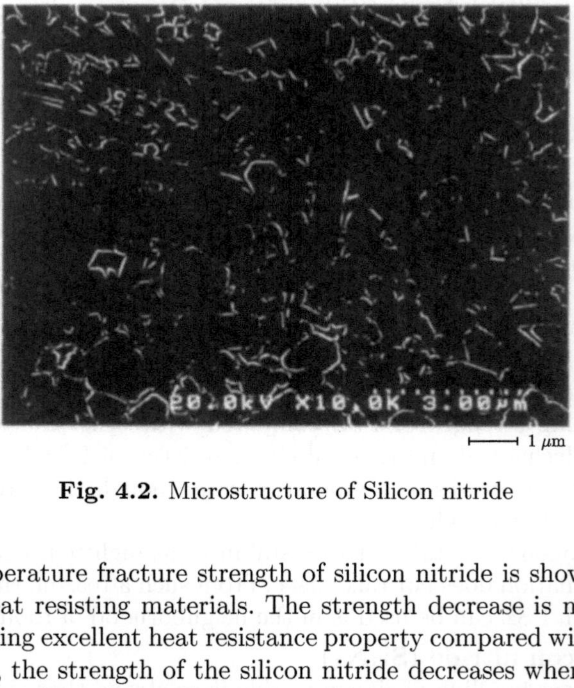

Fig. 4.2. Microstructure of Silicon nitride

High temperature fracture strength of silicon nitride is shown in Fig. 4.1 with other heat resisting materials. The strength decrease is minimal up to 1000°C revealing excellent heat resistance property compared with INCONEL 718. However, the strength of the silicon nitride decreases when the temperature exceeds 1200°C, because Y_2O_3 and Al_2O_3 become glass phase in the grain boundary [11–13].

4.1.2 Strength of Ceramic Spring

Generally, the strength characteristic of PSZ and Silicon nitride keeps its original level when they are formed to the shape of spring. Example of high-temperature strength of PSZ and Silicon nitride coil spring are shown in

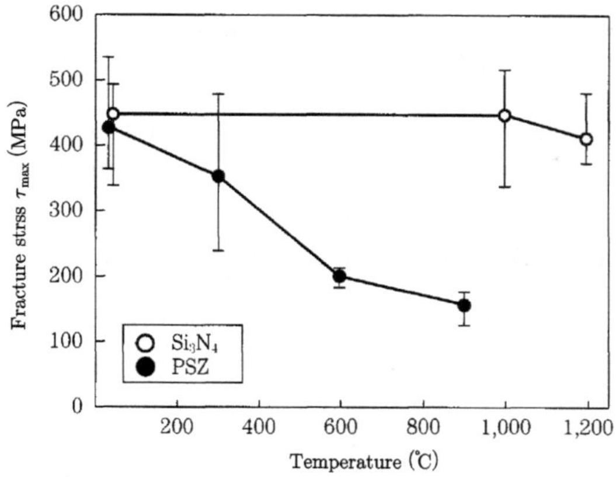

Fig. 4.3. High temperature strength (ceramics coil spring)

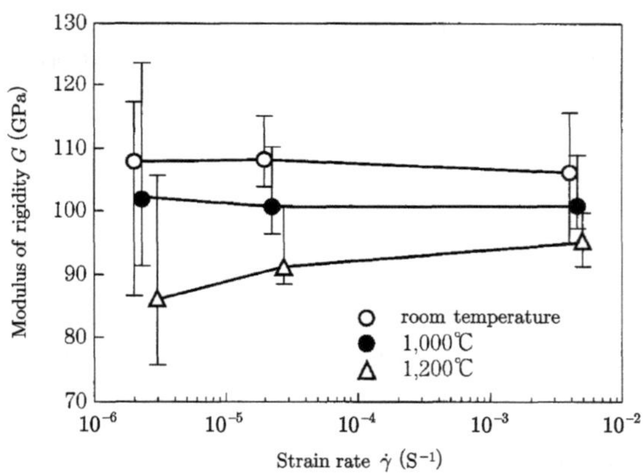

Fig. 4.4. Strain rate and temperature dependence on modulus rigidity (silicon nitride coil spring)

Fig. 4.3. Strength of PSZ decreases along with temperature rise, and it is reduced to the half of its original level when temperature exceeds 600°C. PSZ can not carry out the function of spring over 1000°C temperature range because of a big plastic deformation.

On the other hand, silicon nitride hardly shows the decrease of strength up to 1000°C.

Figure 4.4 shows modulus of rigidity vs. strain rate relationship in different temperature for silicon nitride material. The modulus of rigidity decreases about 6% at 1,000°C compared with the value at room temperature. The modulus of rigidity keeps constant level against strain rate increase in room temperature and 1000°C condition, however, it shows lower level at lower strain rate when temperature goes up to 1200°C.

The reason for this phenomenon is thought to be the presence of plastic flow caused by softened glass phase in the grain boundary in 1200°C temperature condition.

For the reasons described above, it is thought that silicon nitride spring can be used in the temperature range from room temperature to 1000°C.

4.1.3 Operational Temperature of Ceramic Spring

PSZ is categorized as high-strength and high-toughness ceramics. Its strength, however, decreases at higher temperature, since strengthening mechanism is the martensitic transformation. Therefore the operational temperature of PSZ shall be within the neighborhood of room temperature. PSZ has no advantage over metals as far as temperature is concerned. The usage of PSZ may be advantageous where abrasion resistance, corrosion resistance, insulation and non-magnetic characteristic are requested.

On the other hand, silicon nitride is operative up to 1000°C. It has advantage over metals in the temperature region of 700–1000°C where metallic spring cannot be used.

References

1. JSSR: 3rd edition Bane, Maruzen (1982), p. 1
2. O. Shyu: Study of heat-resistant ceramics (Doctoral dissertation, 1994), p. 4
3. T. Yamguchi and H. Yanagida: Engineering ceramics, Gihodo Publishing (1986), pp. 23–24
4. A. G. Evance and A. H. Heuer: Transformation toughening in ceramics, Martensitic transformation in crack-tip stress field, J. Am. Ceram. Soc., 63 (1980), p. 241
5. T.K. Gupta, F.F. Lange and J.H. Bechtold: Effect of stress-induced phase transformation on properties of polycrystalline zirconia containing metastable tetragonal phase, J. of Material Science 13, (1978), p. 1464
6. H. Kobayashi and T. Masaki: High Strength Zirconia Ceramics, Ceramics 17 (1982), p. 427
7. T. Tukuma et al.: Thermal and Mechanical Property of Partially Stabilized Zirconia, Yogyokyokaisi 92 (1984), p. 233
8. K. Komeya et al.: Heat resistant Hyper Composite Material, Pat. No. 759719
9. K. Komeya et al.: Manufacturing Method of Heat resistant Hyper Composite Material, Pat. No. 896931
10. K. Komeya: Development of nitrogen ceramics, Ceramic Bulltin, 63, (1984), p. 1158
11. A. Tuge and Nishida: High strength hot-pressed Si_3N_4 with concurrent Y_2O_3 and Al_2O_3 addition, Am. Ceram. Soc. Bull., 57, (1978), p. 424
12. J.A. Todd and Zhi-Yue Xu: The high temperature creep deformation of Si_3N_4–$6Y_2O_3$–$2Al_2O_3$, J. Mater. Sci., 24, (1989), p. 4443
13. K. Sato et al.: Temperature dependence of anelastic deformation in polycrystalline silicon nitride, J. Am. Ceram., 76, (1993), p. 2043

4.2 RCC (Reinforced Carbon-Carbon)

4.2.1 Properties of RCC

RCC (Reinforced Carbon-Carbon) is a composite material consisting of carbon fiber reinforcement in a matrix of graphite. RCC is also called C/C-composite (Carbon-Carbon composite) for convenience. RCC has high modulus of elasticity and high strength in high temperature environment together with excellent corrosion resistance property.

The strength of various heat resistant materials is shown in Fig. 4.5. RCC has an excellent strength at high temperature compared with other well known materials. Coefficient of linear expansion for various materials is shown in Fig. 4.6. The coefficient of linear expansion for RCC is very small (about 1/16 of metals), so the use of RCC is very advantageous in high temperature environment.

The thermal conductivities of various materials are shown in Fig. 4.7. While copper and aluminum are highly thermally conductive, stainless steel and alumina are less thermally conductive. On the other hand, the thermal conductivities of RCC varies in the range of 30–600 (W/mK), and it is possible to control the value by selecting carbon fiber and baking condition.

4.2.2 Manufacturing Method and Material Characteristics

C/C composite are manufactured by PY (preformed yarn) method. PY is flexible yarn bundle dipped in matrix resin of coke and pitch binder, then encased in a flexible thermoplastic sleeve. C/C spring is manufactured by graphitization at 2,000°C in inert atmospheric environment after the PY was wrapped to a metallic mandrel and hot-pressed. In the standard process of PY method,

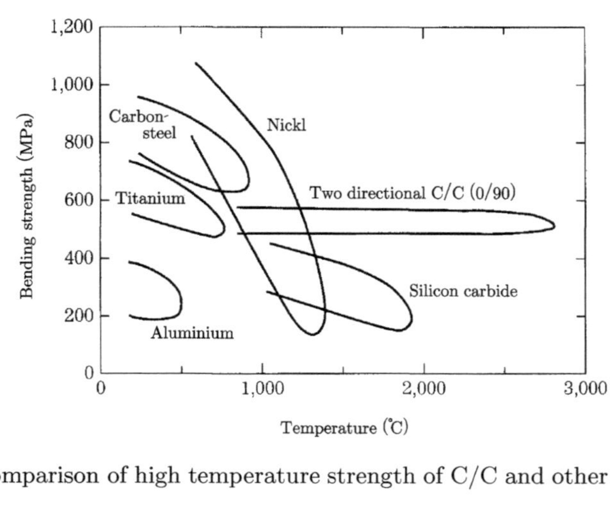

Fig. 4.5. Comparison of high temperature strength of C/C and other heat resisting materials

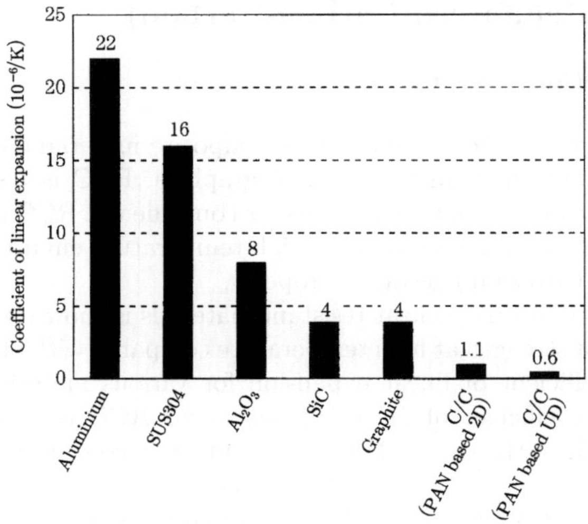

Fig. 4.6. Coefficients of linear expansion

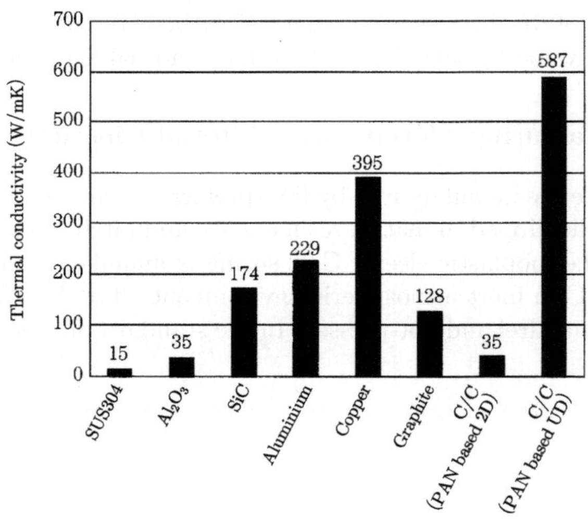

Fig. 4.7. Thermal conductivity

the volume content of carbon fiber before heating is 40–50 vol.%. However, the carbon composition of C/C spring after graphitization becomes 100%, since other resin elements are burnt or vaporized during high temperature process of 2000°C. The characteristics of C/C are shown in Table 4.2.

4.2.3 Shapes and Properties of C/C Spring

Standard shape and specification of C/C spring is shown in Fig. 4.8 and Table 4.3.

Table 4.2. Physical properties of C/C

Specific gravity			1.7
Bending strength		(MPa)	490
Compression strength		(MPa)	235
Charpy impact value		(J/m^2)	52,000
Coefficient of linear expansion	longitudinal	(10^{-6}/K)	0.6
	lateral	(10^{-6}/K)	8.2
Thermal conductivity	longitudinal	(W/(m/K))	68.6
	lateral	(W/(m/K))	10.5
Specific heat	at 20°C	(KJ/kg · K)	0.76
	at 1200°C	(KJ/kg · K)	1.89
Shore hardness			85
Volume resistivity		($\mu\Omega$m)	17

Fig. 4.8. C/C coil springs

Table 4.3. Standard C/C coil spring

Spring rate	(kgf/mm)	2.0	2.0	2.5	3.5
Outside diameter	(mm)	31	50	50	50
Inside diameter	(mm)	19	20	30	20
Free length	(mm)	40	25	27.5	40
Max. deflection	(mm)	7	5	5	7

The comparison of spring materials in high temperature range is shown in Fig. 4.9. Figure 4.10 shows the stability of C/C spring after long service period in 1100°C environment.

It can be said in general from those figures that a C/C spring has an excellent performance in high temperature condition.

In addition to above, there is an advantageous aspect which is unique to C/C composite. Fiber reinforced material is an anisotropic material distinguished from isotropic material such as metals and ceramics. So, springs with

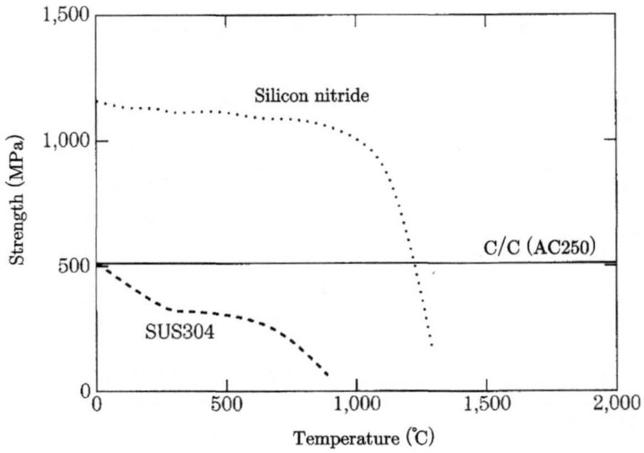

Fig. 4.9. Temperature curve of spring material strength

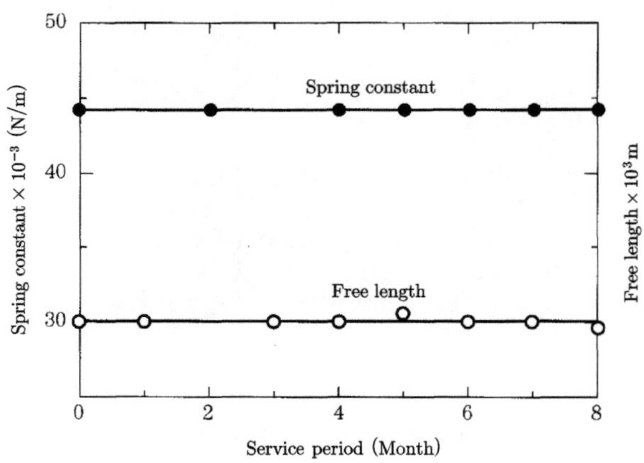

Fig. 4.10. Properties of C/C spring service in high temperature (1,100° C)

various properties can be designed by controlling the orientation and volume percentage of fiber, the shear strength between fiber and matrix resin. Moreover corrosion resistance of C/C is also excellent since it consists of carbon. For instance C/C springs can be used even under corrosive environments such as sulfuric acid (20°C, at the 50% of concentration), hydrochloric acid (60°C, at the 50% of concentration) and nitric acid (50°C, at the 65% of concentration). It has also an excellent performance under high temperature corrosion conditions.

As described above, C/C is the unique spring material that has superior heat resistance and corrosion resistance.

In order to cope with customer's demand for high performance materials, the development of new composite materials like the combination of C/C and ceramics, or C/C and metals can be challenging tasks.

5

Appendix

5.1 Price of Spring Material

Figure 5.1 and 5.2 show the relative price of spring material. The Fig. 5.1 is for wires and the Fig. 5.2 is for sheets or strips. The prices shown in these figures are the relative values for each type of steel. There are no set prices or standard prices for spring material, and prices fluctuate within a considerable range depending on the quantity purchased, the marketing channel, the packaging, the market conditions and so on. Accordingly, the price index in these figures should be used only as a reference guide for comparing various types with one another. The actual price should be confirmed by means of estimates and the like.

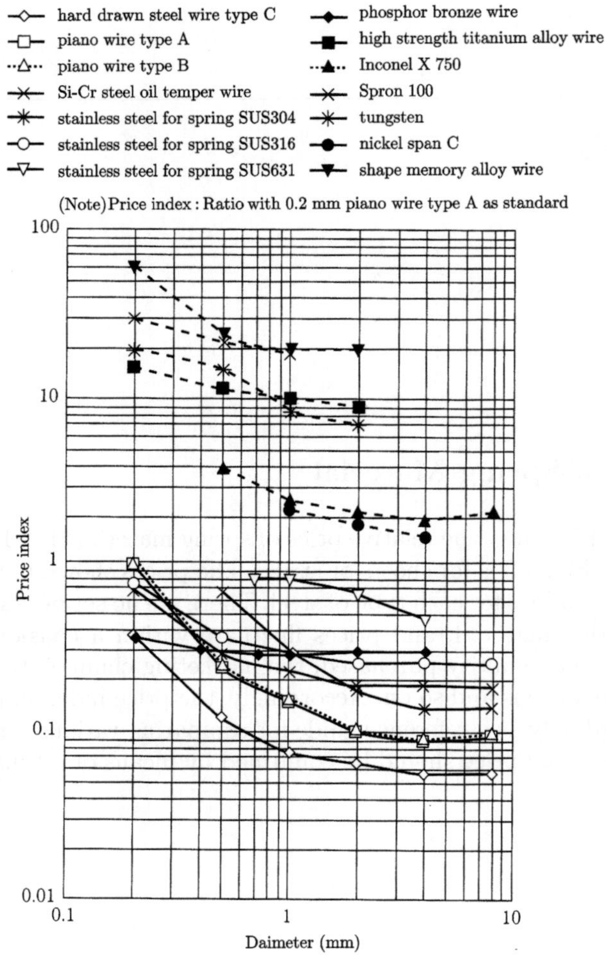

Fig. 5.1. Price of wire for springs

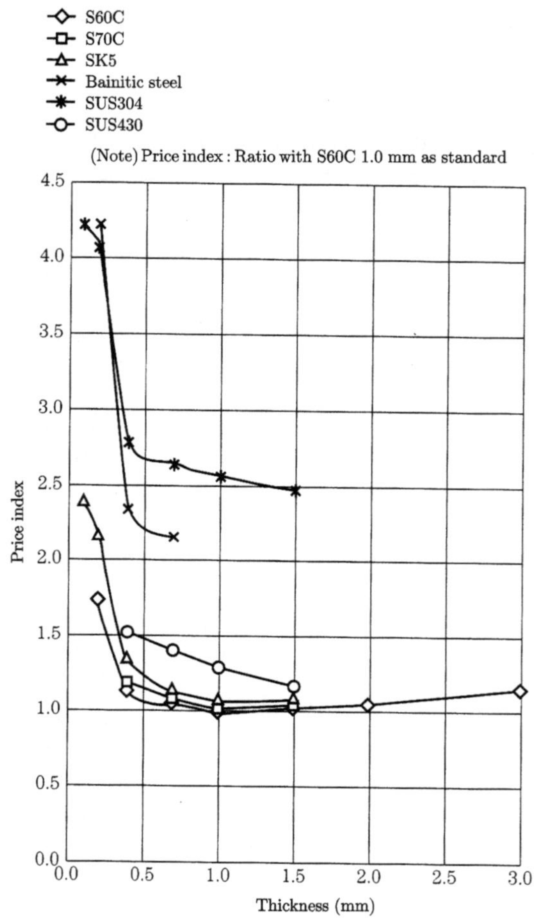

Fig. 5.2. Price of sheets or strips for spring

5.2 Comparison between JIS and International Standards

Table 5.1. Wire for springs (1/2)

Japan Industrial Standard			Equivalent Foreign Standard	
Standard Number	Standard Name	Symbols	Standard Number	Grade
JIS G 3521	Hard Drawn Steel Wires	SW-A	ASTM A 227/A 227 M-93	Hard drawn steel wire for mechanical springs
		SW-B	ASTM A 407-93	Hard drawn steel wire for coil springs
		SW-C	ASTM A 417-93	Hard drawn steel wire for zigzags, squares and wave-like furniture
			ASTM A 510-94	Carbon steel wire rod and wire for general use
			ASTM A 510M-94	Carbon steel wire rod and wire for general use (metric system)
			ASTM A 679/A 679 M-93	High tensile hard drawn steel wire for mechanical springs
			BS 2453-73 (90)	Wire for spokes
			BS 2763-82 (88)	Wire for wire rope
			BS 4637-70 (90)	Carbon steel wire for coil springs for beds and seats
			BS 4638-70 (90)	Carbon steel wire for zigzags and square springs
			BS 5216-91	Cold drawn carbon steel wire for mechanical springs
			DIN 177-88	Cold drawn steel wire (size and weight)
			DIN 2076-84	Wire for spring (size and weight)
			DIN 2078-90	Wire for wire rope
			DIN 17223/1-84	Cold drawn carbon steel wire for springs
			ISO 2232-90	Cold drawn non-alloy steel wire for wire rope
			ISO 6984-90	Non-alloy steel wire for mining winding wire rope
			ISO 8458-1	General requirements of wire for spring
			ISO 8458-2	Cold drawn steel wire for spring wire
JIS G 3522	Piano Wires	SWP-A	ASTM A 228/A 228 M-93	Wire for music springs
		SWP-B	BS 5216-91	Cold drawn carbon steel wire for mechanical springs
		SWP-V	DIN 2076-84	Wire for springs (size and weight)
			DIN 17223/1-84	Cold drawn carbon steel wire for springs
			ISO 8458-1	Steel wire for mechanical springs—Part 1: General requirements
			ISO 8458-2	Part 2: Cold drawn carbon steel wires

Table 5.1. Wire for springs (2/2)

Japan Industrial Standard			Equivalent Foreign Standard	
Standard Number	Standard Name	Symbols	Standard Number	Grade
JIS G 3560	Oil tempered wire for mechanical springs	SWO-A, B	ASTM A 229/A 229 M-93	Oil tempered wire for mechanical springs
		SWOSMA to C	BS 2803-80 (86)	Oil tempered wire for springs
		SWOSC-B	DIN 17223/2-90	Oil tempered wire for springs
			ISO 8458-1	Steel wire for mechanical springs—Part 1: General requirement
			ISO 8458-2	Part 2: Cold-drawn carbon steel wire
JIS G 3561	Oil tempered wire for valve springs	SWO-V	ASTM A 230-A 230 M-93	Oil tempered carbon seel wire for valve springs
			BS 2803-80 (86)	Oil tempered wire for springs
			DIN 17223/2-90	Oil tempered carbon seel wire for mechanical springs
		SWOCV-V	ASTM A 231/A 231 M-93	Cr–V alloy steel spring wire
			ASTM A 232/A 232 M-93	Cr–V alloy steel valve spring quality wire
		SWOSC-V	ASTM A 401/A 401 M-93	Cr–Si alloy steel wire for springs
			ISO 8458-1	Steel wire for mechanical springs—Part 2: General requirements
			ISO 8458-3	Part 3: Oil-harded and tempered wire

Table 5.2. Types of spring steel

Japan Industrial Standard	Foreign Standard equivalent grades						
Standard number / name	Symbol	ISO	AISI SAE	BS	DIN	NF	ГОСТ
JIS G 4801 spring steel material	SUP 3	8457-2	1075 1078	—	—	—	75 80 85
	SUP 6	59 Si 7 683-14	—	—	—	—	60 C 2
	SUP 7	59 Si 7 683-14	9260				60 C 2 Г
	SUP 9	55 Cr 3 683-14	5155		55 Cr 3	55 C 3	—
	SUP 9A	683-14	5160	—	—	—	—
	SUP 10	51 CrV 4 683-14	6150	735 A 51 735 H 51	50 CrV 4	50 CV 4	XφA 50 X ГφA
	SUP 11A	60 CrB 3 683-14	51 B 60	—	—	—	50 X Г P
	SUP 12	55 SiCr 63 683-14	9254	685 A 57 685 H 57	54 SiCr 6	—	—
	SUP 13	60 CrMo 33 683-14	4161	705 A 60 705 H 60	—	—	—

Table 5.3. Types of stainless steel

Japan Industrial Standard		International Standard	U.S.A.		U.K.	Germany	France	Former Soviet Union	European Standard
Standard number / name	JIS	ISO	UNS	AISI	BS	DIN	NF	ГОСТ	EN
JIS G 4313 Cold rolled stainless steel strip for springs	SUS 301	14	S 30100	301	301 S 21	X 12 CrNi 17–7	Z 11 CN 17–08	07 X 16 H 6	
	SUS 304	11	S 30400	304	304 S 31	X 5 CrNi 18–10	Z 7 CN 18–09	08 X 18 H 10	X 4 CrNi 8–10
	SUS 420 J 2	5	S 42000	420	420 S 37	X 20 Cr 13	Z 33 C 13	30 X 13	X 30 Cr 13
	SUS 631	2 (3)	S 177700	S 177700		X 7 CrNiAl 17–7	Z 9 CAN 17–07	09 X 17 H 7 IO	X 7 CrNiAl 17–7
JIS G 4314 stainless steel wires for springs	SUS 302	12, 10 (1)	S 30200	302	302 S 25	—	Z 12 CN 18–09	12 X 18 H 9	—
	SUS 304	11	S 30400	304	304 S 31	X 5 CrNi 18–10	Z 7 CN 18–09	08 X 18 H 10	X 4 CrNi 18–10
	SUS 304 N 1	—	S 30451	304 N	—	—	Z 6 CN 19–09 Az	—	—
	SUS 316	20, 20a	S 31600	316	316 S 31	X 5 CrNiMo 17–12–2	Z 7 CND 17–12–02	—	X 4 CrNiMo 17–12–2
	SUS 631	2(3)	S 177700	S 177700		X 7 CrNiAl 17–7	Z 9 CNA 17–07	09 X 17 H 7 IO	X 7 CrNiAl 17–7

Notes 1. ISO refered to ISO 633/13. But (1) refered to ISO 4955 and (3) to ISO 683/16.
2. U.S.A. was based on UNS resistration number and the AISI steel products manual.
3. Each European country was based on BS, DIN and NF. Also based on reports submitted by each country to ISO.
4. European standards refered to EN 10008-1.
5. ГОСТ quoted by 5632.

Table 5.4. Copper alloy material (sheets, plates and strips)

	Japan Industrial Standard	Similar Foreign Standard				
		U.S.A.		U.K.	Germany	International Standard
	JIS	CDA	ASTM	BS	DIN	ISO
Standard number	H 3110–06	—	—	2870–80 or 2875–69	17670–83	
Material	Phospher bronze and nickel silver	—	—	Copper and copper alloy	Copper and copper alloy	
Material symbol	C5111 P C5111 R	C51100	C51100 (B 103–93)	PB 101	CuSn 4	
	C5102 P C5102 R	C51000	C5100 (B 103–93)	PB 102	—	
	C5191 P C5191 R	C51900	—	PB 103	CuSn 6	
	C5212 P C5212 R	C52100	C52100 (B 103–93)	NS 104	CuSn 8	
	C7351 P C7351 R	C73500	C73500 (BJ 22–92)	—	—	
	C7451 P C7451 R	C74500	C74500 (B 122–92)	NS 103	CuNi 12 Zn 24	
	C7521 P C7521 R	C75200	C75200 (B 122–92)	NS 106	CuNi 18 Zn 20	
	C7541 P C7541 R	C75400	—	NS 105	CuNi 18 Zn 24	
Standard number	H3130–92	—	—	2870–80 or 2875–69	17670–83	
Material	Copper beryllium, copper titanium, phospher bronze, copper-nickel-tin and nickel silver	—	—	Copper and copper alloy	Copper and copper alloy (mechanical properties)	
Material symbol	C1700 P C1700 R	C17000	C17000 (B 194–92)	—	CuBe 1.7	
	C1720 P C1720 R	C17200	C17200 (B 194–92)	—	CuBe 2	
	C5210 P C5210 R	C52100	C52100 (B 103–93)	NS 104	CuSn 8	
	C7701 P C7701 R	C77000	C77000 (B 122–92)	NS 107	CuNi 18 Zn 27	

5.3 Data Books for Spring Material

Name of book	Author	Publisher	Address	Price (yen)
Basics and Applications of phosphor bronze	Mikio Nishihata Hideo Harada	Nikkan Kogyo Shimbun Ltd.	Kudankita 1-8-10, Chiyoda, Tokyo, Japan TEL: +81-03-3222-7131	3,200
Databook of elongated copper product	Japan Copper & Brass Association The databook committee of wrought copper products	Japan Copper & Brass Association	Tsukiji 1-12-22, Chuou, Tokyo, Japan TEL: +81-03-3542-6551	6,000
Handbook of Iron and Steel (CD-R) (2002)	Iron and Steel Inst. of Japan	Iron and Steel Inst. of Japan	Niikura-Bld. Kandatsuki-cho, Chiyoda-ku, Tokyo, Japan	38,000
The Metals Databook	Iron and Steel Inst. of Japan	Maruzen	Nihonbashi 3-9-2, Chuou, Tokyo, Japan	4,000
Heat-treat of steel	Iron and Steel Inst. of Japan	Maruzen	Nihonbashi 3-9-2, Chuou, Tokyo, Japan	16,000
Handbook of Stainless Steel 3rd edition	Japan Stainless steel Association	Nikkan Kogyo Shimbun Ltd.	Kudankita 1-8-10, Chiyoda, Tokyo, Japan TEL: +81-03-3222-7131	53,000
High temperature strength data of spring steel for heat resistance	Spring technology research association	Japan Spring Manufactures Association	Kanda Mikuracho 12, Chiyoda, Tokyo, Japan TEL: +81-03-3251-5234	—
SMI. Spring Material Cross Reference Part 1 Ferrous Materials, Part 2 Nonferrous Materials	Spring Manufactures Institute		2001 Midwest Road. Suite 106. Oak Brook, IL 60521, USA	
Spring materials selector 2nd Edition		Institute of spring technology Ltd. (SRAMA)	Henry Street, Sheffield S3 7EQ, UK TEL +44(0)114-2760771 FAX +44(0)114-2726344 e-mail ist@ist.org.uk	$192
JIS Handbook 1 Steel I (2007)	Japanese Standards Association	Japanese Standards Association	Akasaka 4 chome 1-24, Shibuya, Tokyo, Japan TEL: +81-3-3583-8007	8,925
JIS Handbook 2 Steel II (2007)	Japanese Standards Association	Japanese Standards Association	Akasaka 4 chome 1-24, Shibuya, Tokyo, Japan TEL: +81-3-3583-8007	8,400
JIS Handbook 3 Non-ferrous metals (2007)	Japanese Standards Association	Japanese Standards Association	Akasaka 4 chome 1-24, Shibuya, Tokyo, Japan TEL: +81-3-3583-8007	8,610
JIS Handbook 4-2 Screws 2 (2006)	Japanese Standards Association	Japanese Standards Association	Akasaka 4 chome 1-24, Shibuya, Tokyo, Japan TEL: +81-3-3583-8007	5,775
JIS Handbook 7 Mechanical Elements (2007)	Japanese Standards Association	Japanese Standards Association	Akasaka 4 chome 1-24, Shibuya, Tokyo, Japan TEL: +81-3-3583-8007	10,710
JIS Handbook 26 Plastic 1, Test (2007)	Japanese Standards Association	Japanese Standards Association	Akasaka 4 chome 1-24, Shibuya, Tokyo, Japan TEL: +81-3-3583-8007	11,340
JIS Hand Book 27 Plastic 2, Materials (2007)	Japanese Standards Association	Japanese Standards Association	Akasaka 4 chome 1-24, Shibuya, Tokyo, Japan TEL: +81-3-3583-8007	9,450

Name of book	Author	Publisher	Address	Price (yen)
JIS Handbook 19&20 Rubber 1&2 (2007)	Japanese Standards Association	Japanese Standards Association	Akasaka 4 chome 1-24, Shibuya, Tokyo, Japan TEL: +81-3-3583-8007	6,300 each
JIS Handbook 36 Ceramics (2007)	Japanese Standards Association	Japanese Standards Association	Akasaka 4 chome 1-24, Shibuya, Tokyo, Japan TEL: +81-3-3583-8007	7,700
JIS Handbook 4-2 Heat treatment (2006)	Japanese Standards Association	Japanese Standards Association	Akasaka 4 chome 1-24, Shibuya, Tokyo, Japan TEL: +81-3-3583-8007	11,025
ASM Handbook, Volume 20 Materials Selection and Design	George E. Dieter	ASM International	Material Park, OH 44073-0002, USA	$192
Metals Design Reports of Fatigue Strength (1) Second Edition	Japan Society of Mechanical Engineers	Japan Society of Mechanical Engineers	Shinanocho Rengakan 5th floor, Shinanocho 35, Shinjuku, Tokyo TEL +81-3-5360-3500	
Metals Design Reports of Fatigue Strength (2) Second Edition	Japan Society of Mechanical Engineers	Japan Society of Mechanical Engineers	Shinanocho Rengakan 5th floor, Shinanocho 35, Shinjuku, Tokyo TEL +81-3-5360-3500	
Metals Design Reports of Fatigue Strength (3)	Japan Society of Mechanical Engineers	Japan Society of Mechanical Engineers	Shinanocho Rengakan 5th floor, Shinanocho 35, Shinjuku, Tokyo TEL +81-3-5360-3500	
Metals Design Reports of Fatigue Strength (4) Low Cycle Fatigue Strength	Japan Society of Mechanical Engineers	Japan Society of Mechanical Engineers	Shinanocho Rengakan 5th floor, Shinanocho 35, Shinjuku, Tokyo TEL +81-3-5360-3500	
Manual of Mechanical Elements: Springs	NHK Spring	Kogyo Chosakai Publishing, Inc.	2-14-7 Hongo, Bunkyo-ku, Tokyo TEL +81-3-3817-4700	2,730

In addition to the above databooks, you can ask JICST copy center for copies of other reports to assist you in your research.
A charge will be made for their service.
Hikarigaoka Post Office Box 10, Asahichou 2-8-18, Nerima, Tokyo, Japan
TEL +81-0120-004-381 FAX +81-3-3979-4781
In addition to the JICST you may also access JOIS data base to search for documents.

5.4 Material Manufacturers and Products

Table 5.5–Table 5.8 show the relation between manufacturers and product lines. Manufacturers are listed in no particular order. As product lines change over time, it is recommended that these tables be used for reference purposes only. For the actual study, the product lines should be confirmed by contacting the manufacturers.

Table 5.5. Steel materials for wire springs and their manufacturers

	Hard Drawn Steel Wire	Piano Wire	Piano wire for valve springs	Stainless steel wire for springs	Oil tempered wire for mechanical springs	Oil tempered wire for valve springs	Maraging steel wire	Spring steels	Rolled flat wire		
									Carbon steel	Low alloy steel	Stainless steel
Suzuki Metal Irdustry Co., Ltd.	○	○	○	○	○	○	○		○	○	○
Shinko Wire Industry Co., Ltd.	○	○		○	○	○	○				
Suncall Co., Ltd.	○	○	○		○	○					
Sumitomo Electric Industry Co., Ltd.	○	○	○	○	○	○			○	○	○
Sugita Wire Co., Ltd.	○	○			○	○			○	○	
Tokusen Kogyo Co., Ltd.	○	○	○		○				○		
Aichi Steel Co., Ltd.								○			
Nippon Steel Co., Ltd.								○			
Sumitomo Metal Industries, Ltd.								○			
Kobe Steel, Ltd.								○			
Daido Stainless Co., Ltd.				○			○				
Daido Steel Co., Ltd.								○			
Nippon Seisen Co., Ltd.				○							○
Rokko Metal Co., Ltd.	○	○							○	○	
Sakai Wire Co., Ltd.				○							
JFE Bars & Shapes Corp.								○			
Nichia Steel Works	○	○			○				○		
Toyo Wire, Ltd.	○										
Nippon Metal Industry Corporation				○							

Table 5.6. Non-steel materials for wire spring and their manufacturers

Material Markers \ Type of Material Sorts	Spring Alloy Wire based on Cobalt	Spring Alloy wire based on Nickel (Monel Alloy 400)	Spring Alloy wire based on Nickel (Monel Alloy K 400)	Spring Alloy wire based on Nickel (Inconel Alloy X 400)	Titanium Alloy Wire	Shape Memory alloy Wire	Phosphor Bronze Wire	Phosphor Bronze Wire for Spring	Beryllium Bronze Wire	Nickel Silver Wire
Nippon Mining & Metals Co., Ltd.							○	○		
Mitsubishi Electric Metals Co., Ltd.							○			
The Furukawa Electric Co., Ltd.						○	○	○		
Kiyomine Metal Industries Co., Ltd.							○	○		
Sumitomo Electric Industries, Ltd.				○	○	○				○
Suzuki Metal-Working Industry Co., Ltd.				○	○		○	○	○	
Daido Steel Co., Ltd.	○	○	○	○	○	○				
Nippon Seisen Co., Ltd.	○			○	○	○				
NGK Insulators, Ltd.									○	○
Nippon Bellparts Co., Ltd.								○	○	

Table 5.7. Steel materials for plate, sheet or strip spring and their manufacturers

	Cold-rolled special steel strip	Cold-rolled steel strips for springs	Hot-rolled steel strip	Cold-rolled stainless steel strip for springs	Quenched and tempered steel strip	Baintic steel strip	Spring steel
Nippon Kinzoku Co., Ltd.	○	○		○	○	○	
Hayashio Kinzoku	○						
Nisshin Steel Co., Ltd.	○	○	○	○		○	
Aichi Steel Corporation							○
Daido Steel Co., Ltd.	○	○	○	○	○	○	
Sumitomo Metal Industries, Ltd.							○
Mitsubishi Steel Mfg. Co., Ltd.							○
Kaishin Kogyo				○			
Nippon Mining & Metals Co., Ltd.				○			
Toyo Seihaku Co., Ltd.		○		○			
H.Kishi					○		
Moritec Steel Co., Ltd.	○	○		○	○	○	
JFE Bars & Shapes Corp.			○				
Yodoko	○	○					
Hokuriku Yoshinaka Co., Ltd.	○						
Takasago Tekko K.K.	○			○			
Nas Stainless Steel Strip MFG. Co., Ltd.				○			
Myodo Metal Co., Ltd.				○			
Nippon Kinzoku Co., Ltd.				○			
Nippon Bellparts Co., Ltd.				○			

Table 5.8. Non-ferrous materials for plate, sheet or strip springs and the manufacturers

Type for Material / Material Manufacture	Phosphor bronze	Phosphor bronze	Brass	Nickel silver	Copper beryllium
Harada Metal Industry Co., Ltd.	○	○			
Nippon Mining & Metals Co., Ltd.	○	○	○		
Mitsubishi Electric Metals Co., Ltd.	○	○			
The Furukawa Electric Co., Ltd.	○	○			
Kiyomine Metal Industries Co., Ltd.	○	○		○	
Ohki Brass & Copper Co., Ltd.	○				
NGK Insulator, Ltd.					○
Dowa Mining Co., Ltd.			○		
Toyo Wire, Ltd.	○	○	○	○	○
Yamaha Metanix Corporation					○

5.5 Unit Conversion Table

Table 5.9. Conversion factor for each unit (1/3)

Quantity	SI Unit		NOT SI Unit				
	Name	Symbol		Name	Symbol	Conversion factor	Conversion factor to SI
Angle Solid Angle	Radian Steradian	rad sr		Degree	°	1	$\pi/180$
				Minute	′	1/60	$\pi/10800$
				Second	″	$1/60^2$	$\pi/648000$
Length	Meter	m		Micron	μ		10^{-6}
				Angstrom	Å		10^{-10}
				X-ray unit	X-Unit		$= 1.00208 \times 10^{-13}$
				Fermi	Fermi, fm		10^{-15}
				nautical mile	n mile, M		1852
			Yard and Pound	Foot	ft	1	0.3048
				Inch	in	1/12	2.54×10^{-2}
				Mile	mile	1760	1609.344
				Chain	chain	22	—
				Yard	yd	1	0.9144
Area	Square Meter	m^2		Are	a	1	100
				Hectare	ha	100	—
				Byrne $(-100\,fm^2)$	b		10^{-28}
			Yard and Pound	Square Foot	ft^2		9.290304×10^{-2}
				Square Inch	in^2		6.4516×10^{-4}
				Square Mile	$mile^2$	640	
				Acre	acre	1	4046.8564
				Square Yard	yd^2	1/4840	
Volume	Cubic Meter	m^3		Liter	l, l, L	1	1/1000
				Deciliter	dl, d, dL	1/10	—
				CC (cm^3)	cc	1/1000	—
			Yard and Pound	Measurement Ton	T	100	—
				Cubic Yard	yd^3	27	—
				Cubic Foot	ft^3	1	$= 2.831685 \times 10^{-2}$
				Cubic Inch	in^3	$1/12^3$	—
			Imperial Gallon	Bushel	bushel	8	—
				Gallon	gal	1	4.5460919×10^{-3}
				Pint	pt	1/8	—
				Ounce for Liquid	fl oz	1/160	—
				Minim	minim	1/76800	—
			U.S. Gallon	Barrel	barrel	42	—
				Barrel for Grain	bbl	336/11	—
				Bushel	bu	107521/11550	—
				Gallon	gal	1	$= 3.785412 \times 10^{-3}$
				Pint for Grain	dry pt	107521/739200	—
				Pint for Liquid	liq pt	1/8	—
				Ounce for Liquid	fl oz	1/128	—
				Minim	minim	1/61440	—
Time	Second	s		Year (365 days)	a, y	8760	
				Day	d	24	
				Hour	h	1	
				Minute	min	1/60	
Frequency	Hertz	Hz		Cycle	s^{-1}		1
Rotational Speed	Revolution per Seconds	s^{-1}		Revolution per Minuts	rpm	1	1/60
				Revolution per Seconds	rps	60	1
Velocity	Meter per Second	m/s		Kilo-meter per Hour	km/h	1	1/3.6
				Not	kn, kt	1,852	1852/3600
				Mile per hour	mile/h		0.44704
				Foot per Hour	ft/h		0.3048

Table 5.9. Conversion factor for each unit (2/3)

Quantity	SI Unit		NOT SI Unit			
	Name	Symbol	Name	Symbol	Conversion factor	Conversion factor to SI
Acceleration	Meter per second per second	m/s^2	Gal	Gal		1/100
			G	G		9.80665
			Foot per second per second	ft/s^2		0.3048
Mass	Kilo gram	kg	Ton	t		1000
			Gram	g		1/1000
			Gamma	g		10^{-9}
			Unified Mass Unit of Atom	u		$= 1.6605655 \times 10^{-27}$
			Carat (−200 mg)	car.ct		2×10^{-4}
			Ton (UK)	ton	2240	—
			Ton (USA)	ton	2000	—
			Slug	slug	32.17405	14.59390
			Pound	lb	1	0.45359237
			Ounce	oz	1/16	—
			Grain	gr	1/7000	—
Density	Kilo gram per Cubic Meter	kg/m^3	Weight Kilo Gram square second per meter per cubic Meter	$kgf \cdot s^2/m^4$		9.80665
			Specific Weight	kgf/m^3		9.80665
			Pound per Cubic Foot	lb/ft^3		16.01846
			Slug per Cubic Foot	$slug/ft^3$		515.3788
Weight	Newton	N	Weight Kilogram	kgf	1	9.80665
			Weight ton	tf	1000	9.80665×10^3
			Dyne	dym	—	
			Weight Pound	lbf		= 4.448222
			Poundal	pdl		= 0.1382550
Stress, Modulus of elasticity	Pascal Newton per Square Meter	Pa (N/m^2)	Weight Kilogram per Square Meter	kgf/m^2	10^{-6}	9.80665
			Weight Kilogram per Square Centi-Meter	kgf/cm^2	10^{-2}	9.80665×10^4
			Weight Kilogram per Square Mili-Meter	kgf/mm^2	1	9.80665×10^6
Stress	Pascal	Pa	Bar	bar	—	10^5
			Atmospheric Pressure	atm	1	1.01325×10^5
			Mercury Mili-Meter	mmHg	1/760	—
			Torr	Torr	1.760	—
			Water Colums Mili-Meter	mmAq, mmH_2O		9.80665
			Weight Pound per Square Inch	lbf/in		= 6894.75
			Weight Pound per Square foot	lbf/ft^2		= 47.88026
Moment of a Force, Torque	Newton Meter	N·m	Weight Kilo Gram Meter	kgf·m		9.80665
			Weight Pound Inch	lbf·in		= 0.1129848
Energy Work Heat quantity Gravity unit	Joule	J N·m W·s	Erg	erg		10^{-7}
			I.T. quantity of heat	cal, cal_{IT}, cal (IT)		4.1868
			Quantity of heat (light weight method)	cal		4.18605
			Fifteen degrees quantity of heat	cal_{15}		4.1855
			Thermochemistry quantity of heat	cal_{th}		4.1840
			Kilowatt-hour	kw·h		3.6×10^6
			Power-hour	PS·h		$= 2.647796 \times 10^8$
			Electron Volt	eV		$= 1.60219 \times 19^{19}$
			British thermal unit	Btu		= 1055.056
			Foot·Weight Pound	ft·lbf		= 1.355818
Output Power Electric Power Heat quantity Transfer	Watt	W J/s N·m/s V·A	Power	PS	75	= 735.4988
			Kilo Calories per hour	kcal/h	—	1.163
			Weight kilo gram meter per second	kfg·m/s	1	9.80665
			British power	hp, HP	550	= 745.6999
			British thermal unit per hour	Btu/h	—	= 0.2930711
			foot weight pound per second	ft·lbf/s	1	= 1.355818

Table 5.9. Conversion factor for each unit (3/3)

Quantity	SI Unit		NOT SI Unit			
	Name	Symbol	Name	Symbol	Conversion factor	Conversion factor to SI
Temperature Thermodynamic temp.	Kelvin	K	Celsius degree, Degree	°C	$T\,\mathrm{K} = t\,°\mathrm{C} + 273.15$	
			Rankine degree	°R	$T\,\mathrm{K} = t\,°\mathrm{R}/1.8$	
			Fahrenheit's temperature	°F	$T\,\mathrm{K} = (t\,°\mathrm{F} + 459.67)/1.8$	
Viscosity (Coefficient of viscosity)	Pascal Second	Pa·s	Poise	P	1	1/10
			Centi-Poise	cP	1/100	1/1000
			Weight kilogram second per square meter	$\mathrm{kgf \cdot s/m^2}$	—	9.80665 —
			Slug per foot per second	$\mathrm{slug/(ft \cdot s)} =$	$\mathrm{lbf \cdot s/ft^2}$	47.88026
			Pound per foot per second	$\mathrm{lb/(ft \cdot s)} =$	$\mathrm{pdl \cdot s/ft^2}$	1.488164
Specific heat Specific entropy Gas constant	Joule per kilo gram per Kelvin	J/(kg·K)	Kilocalorie per weight kilo gram per Kelvin degree	$\mathrm{kcal/(kgf \cdot °K)}$	1	4186.8
			British thermal unit per pound per Rankine degree	$\mathrm{Btu/(lb \cdot °R)}$	1	4186.8
			Weight kilogram meter per weight kilo gram per Kelvin degree	$\mathrm{kgf \cdot m/(kgf \cdot °R)}$		9.80665
			Foot calorie per pound per Rankine degree	$\mathrm{ft \cdot lbf/(lb \cdot °R)}$		= 5.380320
Thermal conductivity	Watt per meter per Kelvin	W/(m·K)	Kilocalorie per meter per hour per degree	$\mathrm{kcal/(m \cdot h \cdot °C)}$		1.163
			British thermal unit per foot per hour per Fahrenheit's degree	$\mathrm{Btu/(ft \cdot h \cdot °F)}$		= 1.730735

References

1. JIS Handbook Steel2-2007, Japan Standards Association
2. JIS Handbook Steel2-2007, Japan Standards Association
3. JIS Handbook Steel2-2007, Japan Standards Association
4. JIS Handbook Steel2-2007, Non-steel-1997, Japan Standards Association
5. JIS Handbook Steel2-2007, Japan Standards Association
6. Mechanical Engineering Handbook, Japan Society of Mechanical Engineers, Maruzen

Index

MIX
Papier aus verantwortungsvollen Quellen
Paper from responsible sources
FSC® C105338

If you have any concerns about our products,
you can contact us on
ProductSafety@springernature.com

In case Publisher is established outside the EU,
the EU authorized representative is:
Springer Nature Customer Service Center GmbH
Europaplatz 3, 69115 Heidelberg, Germany

Printed by Libri Plureos GmbH
in Hamburg, Germany